The Value of Information

Ramanan Laxminarayan • Molly K. Macauley
Editors

The Value of Information

Methodological Frontiers and New Applications in Environment and Health

Springer

Editors
Ramanan Laxminarayan
Center for Disease Dynamics,
Economics & Policy
Washington, DC, USA

Molly K. Macauley
Resources for the Future
Washington, DC, USA

ISBN 978-94-007-4838-5 ISBN 978-94-007-4839-2 (eBook)
DOI 10.1007/978-94-007-4839-2
Springer Dordrecht Heidelberg New York London

Library of Congress Control Number: 2012945538

Printed on acid-free paper

Springer is part of Springer Science+Business Media (www.springer.com)

Acknowledgments

We are deeply grateful for financial support from the US National Aeronautics and Space Administration (NASA), Resources for the Future, and the Center for Disease Dynamics, Economics & Policy. Lawrence Friedl, Director of NASA's Applied Earth Sciences Division, shared our vision for asking and answering how to value information and helped guide our efforts. Also guiding our efforts were other leaders in the public and private sectors: Richard Bernknopf, US Geological Survey; Robert S. Chen, Columbia University; William B. Gail, Microsoft; Kass Green, Kass Green & Associates; William H. Hooke, American Meteorological Society; and Charles F. Kennel, Scripps Institution of Oceanography. At Resources for the Future, Abigail Colson, Griffin LeNoir, Andrea Titus and Daniel McDermott were instrumental in helping to manage our effort. Our deep gratitude also goes to Sally Atwater for her careful copyediting of the manuscript.

Contents

Contributors

Richard L. Bernknopf Department of Economics, University of New Mexico, Albuquerque, NM, USA

Western Geographic Science Center, United States Geologic Survey, Menlo Park, CA, USA

Timothy J. Brennan Public Policy and Economics, University of Maryland–Baltimore County, Baltimore, MD, USA

Resources for the Future, Washington, DC, USA

Yan Chen Bates White, Washington, DC, USA

Jessica L. Cohen Department of Global Health and Population, Harvard School of Public Health, Boston, MA, USA

Roger M. Cooke Resources for the Future, Department of Mathematics, TU Delft, Delft, The Netherlands

William T. Dickens Northeastern University, Boston, MA, USA

Brookings Institution, Washington, DC, USA

Timothy Essam Department of Agricultural and Resource Economics, University of Maryland, College Park, MD, USA

Scott Farrow Department of Economics, University of Maryland–Baltimore County, Baltimore, MD, USA

The Woods Hole Oceanographic Institution, Falmouth, MA, USA

Adam M. Finkel Penn Program on Regulation, University of Pennsylvania Law School and Visiting Scholar, Penn Center for Bioethics, Philadelphia, PA, USA

William M. Forney Western Geographic Science Center, United States Geologic Survey, Menlo Park, CA, USA

Steffen Fritz International Institute for Applied Systems Analysis, Ecosystems Services and Management Program, Laxenburg, Austria

Sabine Fuss International Institute for Applied Systems Analysis, Ecosystems Services and Management Program, Laxenburg, Austria

David M. Hartley Department of Microbiology and Immunology, Georgetown University Medical Center, Washington, DC, USA

Fogarty International Center, National Institutes of Health, Bethesda, MD, USA

Petr Havlík International Institute for Applied Systems Analysis, Ecosystems Services and Management Program, Laxenburg, Austria

Ginger Zhe Jin Department of Economics, University of Maryland, College Park, MD, USA

National Bureau of Economic Research, Cambridge, MA, USA

Jonathan T. Kolstad The Wharton School, University of Pennsylvania, Philadelphia, PA, USA

Carolyn Kousky Resources for the Future, Washington, DC, USA

Ramanan Laxminarayan Center for Disease Dynamics, Economics & Policy, and Research Scholar, Princeton University, Princeton, USA

Kenneth L. Leonard Department of Agricultural and Resource Economics, University of Maryland, College Park, MD, USA

Molly K. Macauley Resources for the Future, Washington, DC, USA

Anup Malani Lee and Brena Freeman Professor of Law, University of Chicago Law School, Chicago, IL, USA

Resources for the Future, Washington, DC, USA

Luther Martin Voltage Security, Inc., Cupertino, CA, USA

Ian McCallum International Institute for Applied Systems Analysis, Ecosystems Services and Management Program, Laxenburg, Austria

Joshua Michaud International Development Program, Johns Hopkins University School of Advanced International Studies, Washington, DC, USA

Shruti K. Mishra Western Geographic Science Center, United States Geologic Survey, Menlo Park, CA, USA

Catherine Shelley Norman Department of Geography and Environmental Engineering, Department of Economics, The Johns Hopkins University, Baltimore, MD, USA

Michael Obersteiner International Institute for Applied Systems Analysis, Ecosystems Services and Management Program, Laxenburg, Austria

Daniel Osgood Lead Financial Instruments Sector Team, International Research Institute for Climate and Society, Columbia University, New York, NY, USA

Mead Over Center for Global Development, Washington, DC, USA

Ronald P. Raunikar Western Geographic Science Center, United States Geologic Survey, Menlo Park, CA, USA

Felicjan Rydzak International Institute for Applied Systems Analysis, Ecosystems Services and Management Program, Laxenburg, Austria

Linda See International Institute for Applied Systems Analysis, Ecosystems Services and Management Program, Laxenburg, Austria

School of Geography, University of Leeds, Leeds, UK

Kenneth E. Shirley AT&T Labs Research, Florham Park, NJ, USA

Jana Szolgayová International Institute for Applied Systems Analysis, Ecosystems Services and Management Program, Laxenburg, Austria

Department of Applied Mathematics and Statistics, Faculty of Mathematics, Physics and Informatics, Comenius University, Bratislava, Slovakia

Michael Toman Development Research Group, World Bank, Washington, DC, USA

Abbreviations (Chapter numbers in parentheses)

ACT	artemisinin combination therapy (antimalarial drugs) (7)
AHA	American Hospital Association (6)
AIC	Akaike information criterion (5)
ARC	Africa Rainfall Climatology (1)
AWiFS	Advance Wide Field Sensor (10)
BAU	business as usual (2, 4)
BBN	Bayesian belief net (2)
BSE	bovine spongiform encephalopathy (9C)
CABG	coronary artery bypass graft (6)
COP	conference of the parties (of UNFCCC) (8)
CRRA	constant relative risk aversion (2)
DICE	model Dynamic Integrated model of Climate and the Economy (2)
EMR	expected mortality rate (6)
EO	Earth observation (4)
EPA	Environmental Protection Agency (10)
ESA	European Space Agency (8)
EVPI	expected value of perfect information (8)
FAO	Food and Agriculture Organization (4, 8, 9)
FeliX	Full of Economic-Environment Linkages and Integration dX/dt (4)
FFS	fee for service (6)
FTC	Federal Trade Commission (5)
GEO	Group on Earth Observation (4)
GEOBENE	Global Earth Observation—Benefit Estimation: Now, Next and Emerging (4)
GEOS	Global Earth Observation System (10)
GEOSS	Global Earth Observation System of Systems (4, 8)
GHG	greenhouse gas (2, 4, 8)
GLOBIOM	Global Biosphere Management Model (8)
GWP	gross world product (4)
HARITA	Horn of Africa Risk Transfer for Adaptation (1)
HMO	health maintenance organization (6)

IAA	integrated assessment approach (10)
IAM	integrated assessment model (2)
IIASA	International Institute for Applied Systems Analysis (8)
ICU	intensive care unit (6)
IFPRI	International Food and Policy Research Institute
IIA	independence of irrelevant alternatives (5C)
ILRI	International Livestock Research Institute (1)
IPCC	Intergovernmental Panel on Climate Change (8)
IPCC	4 Fourth Assessment Report of the Intergovernmental Panel on Climate Change
IRI	International Research Institute for Climate and Society (1)
IT	information technology (3)
IV	instrumental variables (5)
LULC	land use and land cover (10)
LULUCF	land use, land-use change, and forestry
MCL	maximum contamination level (10)
MRLI	moderate-resolution land imagery (10)
MSL	maximum simulated likelihood (5)
NDVI	normalized difference vegetation index (9)
OIE	Office International des Epizooties (9)
OLS	ordinary least squares (5)
OMR	observed mortality rate (6)
PHC4	Pennsylvania Health Care Cost Containment Council (6)
RAMR	risk-adjusted mortality rate (6)
RDT	rapid diagnostic test (for malaria) (7)
REDD	Reducing Emissions from Deforestation and (forest) Degradation (8)
RTE	ready-to-eat (cereal) (5)
RVF	Rift Valley fever (9)
SBA	societal benefit area (4)
SBSTA	Subsidiary Body for Scientific and Technological Advice(8)
USDA	US Department of Agriculture (10)
USGS	US Geological Survey (10)
VaR	value at risk (1, 2)
VOI	value of information
VSL	value of a statistical life (3)
WHO	World Health Organization (9)
WIIET	Weather Index Insurance Education Tool (online software)
WTP	willingness to pay
UNFCCC	United Nations Framework Convention on Climate Change (8)

List of Figures

List of Tables

Introduction

Ramanan Laxminarayan and Molly K. Macauley

Although the idea that information has value in both a statistical and a pragmatic sense dates back at least to the 1950s, in recent years, interest in the economic value of information has taken center stage. Policymakers face the burden of justifying large public investment in data on climate and air quality, public health, ecosystems, water, and other natural and environmental resources. Companies face the problem of what types of information best inform consumers of their products, and how to ascribe value to that information. The chapters in this volume explore innovative methodologies and applications of value of information research.

The chapters focus on applications in two disparate fields linked by the importance of valuing information: public health and space. Our selection of these two topics follows from several opportunities. Researchers in the health field have developed some of the most innovative methodologies for valuing information. These applications range from the value of diagnostics in informing decisions to treat patients to the value of private information on health insurance plans when some information is already publicly available. We seek to highlight these innovations in this volume, but broader challenges remain. How much should developing countries that spend less than $10 per person per year on the health of their citizens invest in gathering information to improve that resource allocation decision? How do individuals choose to respond to information, which in turn has an enormous impact on the value of that information?

Our "space" topic refers to innovations in the technologies that collect information; this is the new information provided by the growing number of Earth-observing satellites that collect data about air quality, freshwater supplies, climate, and other natural and environmental resources that affect our health and our overall quality of life. In this field, recent applications of value-of-information methods are critically important for informing investment in the satellite networks. As of 2010, 79 satellites were circling Earth to observe, measure, and monitor the natural dynamics of freshwater, the oceans, land, the atmosphere, and climate, together with human interactions with these resources and consequent implications for human and environmental health. The satellites represent an investment made by more than 40 countries and totaling an estimated $40 billion.

A common theme in this book is that information is often, although not always, a public good. Once the information has been gathered and provided for people to use, one more person can use the information without reducing its value to others or imposing costs on them. This attribute of information—additional use at very low additional cost, or widespread benefit beyond that accruing to just one individual—also justifies public investment in collection and supply of many types of information. For example, testing individuals for a disease will benefit not only the individuals in caring for their own health, but contributes to public information about disease prevalence and in turn informs public health investment in disease prevention. The information collected by an Earth resource observing satellite may benefit all of society, although it may not be optimal for a single individual to launch the satellite.

As the examples in this book show, investment in information has the potential to deliver valuable societal benefits, including better-informed citizens, companies, and governments. But this value has seldom been measured or expressed in ways to ascertain whether the investment is paying off. Worse, suppliers of the information often pay little attention to the form in which information is communicated to decisionmakers, the information-processing costs that decisionmakers face, or their ability to use the information in a time frame that makes the exercise worthwhile.

The chapters are based on papers commissioned for a workshop hosted by Resources for the Future and held in Washington, DC, on June 28–29, 2010. Workshop participants included 120 people from government, the private sector, universities, and other nongovernmental organizations. A steering committee assisted in planning the workshop. Discussants drawn from the government and academic communities were selected to lead the conversation about each paper; these commentaries follow their respective chapters. The discussant chapters were written based on versions of the papers presented at the workshop. Some of the papers have evolved since the workshop.

What Is Distinctive About This Volume?

This volume is the first to present research by experts in two disparate communities—social scientists and experts in the use of satellite data about natural and environmental resources—to identify and critique state-of-the-practice methods for ascribing value and social benefit to information. The findings offer answers to important questions: What is meant by *value of information?* When does information have value? What are the state-of-the-practice methods to ascribe value to information? The contributors identify five discrete approaches at the frontier of methodological advances: price- and cost-based derivation, Bayesian belief networks, regulatory cost-effectiveness evaluation, econometric modeling and estimation, and simulation modeling and estimation. They advance terms to describe what is meant by “value” (which need not be expressed in monetary terms) and identify steps to ascribe, measure, and communicate value.

Overview of the Chapters

In Chap. 1, Daniel Osgood and Kenneth Shirley describe index insurance, a relatively new approach to providing climate risk protection to low-income farmers in developing countries. The success of index insurance hinges critically on the climate information: What weather pattern can be expected over the long term and with what degree of certainty? Osgood present this concrete component of the value of information by quantifying the value of improved data in the reduction it allows for insurance prices.

Continuing a focus on climate but broadening the discussion, Roger Cooke and Carolyn Kousky in Chap. 2 propose a way for society to manage risk in light of the possibility of cataclysmic damages if climate changes abruptly. This concern shifts the policy question to how much risk of catastrophe society is willing to accept. Using the value-at-risk management approach from the banking and insurance sectors as an analogy for managing the risk of climate change catastrophes, Cooke and Kousky provide rough estimates of the value of improved information in these areas. They offer the important observation that knowing when not to wait for more information is just as important as knowing when to seek additional information.

Luther Martin asks in Chap. 3 whether one can infer the value of business information by observing how much companies spend to protect this information. Many businesses claim that information is their most valuable asset. Martin finds that unless protection is required by regulation or legislation, businesses appear unwilling to invest heavily in technologies to protect information—say, from unauthorized access, use, or disruption. Part of the reason may be that in many cases, huge amounts of information flow through a company, and "at the margin," the value of information is quite low. On this basis, the benefit of increasingly large efforts to protect information may not justify the cost. Another explanation may be that businesses perceive the probability of hacking or other breaches of security to be quite low and thus have low incentives to invest in information protection. Martin explores these and other reasons why efforts to secure information may be imperfect measures of information's value.

Martin's point of view is that of the business sector. In Chap. 4, Michael Obersteiner, Steffen Fritz, Ian McCallum, and Felicjan Rydzak characterize the value of investment in information made by the government sector. In particular, they consider the case of investment by European governments in satellites to observe natural and environmental resources. Different governments have each invested in these satellites, leading to the possible opportunity for global coordination of satellite investment to enhance the value of the information. Obersteiner and his colleagues present an engineering approach drawn from the field of systems dynamics to assess the effect of such integration in improving the value of information.

A recurrent theme in the value of public information is how information alters economic choices and thereby influences societal welfare. Often the value of information is in changing consumer behavior—altering decisions on what health insurance to buy, which hospital to go to, or how energy-efficient a washing machine to buy. The expectation is that publishing information, whether about environmental indicators or hospital infections, will alter demand-side behavior, which in turn will result in higher quality. However, does information truly alter consumer behavior? What is its effect in the case of "experience goods," where a single experience with the good could be important in adoption and habit formation?

In Chap. 5, Yan Chen and Ginger Jin study how consumers choose among brands when they have limited information about the product and its quality (even if they know that the product or brand exists). How does information derived from two distinct sources of information—purchase experience and brand advertising—influence consumer purchase decisions? Using Nielsen Marketscan data, they find that advertising does inform consumers about the existence and quality of a product. Incorporating information about the existence of a product increases price elasticity, they find, because informative advertising increases consumer choice set. We can infer from this that in a range of spheres, public information about choices can improve consumer welfare. So an effort by government to provide information on energy-saving appliances could inform consumers about their choices for lowering their energy bills. Indeed, informative advertising may be underprovided by private firms, if part of the value of this advertising is public. Public information can help consumers make a smarter choice of first-time experience and have a lasting effect on consumer welfare through habit formation.

In Chap. 6, Jonathan Kolstad tackles the related theme of the relative value of public and private information. Public information is often justified in its role to correct market failures generated by asymmetric privately provided information. For instance, a hospital may advertise the need for an expensive surgical procedure even if such a procedure has questionable health value. Kolstad looks at how consumers respond to hospital rankings provided by *U.S. News and World Report* hospital reputation before and after the release of report cards on surgeon quality in Pennsylvania's market for cardiac bypass surgery. Kolstad addresses two questions. First, how much do market-based learning and private information (such as that provided by a physician in choosing a specialist) alter consumer choices in the absence of public reporting? And second, does the prior existence of privately provided public data (such as by *U.S. News*) alter the value of public reporting? He finds that privately reported data are a substitute for publicly provided information. There are two main lessons to take away from this study. Researchers must consider existing sources of information on quality available to consumers when deciding on investments in public information. Not accounting for these existing sources could lead to underestimates of the value of public information. Public information could have important distributional consequences that depend on consumers' *ex ante* knowledge of provider quality. Public information, in this case, could have large effects among those who do not have access to privately provided information.

But how likely is information to be provided by private sources? On the one hand, private sources can charge for the information. On the other hand, their profits could be influenced by the information they provide, if they are attempting to sell a bundled good of both information and a commodity or service. Take the case of small drug sellers in Africa who sell antimalarials, including expensive artemisinin-based combinations (ACTs). Use of ACTs when the patient does not have malaria can lead to wasted resources, loss of an opportunity to treat for the true underlying condition that is making the patient sick, and sometimes drug resistance and side effects because of inappropriate use of the drugs. Given the right incentives, small drug sellers (where the majority of Africans seek treatment for malaria) who make money from ACT sales could also sell rapid diagnostic tests (RDTs) that could quickly inform the patient whether he or she has malaria. In Chap. 7, Jessica Cohen and William Dickens find that with symmetric information about the likelihood that patients have malaria, and no subsidies for treatment, drug shop owners will provide RDTs at a socially optimal level. But if ACTs are subsidized and there are external costs associated with the misuse of antimalarials, drug shops will likely underprovide RDTs. This underprovision can be corrected through a wholesaler level subsidy for RDTs; educating customers about the true prevalence of malaria can also help.

Additional questions about how best to characterize the value of information to inform investment are addressed in the remaining chapters. In Chap. 8, Steffen Fritz, Sabine Fuss, Petr Havlík, Ian McCallum, Michael Obersteiner, and Jana Szolgayová evaluate the value of improved global data about land. They develop and then illustrate a portfolio optimization model to find the optimal mix of mitigation options under different sets of information to estimate the benefit of having an improved land cover data set to evaluate policies influencing land use.

The case for greater investment in satellite data is often made on the back of public health. More precise information on weather patterns could help improve our ability to predict disease outbreaks and thereby reduce loss of life and economic damage by allowing local authorities to take preventive action. Emerging understanding of the spatiotemporal determinants of disease emergence and transmission indicates that such prediction is possible, although there is no guarantee that the information will actually be used to prevent disease. As David Hartley notes in Chap. 9, data on the economic payoff from investments in information aimed at reducing health risk are sparse to nonexistent. He discusses economic returns in the context of Rift Valley fever, a disease for which there has been significant investment in weather-based predictive modeling.

In Chap. 10, Richard L. Bernknopf, William M. Forney, Ronald P. Raunikar, and Shruti K. Mishra use a sample of satellite data to disentangle the cumulative regulatory-induced effects of agricultural production on the environment, a set of effects that at present is confounded by the concurrent implementation of agriculture, energy, and environmental policies. They consider the case of corn production in the Midwest of the United States, where biofuels mandates, land and water protection, and crop production subsidies have both intended and unintended consequences on long-term economic output and environmental resources. By coupling space-derived data on spatial and temporal changes in land use with production and economic data, they develop empirical estimates of the value of satellite data in evaluating decisions on the use of natural resources and agricultural output.

Findings and Results

What the chapters show is the array of methods for ascribing value to information and the desirability of bringing these methods to bear to inform public investment in information collection, dissemination, and use—whether one is a corporation, the government, or a private individual.

What is meant by "value?" In general, the authors agree that value connotes a quantitative measure even if it is not necessarily expressed in monetary terms. In some chapters, the authors derive monetary values. In other chapters, the authors derive nonmonetary values, such as number of lives saved, improvements in environmental quality, or enhanced regulatory efficiency. The choice depends on the context of the problem and the data available for empirical evaluation. By emphasizing a quantitative dimension in expressing the value of information, the authors seek to provide a metric that would be relevant for decisionmaking. In the absence of such measures, it is difficult for a company, the government, or a consumer to gauge the relative usefulness of information, distinguish among types and sources of information that can substitute for one another but may differ in acquisition cost, or inform investment decisions in information collection and use.

When does information have value? All the authors agree on the criteria by which "information" has value, with the corollary that there are circumstances in which information has little or no value. These criteria can guide policymakers, corporate managers, and other leaders in making investments in information collection and in demonstrating the value of the information on behalf of consumers, shareholders, and the public. The criteria are as follows:

- Information has the most value when decisionmakers are more indifferent among their alternatives.
- Information has the most value when action can be taken in response to the information. If action cannot be taken, information has less value.
- Information has the most value when the consequence of making the wrong decision is large.
- Information has the most value when the constraints on using the information are few and the cost of using the information is small.
- The value of "perfect information" may not be commensurate with the cost of its acquisition.
- Information has value even if it introduces more uncertainty. In this case, it reveals that what was thought to be certain may not be.[1]

[1] An example is the value of a second opinion in a medical diagnosis.

- Certain attributes of information may confer more value than other attributes.[2] *What are the state-of-the-art methods to ascribe value to information?* The chapters illustrate five methodological approaches.
- Price- and cost-based derivation. The illustrations of this approach include the use of satellite weather and climate data in weather index insurance in developing countries. In this context, the value of the satellite data are expressed in monetary terms derived directly from the insurance premiums and the value at risk. Another example is the value of information in terms of losses averted from having the information, expressed in terms of economic costs associated with vector-borne disease outbreaks. The avoided loss estimate includes avoided control costs, reduced morbidity and mortality, and averted disruption of international trade.
- Bayesian belief network. This approach uses a Bayesian belief net to derive a monetary value for Earth observation data about expected temperature mean and variability in a changing climate. The Bayesian framework is a conventional statistical approach in which people update their expectations when given new information. The belief net allows other information to be brought to bear by a decisionmaker. In this example, the other information was the economy's output (gross domestic product) and damages associated with climate change. This information "conditions" the value of the Earth observations data. The net also provides an efficient computational approach and a means of visually displaying results to show the determinants of the information value.
- Regulatory cost-effectiveness. The illustration of this approach demonstrates direct cost savings enabled by Earth observation data products in implementing land-use and water quality regulation. Another application of this approach demonstrates people's willingness to pay to avoid the loss of information as a means of informing regulatory decisions to maintain and protect information databases.
- Econometric modeling and estimation. Illustrations of this method use statistical approaches involving econometric estimation of hypothesized relationships between information and people's decisions. In these cases, the coefficient on the explanatory variables in the estimated equations serves as a quantitative measure of the value of information. These econometric equations also allow the researcher to control for, or hold constant, other variables that influence the value attributable to the information. For example, econometric evaluation of the role of diagnostic tests for malaria allows quantitative estimation of the size and statistical significance of the information from the diagnostic tests on behavioral responses of patients in their decision to seek additional treatment. The results show by how much the information (from the diagnostic test) contributes to a patient's decision, and the other explanatory variables show how much other factors (age, income, etc.) contribute to the decision. Other applications of this approach illustrate the effects of information in situations where the value of

[2] Attributes include, for example, timeliness, accuracy, precision, spatial resolution, and spectral resolution.

information is expressed in added years of life expectancy or other quality-of-life dimensions.

- Simulation modeling and estimation. An example of this approach is use of systems engineering to design flow charts characterizing multiple uses of the same information. For example, Earth observation data on land use provide information for land carbon assessment. The value of improved land carbon assessment can then be linked to the prices at which carbon is traded in the European Trading System, for example. The common theme across all of those approaches is the goal of a quantitative expression for the value of information, although the value need not be in monetary terms.

Going Forward

The research presented here shows the dire need for those who invest in information collection to better understand the needs of those who use the information. What attributes of the information are most useful? What quality (how much precision or accuracy) is most useful? What are the barriers in using information? How can the constraints on decisionmakers be lowered, to enable them to make better use of information. For example, can we expand the solution set (i.e., enhance the actions taken in response to information)? Can we ease cognitive constraints (i.e., enlarge the number of people who know about the information, including consumers of the information and policymakers)? Can we change resource constraints (i.e., the budgets governing investment in information) by better demonstrating that information has value and is valued?

If we can begin to answer these questions, we can set priorities for information investment in areas that have the ability to produce the greatest economic and nonmarket value.

Chapter 1
The Value of Information in Index Insurance for Farmers in Africa

Daniel Osgood and Kenneth E. Shirley

Abstract Index insurance is a relatively new approach for providing climate risk protection to low-income farmers in developing countries. Because this insurance is implemented in data-poor environments, information constraints and uncertainty substantially affect the products. Since insurance is a tool that can be used to exchange uncertainty in the market, the level of information available directly alters prices, with insurance protection for climate risk and insurance protection for information uncertainties about climate risks both being components of the final price. Using data, methodologies, and contracts for index insurance applications in Africa, the chapter presents this concrete component of the value of information by quantifying the value of improved data in lowering insurance prices. It provides a brief overview of index insurance in developing countries and discusses the value of remote sensing in informing the index and the role of climate trends.

Keywords Agricultural insurance • Index insurance • Value at risk • Value of information • Rainfall simulation

---

D. Osgood (✉)
Lead Financial Instruments Sector Team, International Research Institute for Climate and Society, Columbia University, New York, NY, USA
e-mail: deo@iri.columbia.edu

K.E. Shirley
AT&T Labs Research, Florham Park, NJ, USA

This author's contributions were made while he was a postdoctoral fellow at The Earth Institute, Columbia University, New York, NY, USA
e-mail: kshirley@research.att.com

R. Laxminarayan and M.K. Macauley (eds.), *The Value of Information: Methodological Frontiers and New Applications in Environment and Health*,
DOI 10.1007/978-94-007-4839-2_1,

1.1 Introduction

Smallholder farmers in Africa are severely hurt by droughts and other climate-related events. Most of the literature about the value of information in agricultural production focuses on the production benefits that can be obtained when farmers use the information to alter their behavior to be more cautious in years that are likely to be bad, and more aggressive in years that are not likely to be bad (see literature review by Meza et al. 2008). In the past, smallholder farmers in Africa have had essentially no access to insurance. New types of insurance, such as index insurance, are becoming available to these farmers. Recent literature extends the work on the value of information in production decisions to include important features of insurance (Carriquiry and Osgood 2008; Osgood et al. 2008; Cabrera et al. 2006). The value of information gains an additional and very concrete component, the effect of information on the cost of insurance.

We describe a very different component in the value of information driven by the availability of insurance, the value of information through reducing the cost of insurance. We present this component through pricing exercises of index insurance projects that we have been involved with. For the discussion we use the actual data, software, and formulas used in the insurance programs to explicitly quantify this component of the value of information.

Traditional loss-based indemnity insurance has been extremely difficult to implement in most of the world, leaving most farmers without coverage. There are several challenges with traditional insurance that have prevented it from being workable in many contexts. Traditional agricultural insurance requires a large amount of information on the probability of losses in order to be applied. In many situations, this information is not available. It requires that an adjuster visit the field where a loss is reported, which becomes prohibitively expensive for farmers with small plots in remote locations who have relatively small amounts insured. Loss-based insurance is also fraught with perverse incentives leading to moral hazard and adverse selection. If a farmer receives a payout when there are crop losses, the farmer has an incentive to let crops die. Similarly, farmers who are more likely to have losses are more likely to purchase insurance. Since insurance companies never have complete information about farmers and their actions, these problems often undermine the viability of the insurance. The value of information differences between strategically acting farmers and insurers is addressed in game theory work on asymmetric information, including specific work on agricultural insurance (Luo et al. 1994; Skees et al. 1999).

Index insurance is a relatively new approach intended to address those problems so that insurance may be made more broadly available. For this type of insurance, the payout is based on measurements of an index that is likely to lead to crop loss. Most commonly, this index is a function of cumulative rainfall during critical parts of the agricultural production season, with payouts triggered during droughts, when the cumulative rainfall falls below a predetermined threshold. Using this strategy, historical rainfall data can be used to determine the probability of a payout. Although this information is very often limited, it is typically much less limited than information on yield losses themselves. In addition, payouts can be made based on the weather measurement without requiring an adjuster to incur the cost of visiting each field. The perverse incentives are addressed because the payout is not based on the farmer's behavior. There is no benefit to the farmer through the strategic reduction in yields.

Those benefits come with a substantial cost. Because the insurance provides payments on the rainfall as opposed to the loss, it cannot totally protect a farmer from loss. Farmers may have losses due to pests, flooding, wind, or even differences in rainfall between the measured amount and what the farmer experiences on her field. The disconnect between payouts and losses is called basis risk, and it is a central theme in the index insurance literature. Because of basis risk, index insurance does not function well as comprehensive insurance protection. Instead, it is better suited for incrementally reducing the risk that a farmer faces in the most cost-effective manner. Policy documents that describe index insurance issues include (Hellmuth et al. 2009; Hazell et al. 2010), and (Barrett et al. 2007) provide a review and synthesis of the academic index insurance literature.

For this discussion, we will look at two recent index insurance pilot projects in Africa in which we have participated. One is the second-year (2006) implementation of index insurance for groundnut farmers in Malawi, led by the World Bank ARMT (formerly CRMG), and the other is the launch year (2009) of the Horn of Africa Risk Transfer for Adaptation (HARITA) index insurance project in Adi Ha Ethiopia, led by Oxfam America. Throughout this work, we will draw from reports and policy documents that we have developed as we have assisted in the design and pricing of these products (Osgood et al. 2007; Chen et al. 2010; Dinku et al. 2009; Hellmuth et al. 2009).

1.2 Value of Information in Insurance Price

The cost of insurance is driven by the expense of the financing necessary to ensure payouts, as estimated by the probabilities of payout events. There are two components to the price. The first is simply the expected payout. The insurance premium must be sufficient to cover the average amount of money being paid out. In addition to the average payout, the insurance company must maintain sufficient capital on hand to cover extreme payouts. Insurance companies will choose (or be required by regulations) to keep sufficient liquidity to be able to pay for the largest event expected with a reasonable frequency. Often, this frequency is set to every 50 or 100 years, which is equivalent to holding enough money to cover the 98th or 99th percentile event.

Commonly, this money is borrowed from the insurance company shareholders, so the interest paid is the return on the shareholders' investment in the company. It is money that is held specifically to manage risk, as opposed to be put into investments (such as agricultural inputs) that would provide returns through production.

This is a fundamental cost of risk management. An individual farmer faces a similar choice whether he purchases insurance, maintains savings for a rainy day (in our case, a drought), or borrows to cover losses after the drought has occurred. It is the basic trade-off of how much money to keep liquid in case there is drought versus the money that is put at risk for higher returns, invested in inputs to a productive activity that may experience a loss.

From a risk financing perspective, the primary difference between the insurance company and the farmer is that the insurance company can build a large portfolio of unrelated (or even negatively correlated) risks, such that the amount of money that must be held to cover the farmer's 99th percentile event is less than the farmer would have to reserve. Premiums received each year by the insurance company can also be used as payouts that year, which reduces the amount of money that must be borrowed.

Information quality affects the fundamental cost of insurance. The amount of money to be reserved for risk protection is driven not only by the risks that are faced, but also by the amount of information about the risks that are faced. If the size of the 99th percentile event is not well known, more money must be held to cover a conservative estimate of 99th percentile events that might occur. Even if the average payout is known with certainty, the premium must reflect the range of average payouts that may occur. Otherwise, the insurance company cannot responsibly commit to honoring the insurance contract. As information improves about the probabilities of payouts, that information can reduce the cost of insurance, so that overly conservative levels of reserves and premiums are not required.

Insurance costs have additional components, including the administrative costs of providing insurance and the delivery costs of registering clients and delivering their payouts. Because those components do not reflect information issues and were not explicitly included in the calculation of the actual Malawi insurance cost from our numerical example, we will ignore them here.[1]

Often the cost of insurance is presented as the "actuarially fair" component, which is the simplest accounting calculation of average payouts plus "loading," all of the risk financing, uncertainty, and other costs. The actuarially fair price is "free" insurance—that is, on average, all the money paid to the insurance company is returned to the client. Loading is often expressed in terms of its percentage relative to the actuarially fair price. Insurance is also commonly presented in terms of percentage of maximum liability. This is the size of the premium in relation to the maximum possible payout size. Although this is a convenient indicator for calculating actual premium values for different maximum liabilities, it provides no information about the value that the client is receiving because it does not reflect the frequency or size of actual payouts and can be strategically manipulated to give the appearance of insurance value.[2]

A standard equation for premium (p) calculations is $p = \mathrm{E[Payout]} + r(\mathrm{VaR} - \mathrm{E[Payout]})$ (Osgood et al. 2007), where r is the effective rate of interest paid on the risk reserve funds. E[Payout] is the size of the average payout (including zero payout years) and VaR is the value at risk, the size of the 99th percentile event.

[1] For the Malawi premium calculation, the interest rate on the money held to be able to pay for the 99th percentile event was increased slightly to reflect administrative and delivery costs.

[2] For example, if a $10 premium is paid for a policy with a maximum liability of $100, the percentage price is 10 % for a contract that provides (full) payouts 10 % of the time (an expected payout of $10, or zero loading) as well as for a contract that provides (full) payouts 1 % of the time (an expected payout of $1, or a loading of 900 %).

Typically, insurance price calculations are proprietary, as they reflect the risk handling specifics of the insurance company. In addition, they are commonly affected by the cost of reinsurance negotiations between the insurance company and reinsurer.[3]

The Malawi transaction is an interesting case study to illustrate value of information issues in insurance. Because it was the first index insurance product of its type in Africa (and one of the first in the world), a great deal of effort was taken to make sure that processes were simple, open, and transparent so that the important features of insurance would be clear to participants and observers. In the Malawi example, smallholder farmers in several villages purchased several thousand contracts for insurance costing approximately $2 to insure microloans for groundnut and maize production. Many of these farmers had no previous access to credit, being mostly outside the cash economy. As with most insurance, for these types of projects, it is important that the premiums reflect the true risk costs as accurately as possible. Otherwise farmers will take actions that are too risky or too conservative when they respond to the insurance price incentives (For more information on the project, see [Osgood et al. 2007; Hellmuth et al. 2009; Bryla and Syroka 2009]).

Because the Malawi insurance product was provided jointly by a consortium of insurers and no reinsurance was used, premium calculations were not proprietary negotiations between insurance interests. Instead, the cost of the premiums was determined by the project partners using publicly circulated formulas. The price was calculated as p = E[Payout] + 0.06 (VaR − E[Payout]). Because the Malawi data set had approximately 50 years of rainfall data, the 98th percentile was used to estimate the VaR.[4]

For the 2006 Malawi transaction, because the historical data set was relatively long, the cost of uncertainty about the payout probabilities was not charged to farmers. The price was calculated by calculating what payouts would have been if the index had been applied to the historical rainfall data, a process often referred to as historical burn pricing. This was selected for the initial years of the pilot because it allowed extremely transparent pricing. The intent was that pricing would become more sophisticated as project partners gained familiarity with the concepts involved.

One concern for future pricing was that it responsibly account for uncertainty in calculations, particularly as sites with much smaller data sets were included. In addition, idiosyncratic prices are generated from using historical burn pricing based on the single historical realization of rainfall, since much of the price was

[3] Reinsurance is purchased by insurance companies from global reinsurance companies, which handle very large events that would overwhelm an individual insurance company. Reinsurance companies address these risks through a global portfolio of varied insurance company clients and other investments.

[4] Following the pricing process, partners decided to use the largest payout year that would have occurred using the approximately 50 years of rainfall data to estimate the 98th percentile. This choice was made to make the explanation of the premium simpler and more transparent for early stages of the project; it did not meaningfully change the cash premiums paid by farmers.

determined by the single largest event in 50 years. In essence, pricing based solely on historical data assumes that any amount of rainfall not exactly observed in the past has a zero probability of occurring. A more robust process would estimate distributions of the index from a large number of realizations.[5]

The World Bank therefore commissioned the International Research Institute for Climate and Society (IRI) to develop rainfall models that could realistically generate large numbers of synthetic 50-year time series that could be combined to estimate an appropriate insurance price. This rainfall model is designed to reflect the uncertainty in the probability of rainfall so as to determine a responsible level for reserves and premiums. Because this analysis is less transparent than historical burn calculations and requires insurance partners to be familiar with the assumptions and weaknesses of the models, it was implemented in the form of an educational software tool. It was packaged along with the other analysis software in the online Weather Index Insurance Educational Tool (WIIET, http://iri.columbia.edu/WIIET) so that implementing groups would have full access to the contract design and pricing analysis tools, and capacity could be built for local design and pricing in these types of projects. We use that software for our illustrative example of the value of information in insurance.

Initial pilot sites were selected based on the availability of data. However, for scaling of the project, typical sites must be considered even if they have fewer data. When correctly including the value of information in the premium, many sites with fewer data may have premiums that are substantially more expensive. However, if these premiums are still workable, then the insurance can still be a valuable product. In the insurance-loan-input package, the insurance cost was approximately \$2, the input cost was approximately \$25, and the interest on the loan was around \$7. The insurance price included 17.5 % tax on the premium. The value of the crops at the end of the season was typically about three times the cost of the inputs (Osgood et al. 2007).

Many farmers were so severely constrained in their input availability prior to the project that doubling or tripling of yields was reported. For this production system, an increase in the premium from \$2 to \$4 or even \$5 may still lead to a useful product if it unlocks dramatic production gains. In addition, the premium calculations will allow the government to understand where the generation of improved information is worthwhile (e.g., investments in recovering and digitizing manual rain gauge recordings). The key is calculating the correct premium so that the prices reflect the true value of information and appropriate trade-offs can be made.

Tables 1.1 and 1.2 are raw output from the WIIET software, applied using the Malawi 2006 implementation data and parameters for a maize contract in the capital city, Lilongwe. To illustrate the typical information problem, we present the full 44-year data set and compare it with only the last 9 years, reflecting what might be available at a marginal scale-up site. Table 1.1 presents the historical burn output. The historical data were used for the index formula to calculate payouts and

[5] In addition to the importance of addressing purely statistical pricing issues, it is worthwhile to address physically based processes, such as climate change. This is discussed in Sect. 1.3.

Table 1.1 Price calculations on historical data

	1961–2005 data	1996–2005 data
prem.cash.98 %	86.366	84.181
prem.frac.98 %	0.086	0.084
v_at_var.98 %	484.840	410.800
mean.pay	60.932	63.333
var.pay	20268.577	24350.000
max.pay	576.000	470.000
num.years	44.000	9.000
num.pays	11.000	2.000

Table 1.2 Price calculations with modeled rainfall accounting for information quality

	1961–2005 data	1996–2005 data
prem.cash.98 %	132.27	175.20
prem.frac.98 %	0.13	0.18
v_at_var.98 %	702.05	785.23
mean.pay	95.90	136.26
var.pay	33641.46	48750.91
max.pay	1000.00	1000.00
num.years	989.00	999.00
num.pays	360.00	452.00

the pricing equation was applied. In this table, the maximum liability is set to 1,000 (Kwacha) for illustrative reasons.[6] The premium is expressed as a cash value as well as percentage of the maximum liability. The estimate of the 98th percentile payment is presented as the v_at_var.98 %, as well as the mean payout, the payout variance, the maximum payout, the number of years in the data set, and the number of nonzero payouts. It can be seen from the table that the simple historical burn pricing does not reflect information differences between the two data sets. That is, even though we should be less certain about the rainfall process from using only 9 years of data as opposed to 44 years, the historical burn analysis isn't sensitive to the amount of data used to price the contracts, so the resulting contracts are very similar.

Table 1.2 presents WIIET software output using the module that statistically models rainfall. This model is designed to address some of the statistical issues inherent in pricing using only a single series of historical data. It estimates parameters of a statistical model for rainfall based on the observed data, and when the model is used to simulate additional realizations of rainfall, the simulations are sensitive to the amount of data used to fit the model in the first place. In other words, the model and its associated rainfall simulations account for (1) the natural variation in rainfall (which would still be substantial even if we had exact knowledge of

[6] In 2006, 145 Kwacha was worth about $1, and typical maximum liabilities were approximately 4,000 Kwacha, depending on the specific input package insured.

the true data-generating process); and (2) the uncertainty associated with the parameters of the model, which adds extra variability to the simulated realizations.[7]

We stress that estimating the parameters of the rainfall model without measuring the uncertainty of their estimates does not lead to sensible simulations. To appropriately account for the amount of information observed, it is necessary to preserve the uncertainty in the estimation of the statistical rainfall model parameters. As with any statistical inference, when the parameters of the rainfall model are estimated, the estimates are accompanied by standard errors, which reflect how confident we are in the accuracy of our estimates. For a short rainfall time series, the standard errors will most likely be larger, reflecting less information. As the number of years of observed rainfall increases, standard errors will tend to decrease, reflecting the higher accuracy of estimation. The standard errors of the parameters therefore reflect the set of possible models that may be the true process, which nevertheless cannot be determined based on the available information. In the context of index insurance contract prices, we are most concerned with standard errors related to the estimate of the average payout and the 99th percentile of the payout distribution.

In summary, to price contracts that are appropriately sensitive to the amount of observed data on which they are based, one uses the standard errors. In essence, one first draws from the error distributions to generate a set of parameters that could describe the rainfall-generating distribution. Then, one draws from the distribution of rainfall itself. In this way, both the variability of the rainfall and the amount of information about the variability of rainfall are captured.

The statistical rainfall model in the WIIET software performs this process, reflecting both the variability of rainfall and the amount of statistical information in its generated realizations using a Bayesian statistical model (see WIIET user guide, at http://iri.columbia.edu/WIIET). In Table 1.2, the difference in information between the short 1996–2005 data series and the longer 1961–2005 data series is evident in the substantially increased variance as well as the higher estimate of the size of the 98th percentile event. The average payout is also higher because payouts are due to rainfall levels in the lower tail of the distribution, which is influenced by the increases in variance due to model uncertainty. The frequency of payouts also increases because a higher proportion of years may have sufficiently extreme rainfall levels to trigger payouts.

According to the calculations, the premiums would have to increase approximately 30 % to reflect the reduced level of information if only the past 9 years had been available, compared with the full 44-year data set. The value of the additional information in the longer data set was approximately 30 % of the premium. Although this is a substantial increase in cost, increasing premiums from approximately \$2 to \$3, it is relatively small compared with the other costs in the production

[7] The simulation is set to generate the number of realizations that would most closely sum to 1,000 years of total years generated. This size was selected for feasibility of computation on a web server in a classroom environment.

package (totaling a little over $30) and very small compared with the improvements in the value of farmers' production.

It is also interesting to compare the two tables to see how failing to account for the cost of less information can lead to artificially low premium levels. The insurance was priced using the 98th percentile value at risk and provided payouts approximately 25 % of the time. Because of this, one might expect substantial uncertainty about the mean and VaR of the payouts even with the full 44 years of rainfall data. One can see that including the value of information in the premium calculations leads to more than a 50 % increase in the insurance price for the 44-year data set and an approximate doubling of the price for the 9-year dataset. Therefore it is clear that information can be very valuable in this context, having a value roughly equal to the insurance purchased by Malawian farmers in 2006.

Improving the quality of information is not the only method available to reduce the costs of uncertainty. It is often possible to make products that are less vulnerable to uncertainty, reducing its costs. If insurance payouts are limited to a maximum liability that occurs frequently in the historical data set, the existing information may be sufficient to characterize the distributions effectively. If this restricted insurance product is valuable, then the farmer may save a substantial amount of money over the product with infrequent large payouts. For products such as life insurance, it is unlikely that this restriction would be workable. However, for insuring the costs of agricultural inputs and associated loans, it may be that full payments approximately 10–20 % of the time might be useful. This might be particularly true in situations for which the 1- in 100-year event would be a catastrophe so severe that massive government intervention might be more appropriate than having low-income smallholder farmers finance their own disaster relief. These are insights that the Malawi implementation has provided to help improve microinsurance projects.

1.3 Satellite Information in Index Insurance

One project informed by the experience of Malawi is the 2009 HARITA insurance pilot in Ethiopia, in the village of Adi Ha. For this project, the goal was to develop an insurance product that could be easily implemented in the typical data-poor context faced across much of Africa. A central goal of the project was to build a robust and transparent process for installing a new rain gauge and phasing insurance in at that gauge, once the information was sufficient to derive workable premiums. A site was selected for which there was no official historical rainfall measurement available. Instead, a set of informal, short-length (7-year) datasets had been collected by local extension personnel. Official datasets existed, ranging from about 10 years of data to nearly 50, but those were for sites that were dozens of kilometers away and may have had different amounts of rainfall. In addition, given climate change, it was important not only to reflect the statistically based value of information in the products, but also to account for uncertainty in physical processes such as long-term anthropogenic climate change and the decadal climate variations characteristic of the region (Hellmuth et al. 2009). It was therefore valuable to increase the capabilities of the rainfall models to make sure that they reflected these additional uncertainties and trends.[8]

One year prior to the insurance transaction, the project installed a new, automated weather station on the site and began collecting data. To design the insurance contract and determine thresholds and approximate prices, it was necessary to have some historical information. A satellite estimate of rainfall was deemed the best representative source of information for design because the other sources at the site had not been operating long enough to capture years that were known to be droughts, and the official rain gauges from other locations were known to have somewhat different seasonal timing and average amounts of rainfall.

The index structure was simplified to reduce its vulnerability to errors in the satellite information and short data sets. The rainfall level required for maximum payments was set such that at least one full payment would have happened within the past 15 years, and that most payments would be substantial compared with the full payment. The contract details were repeatedly verified by an elected farmer design team for agreement with vulnerable times of the year, for drought years, and for the timing of the drought during a dry year.

Remote sensing was used in the index design. Remotely sensed vegetative greenness measures, such as Advanced Very High Resolution Radiometer Normalized Difference Vegetation Index (available since 1981) were used to verify whether drought years were evident. Regional greenness measures have been used as indexes in other insurance projects, such as the 2007 Millennium Village Project insurance transaction in Kenya and Ethiopia, an International Livestock Research Institute livestock oriented project in Kenya in 2009, and an ongoing U.S. Department of Agriculture Risk Management Agency livestock product. The application of these products is challenging, since the vegetation observed by the satellite is often not the crops but instead surrounding trees and grasses. In addition, many crops can be green even when they produce little grain. Also, variations due to solar angle, dust, different sensors, and satellite angle may be as large as variations due to drought. More modern satellites address some of these problems through additional spectral bands. However, because these improved satellites have limited data sets that do not extend very far back in time,[9] they are used for validation and understanding the level of information that they provide, typically in situations for which the contract holder has a broad range of risk management options to address shortcomings in the remotely sense data. For example, in the Millennium Village Project, the remote sensing index was purchased by the development project itself, rather than by the individual farmers. The strategy was for the project to use the index in responding to farmer development issues during drought years.

[8] This is a nontrivial challenge, and the development of formal models is currently still in process.

[9] See the technical annex to Hellmuth et al. (2009) at http://iri.columbia.edu/publications/id=1008

Another satellite product use in index insurance is remotely sensed estimates of rainfall. This output has been used much less extensively than vegetative greenness, typically to validate ground measurements or to assist in index design. This is most often performed by using satellite measurements of the temperature at the top of clouds to estimate rainfall.[10] This technique is much more effective at determining whether rain fell than estimating the actual amount, and the quality of estimates is limited by the quality of the ground information used for calibrating the rainfall prediction models. These measures often have relatively high levels of error for daily rainfall but are much more accurate when used to determine average rainfall over a month or so.[11] This phenomenon occurs for many data sources, including ground measurements at different locations. Although daily rainfall between two points or data sets may differ, over time, the differences average out. The Adi Ha indexes were therefore designed to be simple sums over a 1–2-month period[12].

The National Oceanic and Atmospheric Administration Climate Prediction Center Africa Rainfall Climatology satellite rainfall estimate was used as a starting point for the design of the index. This data set has a historical record of 15 years. Although limited when compared with a 50-year data set, when used with the simplified index strategy, this record was long enough to observe several major payouts, including full payouts. Therefore, the satellite record provided basic information necessary for index design, provided the costs due to the value of information were not prohibitive.

The benchmark index was developed using the ARC data set and compared with farmers' reports. The index was refined to obtain the best agreement. In addition, the official ground rainfall measurements were ranked by year in terms of rainfall during critical times in the growing season, and the vegetative remote sensing measures were ranked by year in terms of greenness following the critical rainfall periods. Indexes were evaluated in terms of how well the payout years were reflected in the relevant lower quantiles of the other data sets. Typical agreement was between 60 and 70 % of the payout years, with some datasets agreeing completely in annual ranking. For the first years of the Adi Ha implementation, this process was performed manually. More rigorous statistical models to combine the information in the different measures and to quantify the level of uncertainty are currently being developed and evaluated.

The HARITA project followed the Malawi implementation by a couple of years, and the sophistication of the agricultural microinsurance industry had grown.

[10] More modern satellites use additional information, but their coverage is limited and does not extend very far back in time.

[11] See http://iri.columbia.edu/publications/id=1008

[12] There was one additional feature to these simple contracts. In order to assure that rainfall must be relatively uniform over the contract period, each ten day period had a cap, above which additional rainfall was not included in the total. In this way, a two month period of drought can still trigger the index payment, in spite of a single large rainfall event at the end of the contract period.

Reinsurance had become standard for index insurance pilots, and the focus on simplicity in pricing as calculated by project partners had shifted toward more accurate prices negotiated between the insurer and reinsurer, informed by data and technical analysis provided by project partners.

The initial plan was to have the satellite-based index priced for the new automated station that had been installed the prior year, based on the data available during the past year. It was assumed that smallholder farmers would not be comfortable with using satellite observations in an insurance product, and that the satellite product might not have the necessary accuracy compared with a ground measurement.

After its pricing analysis, the reinsurance company said that the price of a product based on the ground observations would be extremely high if there was only a single year to link the two datasets because of the excess costs associated with the very limited information. The company reported that an index triggered directly by the ARC satellite estimates of rainfall would have a much lower price because much more information was available. The farmers and project partners discussed the options and relative prices. It was decided that it would be worthwhile to use the satellite-only product in the first year, with a goal of transitioning toward the ground measurements once sufficient ground information had been gathered. This was the first time that smallholder farmers had been directly offered a product based on satellite estimates of rainfall.

Following the 2009 contract period, it was found that the satellite estimates were within a few percentage points of the rainfall measurements at the new station and were also within a few percentage points of the rainfall measurements made manually by the farmers themselves. One benefit of the remote sensing product is that it reflects the average rainfall over the region covered by the insurance, as opposed to the amount that falls only where the official rain gauge is located. The satellite observations are available for a much wider range of sites than the ground-based measurements. Also, the satellite information was less vulnerable to tampering or missing data.[13] In follow-up surveys, approximately 90 % of the farmers reported being comfortable with the satellite product (Peterson 2009). A decision was made to pursue a strategy in which the satellite estimates of rainfall would be the primary source of data for the index so long as the historical satellite rainfall was validated through ground measurements, farmer interviews, and satellite greenness observations.

1.4 Conclusion

We have illustrated a new, concrete value of information to African farmers through its reduction of index insurance premiums. Using the data and software from index insurance implementations in Malawi and Ethiopia, we have provided illustrations of

[13] Some of the 2009 data from the newly installed rain gauge were lost because of equipment failure.

this particular component to the value of information and have discussed the value of multiple information sources, such as remote sensing, in insurance.

Much future work remains. For projects such as HARITA, it may be that instead of transitioning to contracts using the ground-based measurements, secondary contracts could be purchased. These secondary contracts could provide a payment in the case where the satellite estimates differed from the measurements of a particular station by a predefined amount, protecting the farmer from dramatic errors in the satellite estimates without providing the full cost of insurance on the gauge. To develop this transaction, a statistical model is needed to quantify the probabilities that the satellite and rain gauge would differ, as well as the uncertainty in these probabilities. In addition, these models would be valuable in making sure that rainfall and climate uncertainties reflected in other sources but not in the satellite data could be used to increase the level of uncertainty in the rainfall modeling. Similarly, these models might be able to reduce the uncertainty in a relatively short but high-quality satellite data series using information from lower-quality but longer satellite products or ground measurements, leading to lower premiums.

The HARITA choice to rely primarily on the satellite data has raised several new questions related to the value of information. For example, there are efforts underway to have the Ethiopian national meteorological agency use its proprietary rainfall records to arrive at an ARC-like product with improved calibration and a doubling of the length of the historical record, to 30 years. This project requires funding. If the information allows premiums to be reduced substantially, that may itself show the new information to be sufficiently valuable to fund the work.

The value of information in the premiums also affects decisions about the installation of new rain gauges on the ground. If their contribution to the information can be systematically modeled, new rain gauges can be strategically located to have the highest value, and the number of expensive new stations to be installed (and maintained) can be determined. The costs of digitizing paper-based historical records can also be weighed against the value of their information in insurance premiums, as well as their value for other applications. Advances in remote sensing of vegetation can be used to validate information from other sources and flag the areas where remote sensing of rainfall or ground measurements do not adequately reflect vegetative changes.

Additional issues have arisen in the HARITA project. During the 2009 implementation, about one quarter of the insurance price was due to uncertainty about climate change. The reinsurance company observed a nonsignificant negative trend in the rainfall data for the 15 years. Although the 15-year dataset was not sufficient to determine whether this trend was spurious, real, or the result of a natural decadal process, the reinsurer held some additional resources to be able to provide payouts in case the trend was real. Rainfall models that could incorporate the physical factors connected with types of trends could allow for less conservative reserves (and therefore lower-priced contracts).

Finally, there may be scope for additional work on the strategic use of information. Contract theory work on incentive-compatible reporting and auditing may be of value for index insurance and remote sensing. In locations where long ground-based datasets are available, there are still concerns that people might tamper with the rain gauge to obtain a payout. Remote sensing information might be used to audit ground-based information even if not of the same level of accuracy. With the appropriate mechanism, the remote sensing merely needs to be accurate enough to credibly signal that people are likely to be caught if they tamper with the ground observations. Similarly, as more farmer observations are used to validate remote sensing estimates, incentives may arise to distort the information obtained. Truth-telling mechanisms (related to those in Sheriff and Osgood 2010) may provide incentives for farmers to measure and report rainfall as accurately as possible.

1. Commentary: Informational and Institutional Challenges to Providing Index Insurance for Farmers

Michael Toman[14]

Managing weather-related risk has been a long-standing challenge in Africa. Poor farmers are especially vulnerable to unexpected weather-induced crop damages or failures because agricultural output plays such a large role in family consumption, and alternative income generation opportunities are limited. For crop insurance to be effective and affordable, the pool of insured farmers needs to be large and dispersed enough that weather conditions across participating farmers are not highly correlated. Because insurance is not a familiar product, however, initial reluctance to purchase it needs to be overcome, in particular by providing credible guarantees that payouts actually will occur once premiums are paid.

Adding to those challenges are the difficulties that are the focus of this chapter. Because decisions by individual insured farmers on protecting their crop yields are difficult to observe, any insurance contract based on measures of farmer-specific loss would be prone to misrepresentation, moral hazard (farmers would reduce their own protective measures), and adverse selection (those less capable of protecting themselves, and thus more costly to cover, would be more likely to buy the insurance). The chapter highlights how insurance coverage based on movement of a general index of weather conditions correlated with individual farm yields can provide reasonably effective coverage without these problems. The analysis is informed by several innovative, controlled field experiments in two African countries. The discussion of this method of analysis is itself an important contribution of the chapter.

A firm offering weather index based crop insurance still faces the challenges of assessing the risks to which its portfolio of policies is exposed, and pricing the insurance coverage accordingly so as to reduce to a minimal level the probability that large contemporaneous claims could exceed its financial reserve. It is in this context that the chapter explores how strategies to improve information about index insurance risks can have value for both the insurance company and its customers. Important findings of Osgood and Shirley include these:

- High uncertainties about payout probabilities can significantly increase index insurance cost. Such uncertainties are common in the context of drought risks, for example, given limited information and modeling available for predicting their occurrence. This presents a challenge for establishing financially sustainable premiums—low enough to be affordable yet actuarially sound.

[14] The views expressed here are the author's alone and should not be attributed to the World Bank Group or its member countries.

M. Toman (✉)
Development Research Group, World Bank, Washington, DC, USA
e-mail: mtoman@worldbank.org

- Simulation models for assessing risks are an important complement to limited observed data on droughts for assessing payout probabilities. It turns out that assessments based on past patterns alone can be very inaccurate and are not very sensitive to changes in information, since new information can only marginally alter the patterns implied by a historical data set. A modeling approach can be useful for exploring how future risks might be altered by climate change. Satellite-based information also can be very useful to improve confidence in probability estimates.
- Index insurance is not a substitute for reducing vulnerability. Drought index insurance addresses only one component of risk, so in the absence of effective insurance for other risks, such as storm or flood damage, complementary farm-level measures still will be needed for reducing vulnerability. Individual actions to reduce vulnerability to drought risks is a cost-effective complement to insurance coverage. An example is output diversification, including crops and livestock, so that planned allocations of land to different products can be modified based on predicted conditions. Here again, earth observation systems that provide better forecasts for an upcoming growing season, and more timely information about emerging threats, can be very valuable.

Osgood and Shirley are careful to note that even though improved information for assessing risks can make index insurance a better value and thus more easily marketed, it is not a sufficient condition for successful introduction of crop insurance. In light of the persistent difficulties encountered in establishing financially sustainable markets for this insurance, it may be useful to highlight some other important considerations that could even preclude the successful introduction of insurance in some circumstances.

- Constraints on liquidity limit the ability of farmers to purchase the insurance, even with modest premiums. This is an especially important consideration if farmers also have used microloans to help finance their current cultivation activity, in which case premiums to cover both farmer and lender may be considerably higher.
- Risk aversion toward using a novel product can decrease demand for insurance, even if improved probability assessments lower the cost. On the other hand, since index insurance is inherently only partial coverage, it is important that potential customers appreciate this. As illustrated by the field experiments underlying the analysis in the chapter, potential customers may require considerable information and education to evaluate the potential advantages of insurance.
- The prospect of climate change inherently reintroduces "noisy priors" for how crop risks may evolve over time, given the degree of quantitative uncertainty about climate change impacts. If crop loss insurance comes into greater use to reduce impacts of short-term climate variability, what adaptive measures by farmers are needed to reduce vulnerability to effects of climate change over the longer term?
- Public policies can weaken the development of an effective insurance market in several ways. For example, to what extent would expectations that the government will continue to provide disaster assistance reinject moral hazard into the

insurance system? Prospective purchasers also will be concerned about the strength of policies to ensure the creditworthiness of the insurers, a common concern in the financial sector of many developing countries. Ultimately, policy-makers need to consider what portfolios of risk mitigation policies can have the greatest impact for a given resource cost. In addition to improved information about risks, such measures could include reducing institutional barriers to accessing insurance, and supporting measures by farmers to reduce their own vulnerability—which will also provide collective benefit by lowering economy-wide risks.

References

Barrett, C., Barnett, B., Carter, M., Chantarat, S., Hansen, J., Mude, A., Osgood, D., Skees, J., Turvey, C., & Ward, M. (2007). *Poverty traps and climate and weather risk limitations and opportunities of index-based risk financing* (Technical Report 07-03). New York: International Research Institute for Climate and Society (IRI).

Bryla, E., & Syroka, J. (2009). *Experiential briefing note. Micro and meso-level weather risk management: Deficit rainfall in Malawi* (Technical Report). Washington, D C: World Bank ARD.

Cabrera, V., Fraisse, C., Letson, D., Podesta, G., & Novak, J. (2006). Impact of climate information on reducing farm risk by optimizing crop insurance strategy. *Transactions of the American Society of Agricultural and Biological Engineering, 49*, 1223–1233.

Carriquiry, M., & Osgood, D. (2008). *Index insurance, probabilistic climate forecasts, and production* (Technical Report, CARD Working Paper 08-WP 465). Ames, Iowa: Center for Agricultural and Rural Development.

Chen, J., Dinku, T., Holthaus, E., Lyon, B., Madajewicz, M., Mullaly, C., Norton, M., Osgood, D., Pascualini, A., Pant, S., Peterson, N., Saavedra, R., Sharoff, J., Shirley, K. E., Small, C., Stiffel, S., & Teh, T.-L. (2010). *IRI technical report 10-08: HARITA IRI report to Oxfam America. Final report for IRI MIEL planning and technical support for HARITA micro-insurance pilot USA 536 / 09* (Technical Report 10-8). New York: International Research Institute for Climate and Society (IRI). http://iri.columbia.edu/publications/id=1002

Dinku, T., Giannini, A., Hansen, J. W., Holthaus, E., Ines, A., Kaheil, Y., Karnauskas, K., Lyon, B., Madajewicz, M., Mclaurin, M., Mullally, C., Norton, M., Osgood, D., Peterson, N., Robertson, A., Shirley, K., Small, C., & Vicarelli, M. (2009). *Designing index-based weather insurance for farmers: In Adi Ha, Ethiopia* (IRI Technical Report 09-04). New York: International Research Institute for Climate and Society (IRI).

Hazell, P., Anderson, J., Balzer, N., Clemmensen, A. H., Hess, U., & Rispoli, F. (2010). *Potential for scale and sustainability in weather index insurance for agriculture and rural livelihoods.* Rome: International Fund for Agricultural Development and World Food Programme.

Hellmuth, M., Osgood, D., Hess, U., Moorhead, A., & Bhojwani, H. (2009). *Index insurance and climate risk: Prospects for development and disaster management* (Number 2 in Climate and Society). New York: International Research Institute for Climate and Society (IRI), Columbia University.

Luo, H., Skees, J., & Marchant, M. (1994). Weather information and the potential for intertemporal adverse selection in crop insurance. *Review of Agricultural Economics, 16*, 441–451.

Meza, F. J., Hansen, J. W., & Osgood, D. (2008). Economic value of seasonal climate forecasts for agriculture: Review of ex-ante assessments and recommendations for future research. *Journal of Applied Meteorology and Climatology, 47*, 1269–1286. doi:10.1175/2007jamc1540.1.

Osgood, D., McLaurin, M., Carriquiry, M., Mishra, A., Fiondella, F., Hansen, J., Peterson, N., & Ward, N. (2007). *Designing weather insurance contracts for farmers in Malawi, Tanzania, and*

Kenya: Final report to the Commodity Risk Management Group, ARD, World Bank (IRI Technical Report 07-02). New York: International Research Institute for Climate and Society (IRI).

Osgood, D. E., Suarez, P., Hansen, J., Carriquiry, M., & Mishra, A. (2008). *Integrating seasonal forecasts and insurance for adaptation among subsistence farmers: The case of Malawi* (Technical Report WPS4651). The World Bank Development Research Group Sustainable Rural and Urban Development Team. Washington, D.C., USA: World Bank

Peterson, N. (2009). *Livelihoods, coping, and micro insurance in Adi Ha, Tigray state, Ethiopia: A report on surveys and interviews conducted in June and August 2009* (Technical Report). Boston, MA, USA, Report for Oxfam America.

Sheriff, G., & Osgood, D. (2010). Disease forecasts and livestock health disclosure: A shepherd's dilemma. *American Journal of Agricultural Economics*. doi:10.1093/ajae/aap042.

Skees, J., Hazell, P., & Miranda, M. (1999). *New approaches to public/private crop yield insurance* (Technical Report). Washington, DC: World Bank.

Chapter 2
The Value of Information in a Risk Management Approach to Climate Change

Carolyn Kousky and Roger M. Cooke

Abstract The standard economic approach to analyzing the climate change problem has been to search for efficient abatement policies. The massive uncertainties and the possibility for cataclysmic climate damages, however, suggest that a risk management approach is more appropriate. This shifts the policy question to how much risk of catastrophe society is willing to accept. Intuitively, this change in focus may shift our information needs, and the needs should be assessed through a value-of-information analysis. Such calculations should allow for examination of how improved information alters the probability of exceeding a given policy target, incorporate rigorous expert judgment for determining beliefs and quantifying uncertainties, and highlight what scientific information is most valuable for a policymaker attempting to keep the probability of catastrophic climate impacts below a set threshold. We discuss how Bayesian belief nets can be useful tools for this type of analysis.

Keywords Bayesian belief nets • Catastrophic climate change • Climate policy • Integrated assessment • Risk management • Value at risk • Value of information

2.1 Introduction

An early, and still predominant, economic approach to climate change has been to treat it as the challenge of pricing a large, global externality. The focus has been on economic efficiency and determining optimal emissions trajectories using integrated assessment models (IAMs) where the avoided damages of climate change could be

C. Kousky (✉)
Resources for the Future, Washington, DC, USA
e-mail: kousky@rff.org

R.M. Cooke
Resources for the Future, Department of Mathematics, TU Delft, Delft, The Netherlands
e-mail: cooke@rff.org

R. Laxminarayan and M.K. Macauley (eds.), *The Value of Information: Methodological Frontiers and New Applications in Environment and Health*,
DOI 10.1007/978-94-007-4839-2_2, © Springer Science+Business Media Dordrecht 2012

compared with abatement costs. Given the possibility of catastrophic impacts and the myriad uncertainties surrounding climate change, we, like several other authors, argue for a risk management approach, not an efficiency approach.

A risk management approach asks what policy should be, given the large range of possible outcomes from that choice. This is quite distinct from asking what the optimal policy is under different assumptions of our uncertain variables. Drawing an analogy to risk management in the insurance and financial sectors, society may wish to keep the probability of facing catastrophic damages to some determined low level. This change in focus to a risk management paradigm dramatically shifts our information needs. A risk management approach highlights the need for research on the possibility of climate catastrophes, their likelihood under various emissions scenarios, and whether we can detect impending catastrophes soon enough to avert them.

There are many places where we can improve our climate information to improve climate risk management, raising the question of where to spend scarce research dollars and when it is worth waiting for better information. Basic value-of-information models make clear that information will be valuable only when (1) the possible policy options perform quite differently in different states of the world; (2) our current beliefs would lead us to pick an option that is worse than what we would do with better information; and (3) we can undertake some measurement, the result of which would shift our belief substantially enough to change our preferred policy.

Bayesian belief nets (BBNs) can be used to take a distinctly risk management approach to the task of valuing improved climate information. BBNs are graphical models of the dependencies between multiple variables. They can be used to calculate how improved information on one or more parameters would change the estimated probability of meeting a given policy target, as well as how improved knowledge alters welfare estimates. BBNs can also incorporate expert judgment for determining beliefs and quantifying uncertainties and highlight what scientific information is most valuable for a policymaker taking a risk management approach, as opposed to an efficiency approach, to the climate problem.

The next section introduces what we consider to be a risk management approach to climate change. Section 2.3 offers an overview of a basic value-of-information framework. In Sect. 2.4 we move on to discussing how BBNs can be used to conduct a value-of-information analysis for the climate problem while incorporating a distinctly risk management flavor of analysis. Section 2.5 concludes.

2.2 A Risk Management Approach

As is known in finance, an increase in expected rewards usually carries with it an increase in risks. Prudent firms in the banking and insurance industries often manage the risk of insolvency using a value-at-risk (VaR) approach. A firm chooses a target solvency probability (or one is set for it through regulation) and then ensures that the risk of insolvency does not exceed this target, through, for example, building capital reserves or reducing exposure. So too with climate change, the benefits of increased economic growth from a carbon economy carry with them risks of negative climate impacts, some of which could be quite catastrophic. When the uncertainties and nontrivial probability of catastrophic outcomes are recognized, it can change preferred policy choices; in these cases, some amount of abatement in the near term as a hedging strategy becomes optimal (see, e.g., Manne and Richels 1995; Lempert et al. 2000).

Following the VaR approach used in the private sector, society could choose to limit the risk of a climate-induced "insolvency." This would be some form of collapse in social welfare—a worst-case scenario whose probability should be kept beneath a defined tolerable level. The policy questions then become, first, what the worst-case outcome is we wish to avoid, and second, how much risk of such an outcome we are willing to tolerate. Regulations for the banking and insurance industries in the European Union dictate the solvency threshold for firms at around 1-in-200. We are currently taking much larger risks of large-scale climate damages than this.

In a risk management approach, then, fully assessing and clearly communicating the uncertainties become essential for policy. Too many studies conducted under an efficiency approach to the climate problem include the uncertainties as a caveat, and too many policymakers dismiss the uncertainties of modeling as fine print. In a world of climate risk management, the size and nature of these uncertainties and our attitudes toward risk determine the optimal amount of abatement today. This requires undertaking a complete uncertainty analysis with current climate models.

It is also the case that a risk management approach highlights different information needs. The correct discount rate becomes less important than an improved understanding of the nature of catastrophic consequences, their likelihood under differing emissions scenarios, our ability to detect tipping points before a catastrophe materializes, and the time frame for response should we pass such tipping points. Although we have some information on catastrophic impacts—for instance, numerous studies point to catastrophic consequences if global temperature exceeds 5 °C above preindustrial conditions, or even above 2.5 °C (Keller et al. 2008)—in general, we have a fairly poor understanding of the tail of the climate change damage distribution.

Satellite data are critical to this type of research. For instance, satellites can be used to document the trends that could be indicative of climate tipping points, such as melting of ice in Antarctica or the amount of methane in the atmosphere. They can also be used to look at effects as diverse as ocean acidification and desertification. This information, however, is under threat because the number of earth-observing satellites is declining, not increasing. Even rough model calculations of the value such information satellites provide in terms of detecting tipping points to avoid catastrophe could be useful for Congress when lawmakers consider appropriating more money to observation systems.

2.3 Value-of-Information Refresher

It is useful to recall a basic model of the value of information. Assume we can choose one of a set of available policy options, and that each option has a well-defined outcome with well-defined utility in each possible state of the world. If the future state of the world were known, we would simply choose the option that would generate the highest utility. Unfortunately, the state is not known, and so we must quantify our uncertainty and then choose the option with the highest expected utility, given our beliefs about the state of the world.[1] Now, suppose we have the opportunity to perform an observation before choosing a policy, which will produce information to alter our beliefs about the likely state of the world. The observation may incline us to choose a different policy than we would have chosen without the observation. A simple result in decision theory states that it is never disadvantageous to perform a cost-free observation before choosing. That does not mean, however, that it is always worth spending money to obtain more information. The value of information quantifies the expected gain of performing this observation, relative to the given set of policy options.

A simple example clarifies the basic properties of the value of information. Suppose we have to choose between three climate policies: (1) business as usual (BAU) with no abatement; (2) tempered abatement (a little now with the possibility of more later); and (3) maximal abatement now. Suppose for illustration that there are two possible states of the world: climate sensitivity (cs) $= 1.5$ and $cs = 5$. The value $cs = 1.5$ corresponds to the most sanguine value given in the Fourth Assessment Report of the Intergovernmental Panel on Climate Change (IPCC 4), and 5 is a very pessimistic value "which cannot be excluded" (IPCC 2007). BAU produces high utility if $cs = 1.5$, as no money is wasted on unnecessary abatement. It is catastrophic if $cs = 5$. The opposite holds for maximal abatement: it avoids catastrophe if $cs = 5$ but is very wasteful if it turns out that $cs = 1.5$. Tempered abatement is intermediate. Since in this simple example $\text{Prob}\{cs = 1.5\} = 1 - \text{Prob}\{cs = 5\}$, the expected utility of each policy is a linear function of Prob $\{cs = 1.5\}$, as shown in Fig. 2.1.

According to our assessed probability of the event $\{cs = 1.5\}$, one of these three options will be optimal. Figure 2.1 shows that for a belief point $\text{Prob}\{cs = 1.5\} = 0.66$, tempered abatement is optimal, but if the probability were a little higher, say 0.7, the preference would shift to BAU. Of course, we would like to know the true cs before choosing. If we could simply observe this number, then we would obviously choose BAU if $cs = 1.5$ and choose maximal abatement if $cs = 5$. Without knowing the outcome of this hypothetical measurement, we can compute the expected value of observing cs before choosing by drawing the thin dotted line in Fig. 2.1. The difference between this line and the maximum expectation of our policy options at our belief point is called the *value of perfect information*, for this belief point and these policy options.

[1] Much ink is spilled over whether we should choose according to the principle of maximal expected utility. We assume for the present discussion that the decider is a rational agent, in the sense of Savage (1954). A rational agent's preferences can always be decomposed into a unique probability over states of the world and an affine unique utility over consequences such that preferences are modeled as expected utility.

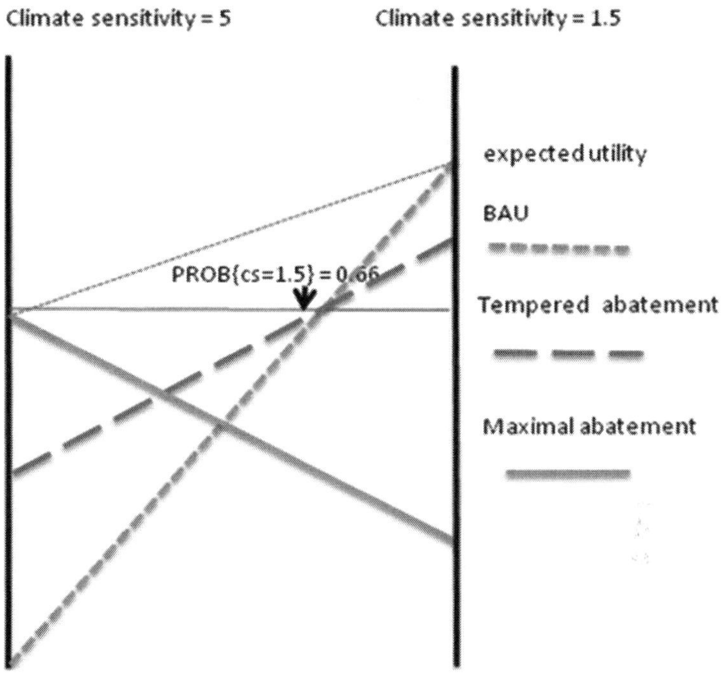

Fig. 2.1 Value of information for simple climate policy choice problem, with value of perfect information

Unfortunately, we are seldom afforded the possibility of performing a perfect observation. The best we can do in practice is find (costly) observations whose possible outcomes would alter our beliefs. Suppose scientists can undertake a study to get better but still imperfect information on the climate sensitivity. Keeping the example simple, suppose the possible outcomes of such a measurement are either HI or LO. Experts agree that if we observe LO, then the probability that $\{cs = 1.5\} = 0.75$, whereas if we observe HI, then the probability of $\{cs = 1.5\} = 0.3$. It is easy to calculate that Prob{outcome = LO} = 0.8.[2] If we observe LO, then we would choose BAU, whereas if we observe HI, then we would still choose tempered abatement, as shown in Fig. 2.2. The expected value of this observation is found by connecting the best choices for each possible outcome by the thin dotted line in Fig. 2.2. The value of this information in this problem is the difference between the thin dotted line and the value of the best option at $\text{Prob}\{cs = 1.5\} = 0.66$. This value is rather small because a HI value doesn't change our initial choice.

[2] The conditional probability $\text{Prob}(cs = 1.5 \mid \text{LO}) = 0.75$, and similarly $\text{Prob}(cs = 1.5 \mid \text{Hi}) = 0.3$. Solve $\text{Prob}(cs = 1.5) = 0.66 = \text{Prob}(cs = 1.5 \mid \text{LO})\text{P(LO)} + \text{Prob}(cs = 1.5 \mid \text{HI})(1 - \text{P(LO)})$ to find $\text{Prob(LO)} = 0.8$.

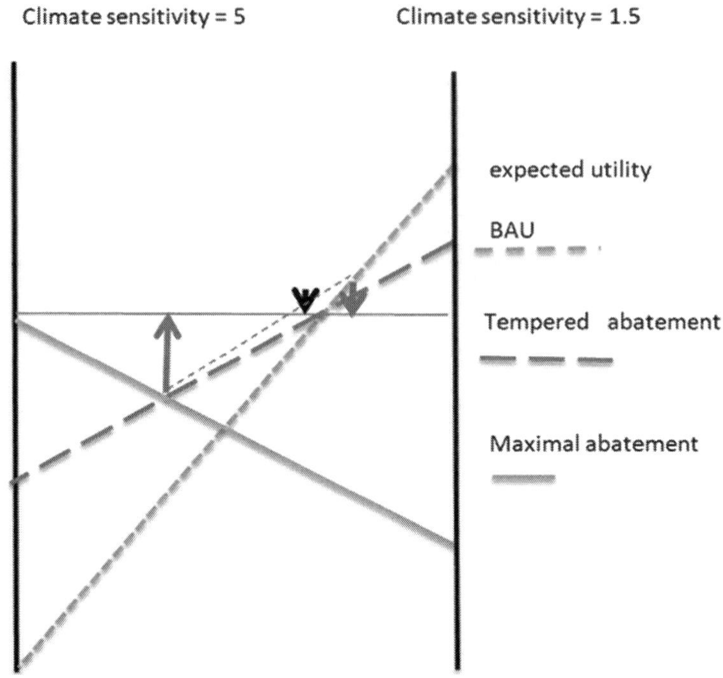

Fig. 2.2 Value of information simple climate policy choice problem with value of imperfect information

Our simple example demonstrates that for the value of information to be important, all of the following must obtain:

1. The set of available options is strongly concave in the sense that it consists of options that are very good in some states of the world and very bad in others, and options that are mediocre in all states of the world.
2. Our belief point leads us to choose an option that is much worse than what we would choose with perfect information.
3. There are observations whose possible outcomes would strongly influence our belief point.

2.4 A Risk Management Approach to the Value of Information

A risk management approach to climate change should translate through to value-of-information calculations. A risk management perspective suggests that value-of-information calculations should allow for examination of how improved information alters our estimate of the odds of meeting a given policy target. Furthermore, they should incorporate rigorous expert judgment for determining beliefs and quantifying uncertainties and should highlight what scientific information is most valuable for a policymaker taking a risk management approach, as opposed to an efficiency approach, to the climate problem.

We argue that Bayesian belief nets are useful tools that meet all three criteria. A BBN is a graphical model representing variables and their conditional probabilities. It allows for quantification of uncertainty in complex models of multiple variables. A simple example based on the IAM of William Nordhaus, DICE, is used, with distributions on three uncertain parameters. We model temperature-induced damages, $\Omega(t)$, at time t as a function of global mean surface Temperature, T(t), with uncertain parameter *dx*:

$$\Omega(t) = 1/[1 + \psi T(t)^{dx}]. \tag{2.1}$$

Temperature is a function of greenhouse gases (GHGs) and the uncertain climate sensitivity parameter, *cs:*

$$T(t) = cs \times \ln(GHG(t)/280)/\ln(2). \tag{2.2}$$

$\Omega(t)$ is a value between zero and one that scales down total output, Q, which is a function of abatement Λ, total factor productivity A (this is a parameter, evolving over time to capture technological change), capital stock K, and labor N, with uncertain Cobb-Douglas parameter gx:

$$Q(t) = \Omega(t)[1 - \Lambda(t)]A(t)K(t)^{gx}N(t)^{1-gx}. \tag{2.3}$$

Different policies are characterized by their GHG emissions: policy 1 involves the lowest emissions at highest abatement cost; policy 10 involves the highest emissions at lowest abatement cost. Greater abatement leads to reduced output.

This model as a BBN is shown in Fig. 2.3. The top three nodes represent uncertain variables in the model: the climate sensitivity *cs*, the exponent in our damage function *dx*, and the exponent in a Cobb-Douglas production function *gx*. We have assigned distributions to each of these variables.[3] One tenet of risk management is that these distributions should be assigned not in an ad hoc fashion by modelers (as we do here simply for purposes of illustration) but in a process of structured expert judgment. This involves transparently choosing a range of experts on the topic, familiarizing them with the study, allowing them to consider the problem and prepare a response, conducting a face-to-face interview, querying experts about measurable variables, querying experts about calibration variables,

[3] This model is meant simply to demonstrate the approach, and in that sense, the distributions chosen are somewhat arbitrary. We model the climate sensitivity as Beta distributed on [1, 15] with parameters (2, 24). The damage and production function exponent are both modeled as uniform variables, the first over [1, 3], and the second over [0.2, 0.4]. Part of the appeal of the BBN approach is that these distributions can be altered and the effects on output examined explicitly.

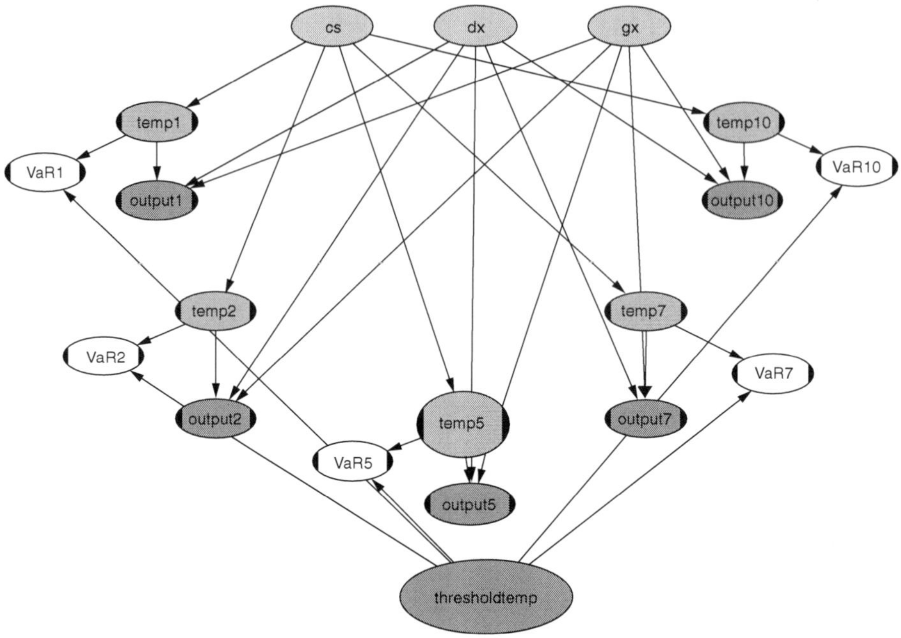

Fig. 2.3 Bayesian belief net for example climate model

and measuring performance on statistical accuracy and informativeness to aggregate judgments (Cooke and Kelly 2010). This process of expert judgment will allow for the best assessment of the uncertainties in the model.

The nodes labeled "output" in Fig. 2.3 represent output over the next 100 years under the five abatement policy options, which are shown in the Temp5 nodes. The arrows connecting the nodes represent defined relationships between those two variables. The thresholdtemp node at the base of the model allows for stipulating a threshold maximum temperature, and the model can then calculate the probability of exceeding this threshold for each policy option. When run, these probabilities would be shown in the White nodes, or the VaR nodes (see Fig. 2.4).

Figure 2.4 shows the result of running the model with temperature threshold set at 3 °C. The expected value of each variable ± the standard deviation is shown at the bottom of each node. The expected value of the VaR nodes is the probability of not exceeding the stipulated threshold. We see that the expected output and expected temperature increase as we move from option 1 to option 10, whereas the probability of staying below the stipulated temperature threshold drops. This reflects the fact, long obvious to investors, that increasing expected gain is coupled with greater risk. The first policy achieves our target 100% of the time, and the second policy achieves it 95.5%. By policy option 3, however, the target is met only about 39% of the time. If we defined 3° as our "collapse" point with a threshold of 5%, then only the first two policies would be deemed viable. We can see that output is higher under the second policy, as would be expected, since there are greater emissions.

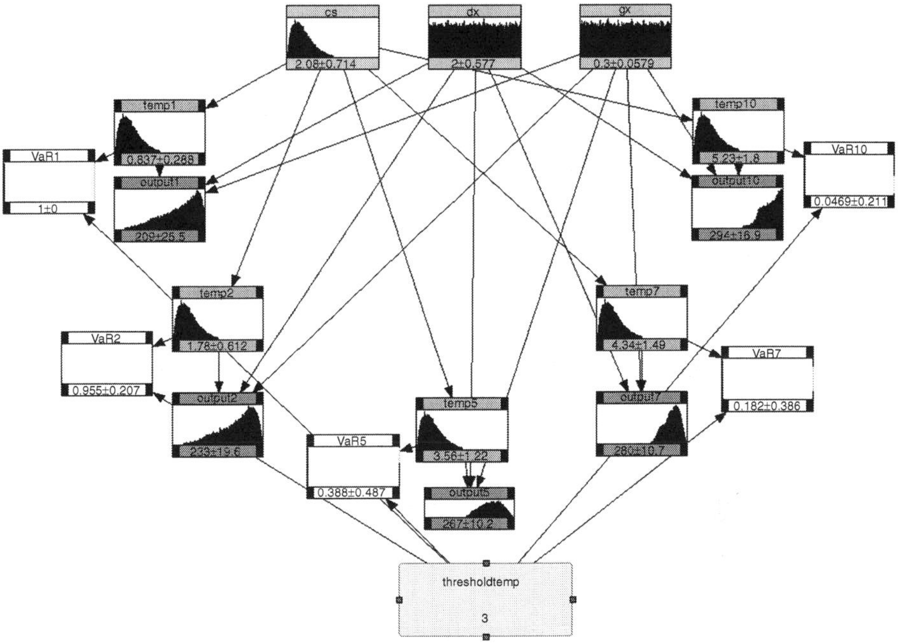

Fig. 2.4 Stipulating temperature increases to not exceed 3°

The BBN thus formalizes our uncertainty over particular parameters and, similar to the simple value-of-information model in the preceding section, will allow us to estimate the value of improved information on any of the uncertain variables. For instance, we can compare the distribution of output under the various policy options when climate sensitivity is modeled as an uncertain random variable and then compare this with the case when it is known with certainty or when its distribution narrows from improved information.

To illustrate, suppose we perform an imperfect observation on climate sensitivity, leading to the distribution shown in Fig. 2.5, with mean lowered from 2.08 to 1.2, and with narrower uncertainty. Now option 3 meets the risk management requirement of holding temperature below 3 °C with probability at least 0.95. The expected output of option 3 is 270. Without performing this observation, our best option meeting the risk management requirement was option 2 with expectation 233. The expected outputs in Fig. 2.5 are a bit higher than in Fig. 2.4, since the lower climate sensitivity leads to reduced damages for all options.

Once defined, the BBN can be sampled. Examining 1,000 such samples, displayed as a cobweb plot in Fig. 2.6, shows, as just one example, the relationship among climate sensitivity, temperature, and output, under the five policy options. The cobweb plot shows clearly that lower climate sensitivity values are associated with lower temperatures and higher output. Although this particular finding is obvious, it demonstrates the way in which a BBN can be used to explore the links among multiple variables. If a measurement could be taken to narrow the possible range for climate

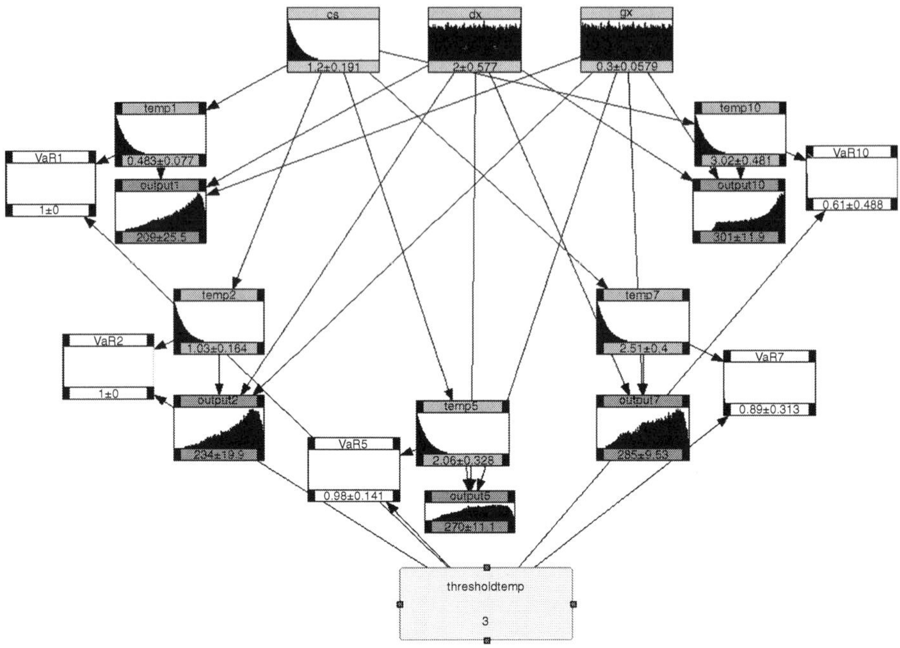

Fig. 2.5 Observation on climate sensitivity leading to shifted distribution

sensitivity, a value of that measurement can be determined by resampling our BBN with the narrowed range comparing output under the range of policies, as well as the probability that various policy options meet our threshold probability.

Finally, the BBN can help risk managers determine what type of information would be most useful and thus where best to direct scarce research dollars, or whether to invest in a particular research project. This can be done by comparing improved information on a variety of multiple uncertain variables. In this simple model, we only have three, but more complicated climate models would include a broader range of the uncertain variables. Our model would then let us uncover which types of information may be useful and which will not be. For instance, if catastrophic tipping points in the climate system are irreversible, detection is impossible, or detection would be too late for society to take action, Weitzman (2007) notes that the option value of waiting for more information would be zero. Thus, knowing when not to wait for more information and when not to invest in learning is just as important as knowing when to do so.

Note that to do this type of analysis effectively, we must clearly determine which uncertain variables are those over which we can undertake measurements to improve our knowledge and those where the uncertainty arises from other sources, such as differing value judgments. For instance, although there is uncertainty over the proper discount rate, this is at base a disagreement of values or opinion and cannot be resolved through better information.

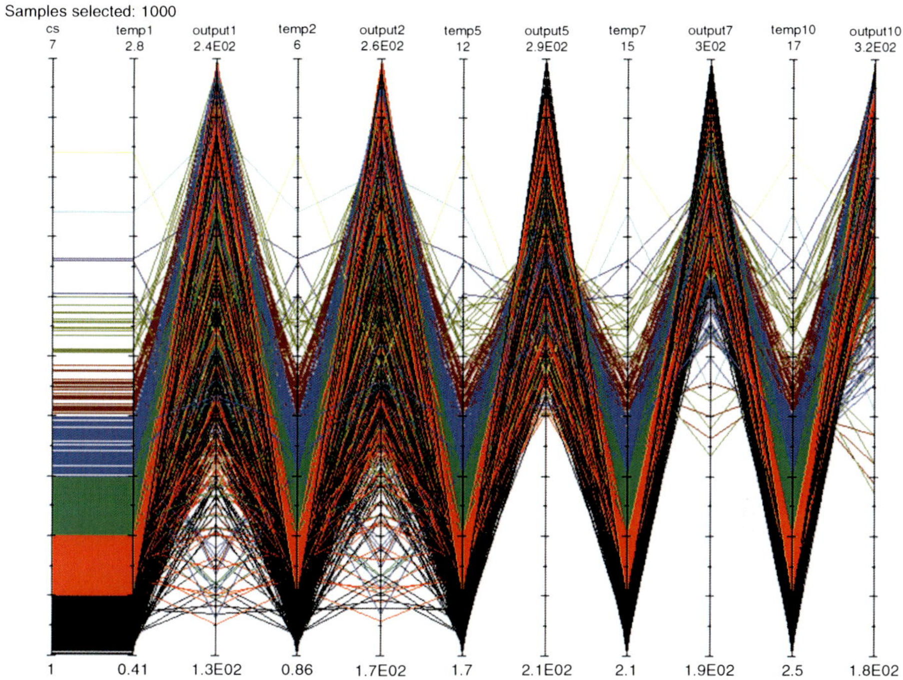

Fig. 2.6 Cobweb plot with 1,000 samples

The simple BBN used here represents just one climate model. It thus makes certain assumptions about the functional form of relationships among variables. From IAM modeling, however, we know that varying these assumptions can produce dramatically different outcomes. Fankhauser and Tol (2005), for example, observe that damages can affect capital depreciation, the utility function, the production function, and population growth. Which is chosen can create profound differences in predicted welfare for various policy choices. These differences can be addressed in one BBN by including different climate damage models.

2.5 Conclusion

Among the most challenging aspects of addressing climate change are the uncertainties and the possibility of truly catastrophic damages should we fail to abate sufficiently. Rather than neglecting these features of the problem, we suggest that a risk management policy approach be pursued, which would aim to keep the probability of reaching catastrophic damages below some tolerable threshold. Within this framework, improved information on some aspects of the climate problem will be more useful than other aspects. A simple model of the value of information suggests when improved information will be helpful—namely, when we have policy options that produce very different outcomes in different states of the world, when our current beliefs lead us to choose a policy we would not choose if we had better information, and when it is possible to learn information that would alter our beliefs substantially. Although these heuristics are useful, more sophisticated analyses of the value of improved climate information based on detailed climate models would help policymakers make improved decisions about where to invest in information, how much to invest, and when more research is even worthwhile.

Such calculations can be performed using BBNs. Translating climate models into this framework creates a visually intuitive model in which it is easy to stipulate risk management thresholds and observe the consequences of improved learning within such targets. We have presented a simple illustration here, but a true analysis would, of course, require a much more detailed model. It would also require the use of expert judgment to adequately characterize the uncertainties, as well as discussion with scientists to discover what uncertainties could be reduced through various investments in research.

2. Commentary: Valuing Information, Ascertaining Risk, and Setting the Target

Timothy J. Brennan[4]

In their contribution to this volume, Kousky and Cooke (this volume) offer a method, Bayesian belief nets (BBNs), as a way to acquire the relevant stochastic information under a risk management approach to climate policy. Under such an approach, the goal is to come up with a policy to meet a given target—for example, that the probability of a given level of warming is less than a tolerable maximum. KC present this in a unified framework, but I want to suggest that the three aspects of the discussion—valuing information, ascertaining risk, and setting the target—are separable issues, in the sense that the merits of the approaches to any of these aspects can be assessed independently of how we regard the others. To support this, I will look at these three aspects in turn, and then conclude with some observations on the use of aggregate expected utility in setting climate policy, whether under a cost-benefit or risk management framework.

2.C.1. The Value of Information: Some General Issues

That information is valuable is obvious; more interesting is why that requires investigation in a way that the value of hamburgers does not. Part of the problem is that markets may not work as well for information as they do for hamburgers. Three aspects of information bring this out. A first is the circular regress in the purchase of information. Applying to information the general principle that one needs to know about a product before one can formulate a willingness to pay for it implies that before one would buy information, one would need to be informed about what the information is (Arrow 1959, 10). Were that true, why would one need it?[5]

---

[4] I thank Molly Macauley for a long series of extensive discussions on the value of information; she should be held blameless for mistakes I persist in making. I am also grateful to Roger Cooke and Carolyn Kousky for discussions of issues raised in their chapter; they too are without responsibility for my errors.

[5] A related issue is the distinction between data, the raw material an instrument gathers, and information, after the data have been turned into useful knowledge through some theoretical interpretation in the chain between initial the gathering of those data and their potential use.

T.J. Brennan (✉)
Public Policy and Economics, University of Maryland–Baltimore County, Baltimore, MD, USA

Resources for the Future, Washington, DC, USA
e-mail: Brennan@rff.org

Two other aspects are intriguing because they are somewhat contradictory. One is that information, apart from any physical medium in which it may be embodied (book, DVD) or communicated (theater, computer), meets the economist's definition of a "public good." In that definition, the consumption or possession of information is "nonrivalrous," meaning that one person's having it does not disable others from having or it as well. Consequently, information should be available to all who place a positive value on it, which generally requires that the price be zero. This is the somewhat technical argument behind the aphorism "information wants to be free" (Brand 1985, 49).

However, if information is free, one has the problem of covering the cost of discovering and providing it. Until recently, information providers recovered these costs through a combination of embedding the information in tangible goods that were relatively costly to duplicate and intellectual property protections against unauthorized copying and resale. The ability to convert information into digital formats detached from tangible goods and easily copied on computers and transmitted through broadband networks has blown a substantial hole in these protections. The good news is that a lot of information is free, but the bad news is that the business models used to support information provision have become unsustainable.

In contrast, although information meets the economic definition of a public good, the value of information often depends on exclusivity—that it is not made available as a public good in practice, even if it could be in principle.[6] One obvious context is business strategy, where the value of information depends on the competitive advantage it conveys over relatively ignorant rivals. Information can be valuable in interactions with buyers or sellers, although in those cases one runs the risk that asymmetry in information between buyers and sellers could cause adverse selection, leading to a collapse of markets to the detriment of all. In such cases, the value of information can be negative: market participants as a whole would be better off without it.[7] But business is not the only context where information is valuable to the extent it is private. Academic researchers strive to limit access to information to protect the priority of discovery on which reputations depend. The second person to write "$E = mc^2$" on a piece of paper is likely unknown, probably forever.

[6] The full quote from Brand (1985, 49) is instructive: "On the one hand, information wants to be expensive, because it is so valuable. The right information in the right place just changes your life. On the other hand, information wants to be free, because the cost of getting it out is getting lower and lower all the time. So you have these two fighting against each other." A quarter of a century later, this is still true.

[7] A virtue of having health insurance purchased through businesses or the government rather than individually is that it creates an information firewall that keeps individual knowledge of health status from creating an adverse selection problem in health insurance markets, where relatively healthy people choose not to purchase insurance, increasing the expected costs of ensuring those who purchase insurance, perhaps to the point where the insurance market disappears altogether. When employees can choose their health care provider, that firewall comes down, and the costs of insurance to all can increase (Cutler and Reber 1998).

2.C.2. *Modeling the Value of Information*

Fortunately and understandably, KC focus not on the strategic value of climate information, but on its public good aspect. Moreover, they get around the paradox by invoking the assumption that the procurer of information knows that the information will resolve a question or reduce uncertainty regarding it, but does not know how that uncertainty will be resolved. They present this graphically, but it is useful to look at the value of information algebraically to see that their formulation applies not just to expected utility maximization but to risk management as well.[8] It also allows us to see that BBNs used to refine uncertainties could in principle be equally useful in either setting.

We face uncertainty about a parameter w reflecting sensitivity of climate to emissions. KC have w equal either to 1.5 or 5, but for notational convenience, we allow w to vary over a connected range W. Before gathering new information I, the likelihood that w takes a particular value is $f(w)$, which as a probability distribution satisfies

$$\int_W f(w)dw = 1.$$

Absent I, a policymaker charged with maximizing expected utility would have to choose a policy x^* to maximize

$$\int_W U(x,w)f(w)dw,$$

where $U(x, w)$ is the utility from policy x if w is realized.

Following KC, assume first that I allows the policymaker to know the value of w before choosing x. If so, she would choose $x(w)$ to maximize $U(x, w)$ given w. Thus, prior to acquiring I, the expected utility the policymaker would achieve with I is

$$\int_W U(x(w),w)f(w)dw.$$

The value of information V at the time one decides to acquire it is the difference between the expected utility with the information and the expected utility without it:[9]

$$V = \int_W [U(x(w),w) - U(x^*,w)]f(w)dw.$$

[8] None of the formal constructions in this section are novel.

[9] Finkel (this volume) points out that one can view this as the expected avoided cost of error or "regret" from choosing x^* instead of $x(w)$.

Figure 2.1 in KC illustrates this formulation in the case where w can take one of only two values.

If V exceeds the expected cost of gathering the information—for example, launching an earth observation satellite, carrying out malaria detection tests, or advertising product characteristics—the investment in information gathering is worth the cost; if V is smaller than that expected cost, the information is not worth obtaining. V will tend to be larger as the difference between the utility of choice knowing the information, $U(x(\mathrm{w}), w)$ exceeds $U(x^*, w)$, over ranges of w that are relatively likely—that is, where $f(w)$ is relatively large. When $U(x(\mathrm{w}), w)$ is not much different from $U(x^*, w)$, either because decisionmakers cannot choose $x(\mathrm{w})$ to be much different from x^* or because the choice does not affect U that much, or when the set of *w values* where the difference matters has a low *ex ante* likelihood of occurring, where $f(w)$ is small, the value of information will not be great and investments to procure it are less likely to be justified on cost-benefit grounds.

It is important to recognize that the value of information here is based on $f(w)$, the likelihood the information will be valuable *before the information is gathered.* Consequently, a study showing that information happened to be valuable, given a particular realization of w after the information was gathered, does not prove that the information was worth gathering in the first place. If the *ex ante* chance of observing that value of w, $f(w)$, was particularly small, the information investment may not have been worth making, even if it happened to turn out to be valuable. Similarly, an information investment could have been worth the cost even if the particular piece of information revealed turned out to have little effect on the decisionmaker's choice or utility, if outcomes that could have been important were sufficiently likely at the time the investment in information gathering was made.[10]

As KC point out in their second example, information may not enable a decisionmaker to identify w; it may give her only better information regarding the underlying distribution of w. To describe this, let the preinformation $f(w)$ be given as the weighted average of a set of probability distributions $p(w, z)$, where the preinformation probability of observing distribution $p(w, \mathrm{z})$ is $g(z)$ taken over a domain Z of distributions.

$$f(w) = \int_Z p(w, z) g(z) dz,$$

[10] This distinction has implications for whether case studies on the value of particular pieces of information really can tell us very much about whether it was valuable to invest in the ability to acquire that information. Such case studies tend to be valuable to the degree that the benefit of the information, $U(x(\mathrm{w}), w) - U(x^*, w)$, is relatively constant over the set of w that one is likely to observe. For example, if (a) one knows that a disease might break out in one out of N areas but one does not know which area will be the one, (b) stopping the disease would have the same benefit regardless of where it broke out, and (c) preventive measures would not be effective absent knowing where the breakout is, then a study showing the benefits of detection will be informative regarding the value of the determining which area will have the outbreak. Whether assumptions along these lines are generally valid could be a useful question to study.

where

$$\int_Z g(z)dz = 1.$$

and for any z,

$$\int_W p(w, z)dw = 1.^{11}$$

Prior to learning z, the policymaker would choose x^* that maximizes expected utility

$$\int_W U(x, w) \int_Z p(w, z)g(z)dzdw.$$

After learning, the policymaker can choose $x(z)$ to maximize expected utility with the distribution $p(z, w)$ that describes the postinformation uncertainty about w:

$$\int_W U(x(z), w)p(w, z)dw.$$

The value of information is, as above, the expected improvement in being able to choose after learning, in this case the distribution $g(z)$ rather than w directly:

$$V = \int_Z \int_W [U(x(z), w) - U(x^*, w)]p(w, z)g(z)dzdw.$$

KC's Fig. 2.2 illustrates this where there are just two potential distributions.

So far we have two uncertainties, the distribution of the climate variable w given the parameter z, and the distribution of the parameter z. Of course, there are other uncertainties as well, associated with how x will affect utility given w, such as how

[11] It may be that $p(w, z)$, the probability of w conditional on z, is learned through a process of Bayesian updating based on $g(z)$, prior beliefs regarding z, and $h(z, w)$, the likelihood that evidence z would be observed were a particular climate observation w valid:

$$p(w, z) = \frac{h(z, w)g(z)}{\int_Z h(z, w)g(z)dz}.$$

well the policy will work and what it will cost. One uncertainty is technological change that may reduce the cost of substituting away from fossil fuel use; another surrounds the ecological costs associated with extreme-case geoengineering. Let θ be a potentially multidimensional parameter reflecting these uncertainties over a range Θ distributed by $h(\theta, w)$, so that the "utility" $U(x, w)$ for any policy under any realization of climate sensitivity is really an expected utility

$$U(x,w) = \int_{\Theta} U(x,w,\theta)h(\theta,w)d\theta$$

where, for any w,

$$\int_{\Theta} h(\theta,w)d\theta = 1$$

Substituting this into the expression for the value of information in the general case where information narrows but does not eliminate uncertainty regarding the climate parameter gives

$$V = \int_Z \int_W \int_{\Theta} [U(x(z),w,\theta) - U(x^*,w,\theta)]p(w,z)g(z)h(\theta,w)d\theta dw dz.$$

We could increase the formal complexity by looking at the value of information about θ that changes the degree of uncertainty regarding the costs and benefits of climate policy. However, this formulation is sufficient to illustrate that the methods for constructing the value of information, the use of Bayesian belief nets, and the choice between expected utility and risk management frameworks, can be regarded as independent issues.

2.C.3. Separating Value of Information, Risk Management, and BBNs

KC propose that climate policy be considered using a risk management framework, as distinguished, presumably, from an expected utility framework. KC (2012, page number tk) characterize the distinction in the following way:

> A risk management approach asks what policy should be, given the large range of possible outcomes from that choice. This is quite distinct from asking what the optimal policy is under different assumptions of our uncertain variables.

Perhaps I'm wearing blinders acquired from decades of being an economist, but the source of "quite distinct" is not obvious. Both seem to be doing the same thing.

The essential difference may be that under risk management, rather than maximizing expected utility, the task is "to keep the probability of facing catastrophic damages to some determined low level," or as said later, a "defined tolerable level." If so, the utility or benefits of the outcome itself are essentially out of the calculation. Maximizing expected utility thus becomes a matter of minimizing the cost of meeting the risk management probability target. In the formulation above, then, we can substitute for expected utility (which was net of cost) the negative of cost.

To put it another way,

$$U(x, w, \theta) = \overline{U} - C(x, w, \theta),$$

where $\overline{U}$ is the utility achieved at the "defined tolerable level" and $C(x, w, \theta)$ is the cost of implementing policy x with climate sensitivity y under circumstances θ. Substituting this for the value of information V above gives that value as the difference between the cost we have to pay for acting before getting the information and the cost we would expect to have to pay after we learned more about the distribution of climate conditions.

$$V = \int\limits_{Z}\int\limits_{W}\int\limits_{\Theta} [C(x^*, w, \theta) - C(x(z), w, \theta)]p(w, z)g(z)h(\theta, w)d\theta dwdz$$

where x^* is the choice of policy that minimizes expected costs prior to acquiring more information about the distribution of climate sensitivity.

In my interpretation of KC, the value of information shows up in the context of a cost-effectiveness test rather than as expected utility maximization. However, as a matter of formalism and, importantly, the underlying uncertainties, the factors determining the value of information are essentially the same. It may be that some values of θ would affect utility but not cost, so the relevant distribution function $h(\theta, w)$ could be less difficult to ascertain. But we still need to know $h(\theta, w)$, $p(w, z)$, and $g(z)$ to determine whether an effort to acquire information to determine z is worth the cost and whether we are operating under an expected utility framework or a risk management framework.

Similar considerations show that the Bayesian belief nets KC advocate would be relevant in an expected utility context. The purpose of a BBN is essentially to use the knowledge of a set of experts to learn the probabilities relevant to a particular policy. For expected utility, one could interpret this as using those experts to learn the distribution of effects on utility instead of costs ($h(\theta, w)$). The expert information can also be used to determine which distribution of climate sensitivity $p(w, z)$ we have, essentially by finding z and eliminating the stochastic step added by $g(z)$. This suggests that the virtues of BBN in improving information regarding risk, and whether it is worth the cost to set one up, are at least qualitatively just as plausible in an expected utility framework as in a risk management framework. The expert survey methods KC propose are equally applicable in both circumstances, so their value is largely separate from whether one maximizes utility or minimizes the cost of reaching a maximum tolerable probability of a climate catastrophe.

2.C.4. Critiques Nonetheless

Establishing that methods for calculating the value of information and Bayesian belief nets could apply equally to expected utility and risk management doesn't mean that risk management and expected utility are equally (in)valid ways of determining policy responses. Neither does applicability to expected utility theory insulate BBNs from critiques of risk management. The point of showing that the concepts are largely independent in principle allows their merits to be assessed largely independently. Some brief observations on risk management, BBNs, and their interrelationship follow. I conclude with a critique of economic approaches to utility maximization in the climate context.

2.C.4.1. Risk Management

One can think of three justifications for using risk management over an expected utility approach. The first two are conventional arguments that justify cost-effectiveness over wealth maximization approaches generally. First, if the benefits of a policy are too difficult to quantify, one might simply evaluate policies in terms of how well they achieve a predetermined policy target. Quantification may be inherently difficult because the data on valuation are highly noisy. For the instructive case of the value of reducing mortality risk, the "statistical value of life," the underlying data on willingness to pay for incremental safety benefits are notoriously variable—but at least there are some markets or behavior trails from which a willingness to pay may be inferred. Since the effects of climate change are global, nonexcludable (and thus outside markets), and future, the present willingness to pay to mitigate it may be impossible to measure with any real confidence. One might be better off making a considered judgment regarding acceptable risk.

The second conventional argument for taking risk management rather than expected utility approach is that ethical considerations as well as economic factors determine the appropriate target, the instant example being the maximum tolerable probability of a sufficiently large-scale climate effect. I conclude below with some observations on the limits of the economic approach and, derivatively, the unavoidability of ethical considerations in the specific context of climate policy. In general, however, when lives or major changes to the social or physical environment hang in the balance, some may argue that policy responses ought not be determined solely by how much people might be willing to pay for them, even if that willingness could be measured accurately. One could set the level of tolerable risk using qualitative assessment, normative judgment, and communal deliberation (Ackerman and Heinzerling 2002) and then manage that risk by gathering information to seek out the most cost-effective programs.

A more recent argument for risk management and against expected utility, cited by KC, comes from Weitzman (2007). As I understand it, Weitzman's argument rests on two premises. The first is that the distributions for climate events that one could statistically infer from the data, the $f(w)$ in the models above, have fatter tails—more weight toward high climate sensitivity—than the underlying distributions might entail. He contrasts the "t-distribution" from the normal distribution in that regard. The second premise is that the appropriate form for modeling the utility of wealth, where wealth is affected by climate, entails assuming constant relative risk aversion (CRRA). These two premises together imply that the expected (dis)utility from climate change is $-\infty$ and that any finite effort to alleviate it is justified.

I lack the expertise to address Weitzman's statistical premise, but the CRRA assumption is unsupported by theory and contradicted by experience. With regard to theory, CRRA is based on a quadratic Taylor series approximation to a utility function to model the willingness to pay to avoid variance in wealth. As such, it is constant only within a small distance around a target wealth level. Nothing suggests that one could extrapolate that approximation far outside such small variations, particularly to catastrophes. Were this so, individuals would regard the loss of life as sufficiently harmful to warrant arbitrarily large expenditures to limit mortality risk. That is, the observed value of a statistical life would be infinite—a prediction violated by almost everyone's behavior almost every day. KC's analysis does not rely on going this far, but it does undercut using Weitzman's argument to justify a risk management rather than expected utility approach.

I was surprised to see KC tout the advantages of "value-at-risk" (VaR) models for risk management. VaR has been taking a beating in the press, where its widespread adoption in assessing derivative portfolio risk has been blamed for the financial meltdown of 2008:

> Given the calamity that has since occurred, there has been a great deal of talk, even in quant circles, that this widespread institutional reliance on VaR was a terrible mistake. At the very least, the risks that VaR measured did not include the biggest risk of all: the possibility of a financial meltdown. "Risk modeling didn't help as much as it should have," says Aaron Brown, a former risk manager at Morgan Stanley who now works at AQR, a big quant-oriented hedge fund. A risk consultant named Marc Groz says, "VaR is a very limited tool." David Einhorn, who founded Greenlight Capital, a prominent hedge fund, wrote not long ago that VaR was "relatively useless as a risk-management tool and potentially catastrophic when its use creates a false sense of security among senior managers and watchdogs. This is like an air bag that works all the time, except when you have a car accident." Nassim Nicholas Taleb, the best-selling author of "The Black Swan," has crusaded against VaR for more than a decade. He calls it, flatly, "a fraud." (Nocera 2009)
>
> VaR, by the way, is a horrible way to measure risk, as has been said again and again by economists, because it calculates the risk for only 99% of the time. As [Simon Johnson, a professor at Sloan School of Management at MIT] says, "VaR misses everything that matters when it matters." Indeed, the VaR metrics obviously missed what led to what now has been dubbed the Great Recession. (Sorkin 2009)

I am not an expert in assessing financial risk, but I find one potential flaw of VaR, a shortcoming it shares with risk management generally: once the level of acceptable risk is determined, factors that might make the costs of unlikely events outside the range of acceptability become irrelevant to decisions, rather than ratcheting down acceptable probabilities as would happen with an expected utility approach. However, a reasonable response may be that the financial meltdown was the fault of not VaR but its application. To paraphrase the National Rifle Association, one could say, "Models don't kill economies, bankers kill economies."

2.C.4.2. Bayesian Belief Nets

My initial reaction to KC's BBN proposal was skepticism. My epistemological predisposition is that knowledge is something an individual acquires by examining the evidence and analyzing relevant theories, not by what appears from taking a poll. Of course, as KC go to some length to point out, BBN is more than mere poll taking. Underlying it is a process, not described in detail in their chapter, for treating expert opinions themselves as data amenable to Bayesian updating and maximum likelihood testing to reduce the uncertainty associated with a particular potential phenomenon, such as climate sensitivity or the costs of various policies.

Along with recognizing the statistical aspect of BBNs is the point that knowledge is routinely combined through institutional mechanisms to arrive at better estimates. Markets, certainly since von Hayek (1945), can be seen as information media in which disaggregated estimates of costs and value are combined to provide prices, the best estimates possible of marginal benefits and marginal costs. Asset markets—stocks and bonds, commodity futures and options, derivatives—similarly combine information regarding expectations of those benefits and costs to estimate their present values. The efficient market hypothesis is that those markets cannot be systematically beaten without unique information (Malkiel 2003), a view currently under fire since the 2008 financial market meltdown (Wighton 2009). Variations on this theme involve prediction markets (Iowa Electronic Markets, Intrade) in which assets are created with payoffs based on the outcome of elections, legislation, wars, or other events. Such markets have been proposed, albeit controversially, to predict the likelihood of catastrophes or terrorist attacks (Hanson 2007).

We, or at least I, should also keep in mind a couple of things about the individual character of knowledge. Academically, material does not become accepted, nor is a lot of research funded, without peer review, itself a kind of collective expert assessment. More broadly, what all of us believe we know is far, far greater than what we have individually found out. From childhood to the present, we depend on teachers, books, colleagues, journals, libraries, and maybe even Wikipedia, to tell us what we believe we "know." Each of those sources is fundamentally a sort of "belief net." That they lack the statistical foundation of BBNs is in BBNs' favor; the crucial point is that resistance to BBNs on the basis of their being collective rather than individual may be misplaced.

It does remain the case that a BBN need not be restricted to risk management settings. As noted above, a BBN can reduce uncertainties to improve policy choices under expected utility maximization as well. We do need to be careful to keep in mind that the value of a BBN is something that needs to be known *ex ante*. KC's portrayal gives the impression that the value of a BBN is realized *ex post*. We need the assorted probability distributions described above to determine whether the costs of a BBN are worth undertaking. More on how one would make this *ex ante* assessment would be useful. One could imagine a staggered set of BBNs, where one undertakes a relatively low cost survey of a small number of experts to determine whether a full-blown BBN would be worth the costs.

2.C.4.3. Putting Them Together

The theme here is that value-of-information calculations, risk management versus expected utility, and the role of BBNs can all be assessed independently. Although that seems largely true, there is one sense in which BBNs combined with risk management could be problematic. As KC say, the goal under risk management is to find the best way to deal with a "defined tolerable level" of risk. This leaves open the question of how "tolerable level" comes to be defined. Under risk management, this is specifically not a matter of ascertaining individual willingness to pay, whatever that may be, and defining tolerable level as the point at which the revealed willingness to pay for further reductions in probability no longer exceeds the cost of those reductions.

If not, then how do we define the tolerable level? If the preferences of the general public cannot be ascertained or relied on, then the question of who gets to choose becomes unavoidable. If so, the BBN framework raises the possibility that the experts may interpret questions to be not about their best guess of a relevant value or probability distribution, but about their interpretation of "tolerable level." This may happen unintentionally but could be problematic nonetheless.

For example, one might find a climate expert who is an avid bicyclist and vegan who finds air-conditioning oppressive. Another expert may love his Hummer and steaks and believe that air-conditioning is great, especially when one keeps the windows open so the house doesn't smell musty. Those experts' assessments regarding relevant probabilities of climate sensitivity and thus justification for incurring costs of policies may become difficult to disentangle from their own judgments about what costs are worth incurring in a society. Perhaps this is just a predictable observation from a paternalism-averse economist, but we probably need to be careful in framing BBN surveys to minimize the degree to which they become a forum for elites' desires trumping the preferences of the public. Even if one has few qualms about letting elites make policy decisions, the political legitimacy of BBNs as an advisory tool may depend on limiting them to "just the facts."

2.C.5. Expected Utility, Risk Management, and Climate Policy

We can conclude with an observation (Brennan 2010) that could cut in favor of the risk management approach, and less obviously but possibly Bayesian belief nets as well. When economists employ utility maximization models for policy, they—we—typically interpret "utility" as in consumer surplus terms—that is, aggregate willingness to pay. The standard normative critique of this approach is that in aggregating surplus across everyone in an economy, it treats each dollar of net benefit as equal, whether the recipient is homeless or a billionaire. The justification for ignoring the normative consequences of the distribution of net benefits is that the winners could in principle compensate the losers, creating a situation where everyone is better off—a generally uncontroversial improvement. Someone other than the policy evaluator can decide explicitly or implicitly whether some other distribution of benefits is preferable.

However, there may be no compensation-in-principle in the case of climate change. If the benefits of climate policy are realized only by generations far in the future, this compensation would require that those in the future pay us for the sacrifices we make on their behalf. Since future output cannot be put into a time machine for delivery to us, such compensation may be impossible. If that is the case, the standard economic approach does not suffice; explicit moral assessment of our obligations regarding the welfare of future generations and environmental protections becomes paramount.

The inevitability of an explicitly normative dimension beyond efficiency to climate mitigation policy suggests we ought to focus not on maximizing expected utility but on determining an ethically tolerable level of risk of severe climate change to be met at least cost—the risk management approach that KC advocate. This would entail gathering information to reduce the relevant uncertainties associated with the costs of various means of meeting that objective. KC provide a framework to accomplish this, making their contribution to the climate policy community important.

References

Ackerman, F., & Heinzerling, L. (2002). Pricing the priceless: Cost-benefit analysis of environmental protection. *University of Pennsylvania Law Review, 150*, 1553–1584.

Arrow, K. (1959). *Economic welfare and the allocation of resources for innovation* (P-1856-RC). Santa Monica: RAND Corporation.

Brand, S. (1985, May). Keep designing: Discussions from the Hackers' conference, November 1984. *Whole Earth Review, 46*, 44–55.

Brennan, T. (2010). The challenges of climate policy. *The Australian Economic Review, 43*, 225–239.

Cooke, R. M., & Kelly, G. N. (2010). *Climate change uncertainty quantification: Lessons learned from the joint EU-USNRC project on uncertainty analysis of probabilistic accident consequence code* (Resources for the Future Discussion Paper 10-29). Washington, DC: Resources for the Future.

Cutler, D., & Reber, S. (1998). Paying for health insurance: The trade-off between competition and adverse selection. *Quarterly Journal of Economics, 113*, 433–466.

Fankhauser, S., & Tol, R. S. J. (2005). On climate change and economic growth. *Resource and Energy Economics, 27*(1), 1–17.

Hanson, R. (2007). The policy analysis market: A thwarted experiment in the use of prediction markets for public policy. *Innovations, 2*(Summer), 73–88.

IPCC. (2007). *Climate change 2007: Synthesis report*. Contribution of Working Groups I, II and III to the fourth assessment report of the Intergovernmental Panel on Climate Change. Geneva: Intergovernmental Panel on Climate Change.

Keller, K., Yohe, G., & Schlesinger, M. (2008). Managing the risks of climate thresholds: Uncertainties and information needs. *Climatic Change, 91*, 5–10.

Lempert, R. J., Schlesinger, M. E., Bankes, S. C., & Andronova, N. G. (2000). The impacts of climate variability on near-term policy choices and the value of information. *Climatic Change, 45*, 129–161.

Malkiel, B. (2003). The efficient market hypothesis and its critics. *Journal of Economic Perspectives, 17*, 59–82.

Manne, A., & Richels, R. (1995). The greenhouse debate: Economic efficiency, burden sharing and hedging strategies. *Energy Journal, 16*(4), 1.

Nocera, J. (2009, January 2). Risk mismanagement. *New York Times Magazine*. http://www.nytimes.com/2009/01/04/magazine/04risk-t.html. Accessed 1 July 2011.

Savage, L. J. (1954). *Foundations of statistics*. New York: Wiley.

Sorkin, A. R. (2009, September 16). Taking a chance on risk, again. *New York Times*. http://dealbook.nytimes.com/2009/09/17/taking-a-chance-on-risk-again/. Accessed 1 July 2011.

von Hayek, F. (1945). The use of knowledge in society. *American Economic Review, 35*, 519–530.

Weitzman, M. (2007). A review of the stern review on the economics of climate change. *Journal of Economic Literature, 45*, 703–724.

Wighton, D. (2009, January 29). Efficient market hypothesis is dead—for now: Business editor's Davos commentary. *The (London) Sunday Times*.

Chapter 3
Understanding the Value of Business Information

Luther Martin

Abstract Many businesses seem to be of two minds when it comes to understanding the value of their information. Firms say that it is their most valuable asset, yet they seem unwilling to invest in information security technologies to protect it. A closer look at a few different ways to understand the value of information suggests how to resolve this apparent paradox.

Keywords Experience goods • Information security • Risk management • Subjective probabilities • Utility versus value

3.1 Introduction

The field of information security encompasses protecting information from unauthorized access, use, disclosure, disruption, modification, or destruction. At its core is the cost-benefit trade-off that businesses need to make when they decide how to invest in information security technologies. Applying traditional cost-benefit analyses to information security, however, can sometimes lead to very puzzling results.

In his doctoral dissertation at Stanford University, "How Much is Enough? A Risk Management Approach to Computer Security," Soo Hoo (2002) performed a careful cost-benefit analysis of several information security technologies, but his results do not necessarily predict how widespread the use of the technologies are. He summarized his findings as follows: "Unless the costs and consequences of computer security breaches are radically erroneous, the optimal solution for managing computer

L. Martin (✉)
Voltage Security, Inc., Cupertino, CA, USA
e-mail: martin@voltage.com

R. Laxminarayan and M.K. Macauley (eds.), *The Value of Information: Methodological Frontiers and New Applications in Environment and Health*,
DOI 10.1007/978-94-007-4839-2_3, © Springer Science+Business Media Dordrecht 2012

security risks calls for very minimal security measures. Thus, the reluctance of both private and government organizations to pursue computer security aggressively may be well justified" (Soo Hoo 2002, 67–68).

He found that firewalls, for example, cannot be justified using a cost-benefit model. The cost of operating firewalls is apparently greater than the losses that they prevent. Even armed with this knowledge, however, most businesses would be unwilling to go without a firewall.

Similarly, Soo Hoo's analysis showed that encryption is a good way to protect sensitive data because the cost of using encryption is apparently less than the losses that the technology prevents. Nevertheless, businesses have been very reluctant to use encryption. The adoption of encryption has traditionally been fairly low and has increased recently only because of laws that either require or strongly encourage its use (Ernst & Young 2009). So even though there is apparently a sound business case for encryption, businesses have been relatively unwilling to use it for many applications unless they are forced to.

Situations like those, where the behavior of businesses differs from what the best models predict, suggest two possible explanations: that the model is wrong, or at least sufficiently inaccurate to make its predictions useless, or that businesses are not making decisions as we might expect them to. Because there has been little to no criticism of Soo Hoo's analytical model, it's useful to consider the possibility that businesses are not making decisions as we might expect them to. Fortunately, there are ways to understand the decisionmaking process that businesses use that seem to make their decisions more understandable.

3.2 Marginal Utility

The theory of marginal analysis (Bernoulli 1738) tells us that the marginal utility of a good is determined by its least important use. Böhm-Bawerk's discussion of the marginal utility of grain to a farmer in *The Positive Theory of Capital* gives a good example of this (1891, III.IV.9):

> A colonial farmer, whose log hut stands by itself in the primeval forest, far away from the busy haunts of men, has just harvested five sacks of corn. These must serve him till the next autumn. Being a thrifty soul he lays his plans for the employment of these sacks over the year. One sack he absolutely requires for the sustenance of his life till the next harvest. A second he requires to supplement this bare living to the extent of keeping himself hale and vigorous. More corn than this, in the shape of bread and farinaceous food generally, he has no desire for. On the other hand, it would be very desirable to have some animal food, and he sets aside, therefore, a third sack to feed poultry. A fourth sack he destines for the making of coarse spirits. Suppose, now, that his various personal wants have been fully provided for by this apportionment of the four sacks, and that he cannot think of anything better to do with the fifth sack than feed a number of parrots, whose antics amuse him.

In these circumstances, the value of the fifth sack of grain is quite low to the farmer. If he loses one sack out of five, he will not scale back each of his uses for the grain by one-fifth but will stop the use that provides him with the least value—feeding his parrots. On the other hand, if he has only a single sack of grain, the value to him is extremely high, for losing the final sack may mean that he starves to death.

In general, the more grain the farmer has, the less value an additional sack of grain has to him. This can be summarized in the law of diminishing marginal utility: the additional benefit provided by an additional unit of a good tends to decrease as the total amount of the good increases. Applying the law of diminishing marginal utility to information may both provide some useful insights into the behavior of corporate security departments and let us predict some future trends.

The information age has caused an explosion of information, and we should expect a diminishing marginal utility for this information as the total amount of it increases, particularly because this ever-increasing amount of information is often close to indistinguishable. It is currently unfeasible to classify information to any significant granularity; projects that try to classify data based on more than the source of the data usually fail (Strategic Data Management 2008). So current technology might require the same handling of any information that comes from a financial management system, for example, or it might require the same handling of any information that is processed by an email system. Within such broad categories, information is essentially handled in a common way.

The law of diminishing marginal utility tells us that the marginal utility of such information is determined by its least important use, and we should expect corporate information security organizations to protect their information as if this were the case. Even though some information in email may have high value, therefore, we should expect email to be protected as if it were of low value. And because it is currently impractical to classify some data according to the actual value of the data, we should expect to see high-value data often remain unprotected. The slow adoption of security technologies like email encryption or whole-disk encryption, which is meant to render data on a lost or stolen computer unavailable to an unauthorized user, may be due to the low value of some corporate data, and thus to the low marginal value for all data on an enterprise-wide basis.

The steady trend toward outsourcing core business functions extends to those that involve extremely sensitive data. Not many years ago, it was unheard of for any business to outsource functions like accounting or payroll; today these functions are routinely outsourced. More recently, there has been at least one successful business that provides a service that outsources the management of sales data—information that is extremely sensitive and potentially valuable to competitors. Even information security functions are starting to be outsourced.

The trend to outsource more and more critical business functions has coincided with the explosion of information, which the theory of marginal utility seems to predict: we should expect information to have a decreasing marginal utility as the total amount of information increases. Thus the trend toward outsourcing is certainly predicted by marginal utility theory. If a business has a relatively small amount of information, the marginal value of the information is relatively high, and outsourcing is viewed unfavorably because it provides a chance for the loss of valuable information. But when a business has a relatively large amount of information, the marginal value of the information is relatively low, and objections to outsourcing disappear.

Although it is today infeasible to classify data beyond the source of the data, technology is now being developed that may produce better solutions in the not-too-distant future and allow businesses to classify information according to its actual value. When this happens, marginal utility theory predicts, demand will increase for encryption technologies and other products that provide strong protection to the high-value data that future data classification technologies will identify.

So in the future, data will still be protected according to its least important use, but the ability to separate data into different categories will make it possible to more narrowly define these categories. In this case, even the least important uses of valuable data will justify the use of encryption to protect it. Whole-disk encryption and email encryption products that have so far experienced fairly slow adoption rates may become more widely used as it becomes easier to identify exactly what data should be encrypted.

The framework of marginal utility may explain some observed behavior. It seems to provide a good way to understand the reluctance of many businesses to use encryption to protect some sensitive data, as in the case of email. It also seems to predict when encryption will be more commonly used. Social Security numbers, for example, are sensitive data, so the fact that encryption is commonly used to protect fields in a database that includes Social Security numbers is correctly predicted by this framework. Other cases are not as easy to understand. Laptop computers, for example, almost always contain sensitive and valuable data, yet encryption of the data on laptops is far from ubiquitous. Other models are needed to understand cases like these.

3.3 Measuring Risk

In Lord Dunsany's story "Jorkens' Revenge" (Plunkett 1935), the Munchausen-like Jorkens manages to win an unusual wager with his nemesis, Terbut: Jorkens bets him £5 that it is farther from Westminster Bridge to Blackfriars Bridge than it is from Blackfriars Bridge to Westminster Bridge.

The perplexed Terbut then finds that the taxi ride one way is indeed longer than the ride the other way and grudgingly pays Jorkens £5 without fully understanding why he lost the bet. The secret to Jorkens's victory was that the road between the two bridges is semicircular, and driving an arc of a smaller radius gives you a shorter distance than driving an arc with a larger radius. Clearly, how we measure things can be very important. How do we measure the risk associated with sensitive information?

Risk, as it is understood by professional risk managers (Stoneburner et al. 2002), is the expected loss associated with a particular event. If we denote the probability of a loss-causing event by P and the loss that will accompany this event by L, then we can write the risk associated with this event as R, where

$$R = PL$$

So if an event causes \$1 million in loss when it happens, and this event happens with a probability of 0.001, then this event represents

$$(\$1 \text{ million})\ (0.001) = \$1{,}000$$

or \$1,000 of risk. This definition of risk gives us a good way to decide which information security technologies to use: if the decrease in risk that using a particular technology causes is less than the cost of using the technology, then it makes sense to use the technology. Otherwise, we'd be better off not using the technology at all. In most cases, however, neither the probability of security incidents happening nor the damages that result from an incident are known very well. This makes the traditional definition of risk not very useful in this particular context.

Consider the simple case of a web server. Almost all software have security vulnerabilities, some of which have yet to be discovered. In such a situation, what is the probability of that your web server will be compromised by a hacker? And if it is compromised, how do you put a dollar value on the damage? Because it's hard to answer questions like these accurately, it is difficult to apply the classic definition of risk to information security or use common risk management methodologies to manage information security. Other approaches—ones that do not depend on accurate estimates of probabilities—are needed.

3.4 Uncertainty

Knight described the difference between risk and uncertainty in his 1921 book *Risk, Uncertainty and Profit* as follows:

> The essential fact is that "risk" means in some cases a quantity susceptible of measurement, while at other times it is something distinctly not of this character; and there are far-reaching and crucial differences in the bearings of the phenomenon depending on which of the two is really present and operating. There are other ambiguities in the term "risk" as well, which will be pointed out; but this is the most important. It will appear that a *measurable* uncertainty, or "risk" proper, as we shall use the term, is so far different from an *unmeasurable* one that it is not in effect an uncertainty at all. We shall accordingly restrict the term "uncertainty" to cases of the non-quantitative type. It is this "true" uncertainty, and not risk, as has been argued, which forms the basis of a valid theory of profit and accounts for the divergence between actual and theoretical competition. (Knight 1921, I.I.26)

A concise summary of Knight's thinking is that if you know the probability of an event, then you're dealing with risk, and if you don't know the probability, then you're dealing with uncertainty.

It's easy to essentially eliminate the effects of risks by buying insurance, for example. Dealing with uncertainties is more difficult, but there are a few general ways to understand uncertainty that help to explain the way that people make decisions in the face of uncertainty.

3.4.1 Subjective Probabilities

One way to account for the decisions that people make in the face of uncertainty is to assume that people use their best guesses for probabilities in the absence of reliable data. People are not good at estimating probabilities, however. We tend to systematically overestimate low probabilities and underestimate high probabilities. So we might expect people to estimate that an event with a 99.9% chance of happening has only a 90% chance of happening. Or we might expect people to estimate that an event with only a 0.01% chance of happening has a 10% chance of happening. The framework of prospect theory, as pioneered by Kahneman and Tversky (1979) provides a good way to understand why this happens, but here we just observe that these biases exist.

The bias that creeps into such estimates makes estimates of risks based on them inherently inaccurate, and we should be wary of placing much faith in such estimates. Some types of security compromises are fairly rare, and it is likely that their chances may be routinely overestimated; the result is that a firm may spend too much on countermeasures designed to reduce the already-low chances of a security breach.

If a particular breach could cause a loss of $10,000, we should expect a rational business to spend up to $1,000 to eliminate exposure to it if the chances of its occurrence are 10%, but to spend $1 or less to eliminate this exposure if the chances of its occurrence are only 0.01%. So we can expect that inaccurate estimates of probabilities of security compromises to make a difference in the firm's willingness to invest in information security technologies.

It is difficult, however, to get a careful and precise definition for probability. In one case, we can define objective probabilities. These represent how often we expect a particular event to occur. If we flip a fair coin, for example, we expect it to come up heads half of the time, so we say that the probability of this event is one-half.

Another point of view is that a probability is only a subjective estimate for how often a particular event will happen. In this case, we need to update our estimate of a probability every time we get more relevant data. So we might start with the estimate that a coin is fair, but if we flip it several times in a row and it ends up showing heads every time, we might want to use these observations to update our estimate that this particular coin will be heads half the time. The framework of Bayesian inference models such cases. The most notable feature of Bayesian inference is that we commonly replace probabilities with conditional probabilities, or probabilities of the form $P(A)$ with probabilities of the form $P(A|K)$, where $P(A)$ represents the probability of the event A happening while $P(A|K)$ represents the probability of the event A happening given K, the other knowledge that we have.

Such subjective probabilities can model many of the biases that tend to cause people to make errors in estimating probabilities and, in some cases (El-Gamal and Grether 1995), can explain observed behavior with a reasonable degree of accuracy. For example, the information security staff at businesses know that security vendors have an interest in making exaggerated claims about the need for their products. Because of this, the probability that staff will believe a vendor's claim is more accurately represented as $P(C|K)$ instead of $P(C)$, and the fact that we should tend to discount vendors' claims is easily reflected in such a model.

3.4.2 *Email as an Example*

That people tend to think in terms of subjective probabilities instead of objective probabilities may explain how some security threats are perceived. The security of email is probably a good example of this. A hacker who wants to intercept and read email has two ways to do this: he can either intercept an email message while it is being transmitted across a network, or he can get the email from a server where it is stored.

To estimate the relative chances that these types of compromises will happen, we polled 20 information security experts. None of the experts could think of even a single case where email was intercepted and read while being transmitted across the Internet, but every one of the experts could think of a case where email that was stored on a server was accessed and read by a rogue administrator. This is little more than anecdotal evidence, but it certainly suggests that the chances of email's being intercepted and read on the Internet is fairly low while the chances of email's being intercepted and read when it's stored on a server is probably much higher.

It's likely that same general principle applies to other forms of sensitive information. A hacker wanting to obtain credit card numbers, for example, can either intercept them while they're being transmitted across a network or compromise a server where lots of them are stored. If she tries to intercept information, the hacker will probably spend lots of time and recover little sensitive information. A better opportunity is to compromise a server where lots of sensitive information is stored.

When asked by a reporter why he robbed banks, Willie Sutton is said to have replied, "Because that's where the money is." It turns out that Sutton didn't actually say this: a reporter looking for a noteworthy quote actually just made it up (Sutton and Linn 2004, 161). Despite this slight inaccuracy, we might think that Willy Sutton might advise twenty-first-century hackers to target storage systems "because that's where the data are."

Even though sensitive information is rarely compromised while it is being transmitted on the Internet and is more likely to be stolen while it's in storage, the commonly held perception seems to be the opposite. That is, the subjective probabilities of the different types of compromises seem to differ significantly from more objective estimates. If decisions are based on the subjective probabilities, this would seem to explain why people tend to worry about information being transmitted across the Internet and worry less about information in storage

3.5 Adams' Point of View

Adams (1997) divides risks into three general types. In his model, some risks can be directly perceived. A person standing near a busy road, for example, can directly perceive the danger associated with the nearby traffic. Risks that are directly perceived are easily managed. They are so easily managed, in fact, that everyone becomes his or her own risk manager, which can easily cause conflict with organizational policies. Few information security risks are easily perceived and measured.

Other risks can only be perceived with the aid of science or technology. This is the area in which traditional risk management methodologies usually excel. Some information security risks fall into this category. You cannot directly see if your network is under attack, for example, but with the right technology, such attacks become obvious. However, because the probability of many security incidents or the damage resulting from such incidents is difficult to accurately quantify, traditional risk management methodologies can be difficult to apply in such situations.

Other risks are virtual risks: science and technology cannot provide a definitive understanding of the risk. Much of information security may be appropriately thought of as the management of virtual risks. In the case of virtual risks, what people believe depends on the source of their information. People tend to put less trust in experts, who actually have access to the most reliable information, and put more trust in friends and family, who tend to be those with access to the least reliable information. This leads to situations in which some virtual risks are deemed very serious despite any credible evidence.

This may be just an example of Bayesian inference, with people naturally discounting the views of experts because they often have an interest in certain outcomes. Experts on encryption often work for encryption vendors, for example, but their interest in seeing their products sold might make people who know little to nothing about encryption severely discount their expert opinions. The net result is that encryption is used less than it should be. Another explanation is that the nature of information security technologies may actually cause this to happen.

3.6 Nelson's Categories of Goods

Nelson (1970) describes how to divide goods into three types based on how easy it is to verify their manufacturers' claims: search goods, experience goods, and credence goods. Search goods have properties that are easy to check before you consume them. If you are in the market for a red car, for example, it is easy to determine whether a potential purchase is really red. Very few, if any, information security technologies fall into this category.

Experience goods have properties that are not obvious before you buy are easy to verify after you consume them. If you are looking for a car with a certain fuel efficiency, perhaps getting at least 35 miles per gallon under your typical driving conditions, you cannot tell this by looking at the car itself, but you can easily test it. It's probably the case that very few security technologies are experience goods.

Credence goods have properties that cannot easily be checked either before or after they are consumed. Organically grown produce and meat from animals raised in humane conditions are examples of credence goods; it is very difficult to verify these particular properties, even after you consume them. Many medicines also fall in this category because it is difficult to tell whether your recovery was due to the medication, a placebo effect, or your body's self-healing mechanisms.

Most information security technologies are probably credence goods. It requires expensive and uncommon skills to verify that data are really being protected by the use of encryption, for example, and most people cannot easily distinguish between very weak and very strong encryption. And even after you use encryption, you are often never quite sure that it is protecting the data as it is supposed to. It is always possible that an adversary could develop a clever cryptanalytic attack that defeats even the strongest encryption, and the legitimate owner of the data would have absolutely no idea that the adversary had decrypted the information.

3.7 Akerlof and Quality Uncertainty

As described by Akerlof (1970), uncertainty about the quality of products can have unfortunate consequences for markets. In particular, it can cause market failures in which low-quality products succeed while high-quality products fail and prices spiral downward, pushing the quality of products even lower. Akerlof's reasoning, when applied to the market for used cars, gives us the following situation.

Suppose that all used cars are worth $10,000 if they are in good repair, but half of them, the lemons, actually need $2,000 worth of repairs, yet buyers cannot tell the difference between the good cars and the lemons. If buyers expect to have to spend an average of $1,000 on repairs, we should expect them to pay $9,000 for a used car. So the imperfect knowledge of the buyers has set the market price of used cars at $9,000. But at this price, those who have cars that are actually in good repair will not be inclined to sell their cars. After all, their cars are worth $10,000 each, but they can only get $9,000.

This means that all of the cars offered for sale at $9,000 will be the lemons, and the difference in information between the buyers and sellers has resulted in a situation that benefits only those who are selling the lemons. The declining quality of the cars offered for sale will eventually result in the lowering of buyers' expectations, and as the market becomes dominated by lemons, it may even fail altogether.

Similarly, the fact that most information security technologies are credence goods makes it difficult for businesses to judge whether a technology is 99.9% effective, 90% effective, or even 50% effective or less—a situation that invites the sort of problems that Akerlof predicts. And if good security products are indeed driven from the market by their lower-quality competitors, we might expect businesses to be less than enthusiastic about deploying and using the low-quality products that remain.

So it seems that several factors contribute to the low adoption of information security technologies, and it is difficult to explain this by a naïve application of the model of risk as $R = PL$. And although there are models that we can use to get insight into how people will deal with uncertainty, none let us estimate the value of information. A different approach, however, may be able to do just that.

3.8 Utility vs. Value

Another way to model the decisions that people make in the face of uncertainty is to generalize the value of a loss to the utility associated with it, instead of its financial value (von Neumann and Morgenstern 1944). Utility accounts for all of the ways in which the value of something is determined and even includes the effects of irrational preferences or prejudices. And although it may be very difficult to estimate utility a priori, we can often infer it from people's behavior.

3.8.1 The Value of a Statistical Life

Some attempts to estimate utility in cases where people are either unable or unwilling to make reasonable estimates have been fairly successful. The most notable of these may be the value of a statistical life (VSL) concept (Viscusi and Aldy 2003). When writing laws designed to increase public safety, it's important to know how much cost should be incurred to save a life. Unfortunately, it's difficult to get accurate estimates from people for how much they value a life. Such answers as "it's impossible to put a value on a human life" or "a life is of infinite value" are of little value when it comes to determining what costs should reasonably incurred to save a life.

The insight that leads to a way to get useful estimate of the value of a life is that it's possible to infer from people's behavior how much they value a life. Statistics on wages in risky occupations can be useful. For example, if a worker requires a premium of \$1,000 in annual wages to accept a 0.01% chance of being killed on the job each year, we can reasonably estimate that he puts a value of roughly

$$(\$1,000/\text{year})/0.01\ \%/\text{year}) = \$10\ \text{million}$$

on his life. Somewhat surprisingly, estimates for VSL are actually fairly consistent, even when created using data for different occupations and different industries (Viscusi and Aldy 2003).

The same obstacles to getting a good estimate for the value of a life also make it difficult to get a good estimate of the value of personal information. As with questionnaires about the value of a life, we should expect to get answers like "it's impossible to put a value on privacy" or "privacy is of infinite value." But much as the VSL methodology provides a way to infer from their behavior how much people actually value a life, it's possible to use a similar approach that infers from firms' behavior how much they actually value information. There may be enough information about the loss and theft of laptop computers to make a reasonable estimate of this.

There is probably about a 2.5% chance per year that one's laptop computer will be lost or stolen, an estimate based on the fact that \$3,000 of laptop insurance with a \$50 deductible currently costs approximately \$75 per year. If a laptop is lost or stolen but its data are encrypted, the information on the laptop has a reasonable level of protection against being compromised. The cost of laptop encryption software, as measured by the total cost of ownership of the software, is approximately \$150 per year (Checkpoint 2009). This leads to the estimate that the value of the data on the laptop is probably about

$$(\$150/\text{year})/(2.5\ \%/\text{year}) = \$6,000$$

At this cost, businesses seem indifferent to the decision of encrypting laptops or not doing so—roughly 50% of businesses currently use the technology (Johnson 2008), so \$6,000 may be a reasonable estimate for the value of the data on a typical laptop. Even though businesses may say that their information is very valuable, their behavior tells us that this may not actually be the case. A more careful look at the value of information, however, may provide a better explanation for this discrepancy.

3.8.2 The Value of Information in Context

Another way to understand decisionmaking in the face of uncertainty is to generalize the meaning of an event to let us handle cases that might cause different behavior. To understand the value of an umbrella, for example, we could divide days into two classes: rainy days and rain-free days. We can then estimate the value of an umbrella as being its value on a rainy day weighted by the probability of rain plus its value on a rain-free day weighted by the probability that it does not rain. This idea is formalized in decision tree methodology (Breiman et al. 1984), and it turns out that we can use this idea to understand why seemingly valuable information often remains relatively unprotected.

Much as the value of an umbrella depends on the chances of rain, the value of information is highly dependent on its context. Take two random laptop users. One is a sales executive for a software company whose laptop contains a list of important customer contacts and pricing information for the products that she sells. The other is a marketing manager at a plumbing fixtures company whose laptop contains information on the company's plans to change the composition of the bronze alloy that they use in certain pipes and to outsource the production of these pipes to China. If these two laptop users switch laptops, both will end up unhappy. The software sales executive doesn't care about the manufacture of pipes, and the marketing manager doesn't care about the software pricing information. In both cases, the information is very valuable to its owner but of little to no value to most other people.

Surveys of corporate IT departments seem to indicate that this model is fairly reasonable. A recent study (Ponemon 2009) estimated an approximately 0.24% chance that lost or stolen data would have significant value to anyone who might end up with the information. Surveys have also suggested that people estimate the value of the information on their laptop computer to be approximately \$1 million

(Sturgeon 2006). This provides enough information to use decision tree methodology to estimate how the owners of laptops should behave.

Suppose that a laptop owner believes the value of his information is \$1 million to him and a few select others but essentially zero to the remaining 99.76% of the population. If this is the case, then we might expect the owner to behave as if the data on the laptop is worth only \$2,400:

$$(\$1\text{ million})(0.0024) + (\$0)(0.9976) = \$2,400$$

Businesses' enthusiasm for whole-disk encryption is less than information security experts would like it to be. Perhaps the uncertainty about the value of the data on the laptop can explain this behavior. If that's the case, then it is certainly seems possible that the lack of protection of data on laptops is consistent with a valuation of \$1 million for those data.

3.9 Summary

The traditional risk management model, where we quantify a risk as the expected loss associated with an event, seems to fail badly in the case of information security. This should not be surprising since neither the probability of a security breach nor the damage resulting from an incident is known with any significant accuracy. Despite this limitation, other models that require less precise information can be used, but none of the models are ideal.

The theory of marginal utility provides an explanation for some observed behavior but not others. In other cases, if we have a reasonable estimate for one of the two unknowns, either the probability of an event or the loss associated with an event, it's possible to get similarly reasonable estimates for the other unknown based on observed behavior. This may give us estimates for values that are really subjective probabilities instead of objective probabilities, or values that are for utility instead of just financial value.

In any event, it seems that a closer look at how to quantify the risk to information can do a reasonable job of predicting the current situation in which businesses protect their information less thoroughly than information security experts would like them to.

3. Commentary: Harvesting the Ripe Fruit: Why Is It so Hard to Be Well-Informed at the Moment of Decision?

Adam M. Finkel

3.C.1 Introduction

The mindset and the algorithms that enable the systematic appraisal of the value of information (VOI) confer a power that is hard to imagine refusing. And yet, VOI methods remain confusing and underutilized. We devote hundreds of billions of dollars each year to public sector decisions (waging war, protecting health, safety, and the environment, etc.), so the stakes—measured by the potential for wasted costs or harms mistakenly tolerated—are vast. We also spend billions of dollars each year on research ostensibly related to these decisions (i.e., applied research and data collection), and sometimes the tiniest ripple in the realm of information can cause huge waves in the much larger realm of the costs and benefits of decisions. For example, in hindsight it is at least conceivable that a few thousand dollars of additional effort spent resolving the factual controversy over whether Saddam Hussein was trying to obtain uranium from Niger might have changed the decision whether to begin a war that lasted for nearly nine years. In other words, spending on research only a percent or so of what we spend on control may be a foolish way to economize, but even more questionable is our refusal to spend even a percent of our research budget on asking the meta-questions that would optimize the value of that research.

The tools of VOI analysis can help us decide how much we need to know before we should feel ready to make a decision, and can even help channel our efforts toward or away from specific subsets of information collection, yet they are barely used where they are most needed. This essay responds to Luther Martin's chapter about the value of information in data security and then tries to explain why more general concepts of VOI remain curiosities rather than centerpieces, from the perspective of a former federal agency senior executive who has tried to evangelize about VOI methods to environmental, health, and safety agencies (especially the U.S. Environmental Protection Agency) over the past 25 years.

A.M. Finkel (✉)
Penn Program on Regulation, University of Pennsylvania Law School, Philadelphia, PA, USA
e-mail: afinkel@law.upenn.edu

3.C.2 Further Thoughts on the Value of Business Information

Luther Martin uses his deep knowledge about computer security and the slow adoption of inexpensive safeguards by businesses to make some excellent points about how risk analysis can shed light on the value of protecting data. Martin essentially takes a revealed-preference approach to estimating the value of safeguarding laptop data: according to this approach, the demand for encryption software should be a function of the losses incurred if a user who can exploit the information acquires it, multiplied by the probability of this untoward event. However, Martin might have considered one or more of these three refinements to this basic (probability × consequence) approach to estimating the value of preventing a loss:

- *Risk attitude.* The value an individual places on protecting an asset may, of course, be either smaller or larger than the expected monetary consequences of the threat (because of a nonlinear relationship between value and utility), or not fully captured by expected utility (see the literature on decision regret, prospect theory, and other refinements of expected utility, including Bell (1982) and Kahneman and Tversky (1979)).
- *Interindividual variability in preference.* It is possible that the subpopulation of customers who do buy laptop insurance or encryption software are precisely the users who place a higher relative value on their own data.
- *Uncertainty in risk.* The point estimate of 0.0024 for the probability that someone who finds a laptop can exploit valuable data therein seems to imply randomness where non-randomness applies: where valuable business data are known to exist, thieves may not target victims at random, and someone who finds a lost laptop and doesn't know how to exploit its data may be able to find someone who does value the data highly and sell the laptop to him.

Nevertheless, his chapter shows that valuation—a concept that many readers of this collection may think of only in terms of willingness to pay for intangible benefits to longevity, quality of life, or the environment—is applicable to intangible market commodities as well. That the demand for tools that protect stolen data from being used against the victim is weaker than the purveyors of the tools would prefer is a familiar story to regulatory agencies, who often struggle to mobilize public support for protective measures or to catalyze public willingness to take self-protective steps.

3.C.3 Two Kinds of Value of Information

In his opening remarks at the Resources for the Future workshop, Lawrence Friedl recited two lists of technical terms, each used by a particular discipline, to show compellingly that collaboration between (say) geoscientists and economists is made more vexing by the lack of jargon in common. I think Martin's chapter shows in addition that some terms that *appear* to be common to multiple disciplines may be an even bigger impediment to interdisciplinary collaboration because disciplinary specialists *believe* they use the term the same way as those in other fields do. The kind of "value of information" I work with and the kind Martin writes about here have much in common: they both start from concern about losing something of value. But in data security, the information *itself* is the commodity that we value (in the sense of "are afraid to lose"), whereas in more general decision theory, information is something we ascribe value to—it is a means to avoid losing something *else* of value. In the latter context, information—perhaps more precisely called research—has value (in the sense of "efficacy") because armed with it, we can get more of what we really value.

So in Martin's example, the value of information is akin to the value of life in the kind of regulatory decision problem I will sketch out below. By putting a value on laptop data, we can help determine which of the decisions we could make would enable us to best protect the data, in light of the increasing cost of achieving more assurance. But the choice of how assiduously to protect data, like every other important and nontrivial decision each of us will ever make, is complicated by uncertainty. Martin could have extended his chapter, therefore, to ask some value-of-research questions, all flowing from the idea that we might seek information to better protect our data (our information). How uncertain is the assessed probability of the data falling into the hands of someone who could use the data to harm me? How uncertain is the loss I would incur in this eventuality? What could I learn that would reduce my uncertainty about these parameters, and how much would it cost me to learn more? These are the raw materials for assessing the value of information—whether it will be harnessed to protect lives, ecosystems, corporate profits, or in this somewhat confusing mix of two different usages of the same word, to protect *other* information.

3.C.4 The Classic VOI Setup for Risk Regulatory Decisions

Information has value only insofar as it reduces potential losses that follow from suboptimal decisions (Finkel and Evans 1987; Yukota and Thompson 2004). That bold statement already excludes some of the most important aspects of how we colloquially treat information—in the immortal words of the Faber College motto in the 1978 movie *Animal House*, "Knowledge is Good," after all—but the tight link between the performance of decisions and the salutary power of information is what enables quantitative estimates of VOI and ordinal comparisons among possible research strategies. To set up a VOI inquiry, therefore, the involved protagonists have to be willing to answer certain preliminary questions (here I pose them generally, but they are also specific to the kind of regulatory cost-benefit examples I work in):

- What are we trying to achieve? (In environmental, health, and safety regulation, to reduce risk net of the cost of control, otherwise known as "maximize net benefit").
- What choices do we have? (Here, either do nothing, or implement one or more control options whose costs and benefits can be estimated).

- What don't we already know perfectly? (Although I have written extensively about inattention to uncertainty in regulatory cost (Finkel et al. 2006; Finkel 2010), assume for simplicity here that only the risk is uncertain).
- Which option would outperform all others, for each possible value of the uncertain quantity? (If we knew exactly how large the risk was, how tightly would we control it to avoid errors of overspending and underspending?)

Those questions set the stage for the "VOI question." The real power of this method is that it encourages those involved to try a leap of insight—to imagine that they have already made a decision and can look back with pride or regret on what they did or might have done.[1] The VOI question therefore is: How much do we stand to lose if we decide now and later come to wish we had chosen otherwise?" The fundamental assertion of VOI analysis is that perfect information is worth exactly as much as the expected losses we stand to incur by doing the best we can now, within the shadow of uncertainty. This leads directly to the fundamental corollary: information that costs[2] less than it is worth should be pursued, while information that costs more than the benefits it delivers should be shunned.

The following example, adapted from my 1987 paper with John Evans, shows the relationship among choices, uncertainty, and information value. Assume that we face an uncertain risk to human health that, if left uncontrolled, will kill R people every year, and assume that we value a statistical life at $1 million (this estimate was less appallingly low when we developed this example for illustrative purposes 25 years ago). The agency charged with regulating the risk has three possible choices: (1) do nothing; (2) require polluters to spend a total of $10 million every year on controls that will reduce the risk by 80%; or (3) require polluters to spend $20 million per year on more efficient controls that will reduce the risk by 96%.

The total cost (TC) of each option, the control costs plus the monetized health harms left behind, is a function of only one unknown (R), and the values of TC (in $million) for each decision option are (1) R; (2) 10 + 0.2R; and (3) 20 + 0.04R. Simple algebra shows that for $R < 12.5$, TC is least when Option 1 is chosen, and that for $R > 62.5$, TC is least when Option 3 is chosen; for any intermediate value of R, Option 2 has the least cost. Figure 3.1 shows the TC of each option; the dotted line demarcates the least-cost frontier as a function of R.

---

[1] A great book was made into an inscrutable movie about this very sort of insight. The text of *2001: A Space Odyssey* makes clear that while "Thus Spake Zarathustra" was blaring through the speakers, the alien monolith was implanting in the prehominids a vision of how using rudimentary tools could enable them to eat in safety, away from the danger posed by predators. According to author Arthur C. Clarke, this training helped humans not only to survive but also to evolve habits of mind (envisioning a future where what one is about to decide has become part of history) that have served us well.

[2] To oversimplify, the cost of obtaining information consists of direct resource costs plus any harms that mount up because information takes time to develop, and because the resulting delay may have its own consequences.

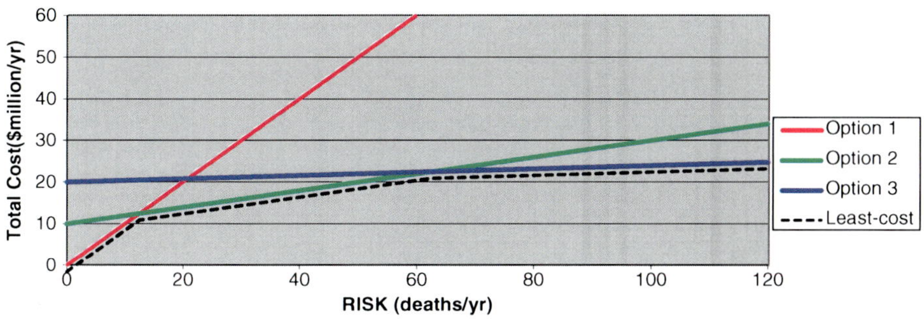

Fig. 3.1 The total cost (control costs plus the monetized harms of risks not controlled) of 3 hypothetical decision options, as a function of the uncertain baseline risk

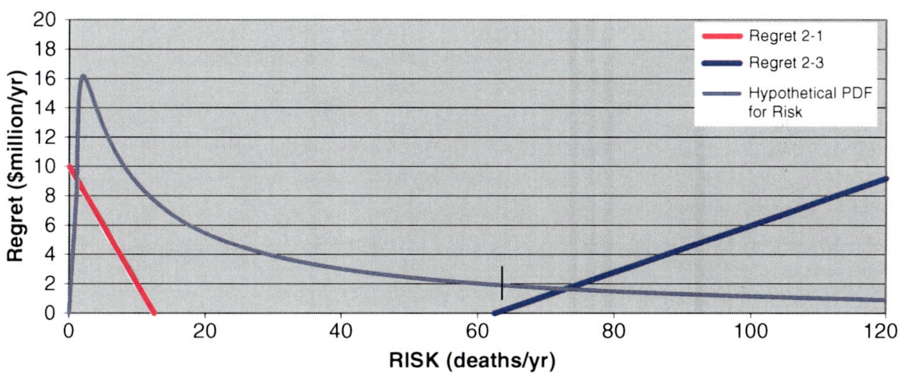

Fig. 3.2 The additional cost of choosing Option 2 as compared to Option 1 (the red line) or as compared to Option 3 (the blue line), as a function of the uncertain baseline risk

Now assume that R is uncertain—because if it isn't, we already know what to do and no additional information is germane, nor worth anything to obtain. Suppose R has an expected value of 29.6 but is lognormally distributed (about a median value of exactly 4) with a logarithmic standard deviation of 2 (i.e., the natural log of R is normally distributed with a standard deviation of 2). In this case, it turns out there is about a 72% chance that R is less than 12.5, about a 21% chance it is between 12.5 and 62.5, and about a 7% chance R exceeds 62.5. But again, on average R is 29.6, and the *expected* cost of Option 2 is still less than that of either of the other two choices.

So if we have to live with the uncertainty, Option 2 is the best we can do. But with 79% probability (72 + 7), we will someday look back at that choice with regret: if we overestimated R, we could have saved $10 million per year (imposed no controls) and accepted a small amount of risk, whereas if we understated R, we could have spent $10 million per year more and reduced a very large risk more thoroughly (thereby saving lives worth more in total than $10 million per year). Figure 3.2 shows the *regret* of choosing Option 2 as a function of what we might learn R to actually be; superimposed on Fig. 3.2 is the (lognormal) uncertainty in R that we might choose to eliminate with more information. The regret of choosing

Option 2 when Option 1 was wiser follows the line (10 – 0.8 R); the regret of choosing Option 2 when Option 3 was wiser follows the line (0.16 R – 10). By integrating the expressions

$$\int_{0}^{12.5} (10 - 0.8R)f(R)\, dR \quad \text{and} \quad \int_{62.5}^{\infty} (0.16R - 10)f(R)\, dR,$$

where f(R) denotes the probability density function for the uncertain risk, one can calculate the expected regret of choosing Option 2 without gathering more information; in this example, it amounts to roughly \$8 million per year. So VOI theory dictates that perfect knowledge about the exact value of R, which would allow us to choose an option with perfect confidence that it was the best available one, is worth about \$8 million (per year, or converted to net present value using an appropriate discount rate).

Note that this sum is fairly large relative to the general stakes of this decision. On average, we expect to spend \$10 million and incur (29.6 × 0.2) = \$5.9 million in cost attributable to "lives not saved," so is worth roughly half this \$15.9 million to eliminate the uncertainty.[3] But this ratio is large because the uncertainty in R is quite large, and because the best a priori decision is superior only 21% of the time. *When the best decision is this precarious, knowledge is more than "good," it is valuable*. But by the other side of the same coin, when uncertainty is small and/or when it would take a large misestimation to make a different decision better than the one about to be chosen, knowledge could have little extra value, and quixotic attempts to obtain it may cost much more than they help.

The example above may help elucidate some practical rules-of-thumb about VOI:

- To a rough first approximation, bigger decisions justify more extensive research, as do larger uncertainties. Obviously, it makes little sense to buy a \$5 racing magazine to help decide which horse to place a \$1 bet on.
- The converse of this advice, however, is more important: not all big decisions justify extensive research. The lesson of Fig. 3.1 (although lessons are made to be challenged) is that once we've learned enough to be very (completely) confident

[3] On average, therefore, the total cost of the best *a priori* decision is roughly \$16 million, but it would be incorrect to assume that perfect information is worth nearly as much as the full \$16 million. One must remember that even with perfect information, there will be costs. In fact, if we knew the exact value of R, we might be able to get by with Option 1 and incur costs of roughly \$6.25 million (if R turned out to be between 0 and 12.5), or we might learn that Option 3 is needed, with associated costs of \$22.5 million (if we learned that R was exactly 62.5) or more than \$28 million (if we learned that R was 200 or greater). On average, we could *decrease* our expected costs from \$15.9 million to about \$8 million if we learned R exactly: that difference is mathematically the same as the expected regret of deciding now, and both are tantamount to the value of perfect information.

that R must lie between 12.5 and 62.5 deaths per year, further information has little (zero) value. With one important caveat (see the last paragraph of this essay), if nothing you can learn can make you want to change your mind, it's time to stop dithering and act.

- When it's clear we do need to know more, VOI theory says that not all uncertainty reductions have equal value, and that small targeted reductions can be much more useful than large reductions achieved by brute force. Although this is easier said than done, the goal of reducing uncertainty in this context should be to end up with an uncertainty distribution completely contained within one of the regions in a schematic like Fig. 3.1, where one particular decision dominates all others. In practice, this sometimes means focusing attention on one or both tails of the current uncertainty distribution; if you can rule out the tails, you can "rule in" the best course of action. In the general case where uncertainty stems from several separable components, one can simulate the results of a research investigation before conducting it, to see what effects it would have on the tails (or on any part of the uncertainty distribution that straddles more than one region where a particular decision is optimal). In human health risk assessment, both the extent of exposure to the stressor and its potency (the probability of harm per unit of exposure) are always uncertain—so it is always possible to envision what the uncertainty distribution would look like after resources were expended to obtain N more environmental samples, or instead to conduct dose-response experiments on N more laboratory animals or epidemiologic investigations on N more exposed humans.[4]

3.C.5 *Whence the Resistance?*

As a naive graduate student in the mid-1980s, I went with my mentor to several program offices of the U.S. Environmental Protection Agency (EPA), full of enthusiasm for a set of tools that could shed light on how much research is too much, but especially on how pound-foolish it is to make billion-dollar decisions with million-dollar (or smaller) research programs behind them. Between our salesmanship and the agency's receptivity, little was accomplished. More than 20 years later, I found myself on the Board of Scientific Counselors advising EPA's Office of Research and Development on how it could develop strategic plans for environmental research in support of EPA's program offices, and I found that VOI thinking had advanced scarcely at all in the intervening decades. I offer several reasons for the slow adoption of VOI methods.

[4] It is possible, as was mentioned at the conference, that new information can appear to increase uncertainty. I believe that this view is illogical, and I prefer to explain it as "more information can sometimes reveal that there was more uncertainty than you realized at the time: the uncertainty is smaller now, though perhaps larger than the overconfident view you held previously."

First, agencies are often risk averse and populated by risk-averse individuals. A tool and a mindset that could reveal the general need for substantial increases in applied research—but could also suggest specific instances where additional research would be superfluous—may be seen as a mixed blessing at best.

Second, because VOI is in essence the value of uncertainty reduction, it presupposes the willingness and the capability to estimate how uncertain are the key parameters (risk, cost, efficiency of controls, etc.). Agencies may be resisting this step rather than the VOI mindset *per se*. However, I think this becoming a less likely explanation, as EPA and the other agencies have made tremendous strides in making quantitative uncertainty analysis of risk routine and advancing new methods for it (see, e.g., NRC 2009), although without commensurate attention to uncertainty in cost, these advances may promise more than they can deliver.

Resistance to VOI may also be a symptom of resistance to more general methods of quantitative decision analysis. In my experience, people elected or appointed to positions of decisionmaking responsibility sometimes believe, overtly or tacitly, that they must be good decisionmakers—that their innate skill (or their well-developed gut feelings) surpasses any formal method.

Moreover, since the organizational goal of VOI analysis is to harness research plans to improve decisionmaking, well-intentioned research managers may believe they have already made that leap when they take a baby step towards it without having used any VOI methods. I have seen several research programs highlight the fact that they are now beginning to link their research agenda to "serve the needs of decisionmakers"—but by this they often mean that they ask the program offices for clues as to what problems are most important to them and try to focus more of their research efforts on the A-list problems. This is assuredly an improvement over any less interactive method of setting research priorities, but of course it never considers decision regret (the sine qua non of quantitative valuation of information), simply because no decisions are ever mentioned. Just as big uncertainties do not necessarily imply valuable research, big problems are an even less reliable indicator of critical knowledge gaps. Big problems with clearly optimal (even if costly) solutions don't demand extensive research, nor do big problems with intractable uncertainties. But any dialogue across the research-program divide may tend to foment the sense that all the desired conceptual linkages have also been forged.

That leads to perhaps the most fundamental problem of all. EPA and many other agencies operate under a linear research-analysis-decision paradigm, probably first codified in the landmark 1983 National Academy of Sciences report *Risk Assessment in the Federal Government: Managing the Process* ("The Red Book"), in which little thought is given to solutions until the problems are analyzed *ad nauseam*. Statutory design sometimes dictates such a process: for example, Congress has told EPA to refine its estimates of the risks of criteria air pollutants every 5 years, but the National Ambient Air Quality Standards that EPA sets are aspirational only and dictate no specific actions of any kind. In other situations, EPA chooses to study individual substances and set emissions or concentration standards, rather than to compare any actual controls—and yet arguably, the nation does not have a "dioxin problem" but a series of product and technological choices that each contribute to an unacceptable total dioxin load in the environment. The dogma that risk assessment must precede and inform risk management is actually diametrically counter to decision theory, which starts from the premise that assessment exists to help discriminate among choices, not to exhaust itself and only then pass forward (incomplete) understanding to those responsible for thinking about solutions. If information has value only insofar as it sheds light on choices, and no one thinks hard about choices until too late, then all the resources previously devoted to information collection will have been aimless, and the urgency of doing anything at all, rather than "calling for further study," may be irresistible.

So in part out of concern for a process that does not harness research to reduce decision regret, but more out of a larger concern that we are becoming too good at doing cost-benefit analysis and yet are not solving the problems we study, I have proposed that we consider a new policy paradigm I've termed solution-focused risk assessment (SFRA) (Finkel 2011; NRC 2009, Chapter 8). By asserting that cost-benefit analysis should not begin in earnest until after agencies and their affected stakeholders have given some concrete thought to solutions to be analyzed, SFRA would also provide a template for VOI methods to flourish. Perhaps even more significantly, it could enable the beginnings of a feedback from the study of problems to the study of solutions. One conundrum of VOI theory is that more and better choices can sometimes increase the value of information—you may not regret flipping a coin and picking one of two lousy choices, until someone suggests a third alternative for which better information could truly be a life-saver. A new relationship between analysis and action that encourages the analyst to say, "This research would help you make a better choice, but here's another choice that might be even better," is, in my view, the true validation of the value of wisdom, of which VOI is the price of admission.

References

Adams, J. (1997). *Cars, cholera and cows: Virtual risk and the management of uncertainty*. Manchester: Manchester Statistical Society.

Akerlof, G. A. (1970). The market for 'lemons': Quality uncertainty and the market mechanism. *Quarterly Journal of Economics, 84*(3), 488–500.

Bell, D. E. (1982). Regret in decision making under uncertainty. *Operations Research, 30*(5), 961–981.

Bernoulli, D. (1738). Specimen theoriae novae de mensura sortis. *Commentarii Academiae Scientiarum Imperialis Petropolitanae, 5*, 175–192. (Reprinted in translation as Exposition of a new theory on the measurement of risk. *Econometrica, 22*, 123–136, 1954)

Böhm-Bawerk, E. (1891). *The positive theory of capital* (William A. Smart, Trans.). London: Macmillan and Co.

Breiman, L., et al. (1984). *Classification and regression trees*. Monterey: Wadsworth.

Checkpoint. (2009, October 21). *Guide to the TCO of Encryption* (Checkpoint Software Technologies) Redwood City, California, available at http://security.networksasia.net/content/guide-tco-encryption accessed July 17, 2012.

El-Gamal, M. A., & Grether, D. M. (December 1995). Are people Bayesian? Uncovering behavioral strategies. *Journal of the American Statistical Association, 90*(432), 1137–1145.

Ernst & Young LLP. (2009). *Outpacing change: Ernst & Young's 12th annual global information security survey* at http://www.ey.com/Publication/vwLUAssets/12th_annual_GISS/$FILE/12th_annual_GISS.pdf.

Finkel, A. M. (2010, October 20). *Out of balance: (Why) are risk assessors more interested in uncertainty and variability than regulatory economists are?* Presentation at the Society for Benefit-Cost Analysis annual meeting, Washington, DC.

Finkel, A. M. (2011). Solution-focused risk assessment: A proposal for the fusion of environmental analysis and action. *Human and Ecological Risk Assessment, 17*(4): 754–787. See also five commentaries on this article in the same issue of the journal, 788–812.

Finkel, A. M., & Evans, J. S. (1987). Evaluating the benefits of uncertainty reduction in environmental health risk management. *Journal of the Air Pollution Control Association, 37*(10), 1164–1171.

Finkel, A M., Shafir, E., Ferson, S., Harrington, W., et al. (2006). *Transferring to regulatory economics the risk-analysis approaches to uncertainty, interindividual variability, and other phenomena* (National Science Foundation Grant #0756539). Decision, Risk, and Uncertainty program, (Human and Social Dynamics of Change competition).

Johnson, C. (2008, October 15). The global state of information security. *CIO Magazine.*

Kahneman, D., & Tversky, A. (1979). Prospect theory: An analysis of decision under risk. *Econometrica, XLVII*, 263–291.

Knight, F. (1921). *Risk, uncertainty and profit.* Boston: Hart, Schaffner & Marx/Houghton Mifflin Co.

National Research Council. (1983). *Risk assessment in the federal government: Managing the process (the "Red Book").* Washington, DC: National Academy Press.

National Research Council. (2009). *Science and decisions: Advancing risk assessment.* Washington, DC: National Academy Press.

Nelson, P. (1970). Information and consumer behavior. *Journal of Political Economy, 78*(2), 311–329.

Plunkett, E. (Lord Dunsany). (1935, January 28). Jorkens' Revenge. *The (London) Evening Standard.* (Reprinted in *The collected Jorkens* (Vol. 2). San Francisco: Night Shade Press, 2005)

Ponemon, L. (2009, April 22). *The cost of a lost laptop.* The Ponemon Institute. Michigan: Traverse City

Soo Hoo, K. (2002). *How much is enough: A risk management approach to computer security.* Ph. D. dissertation, Department of Management Science and Engineering, Stanford University, Stanford.

Stoneburner, G., Goguen, A., & Feringa, A. (2002, July). *Risk management guide for information technology systems* (NIST Special Publication 800-30). Gaithersburg: National Institute of Standards and Technology.

Strategic Data Management. (2008, April 21). *Effective data management: Imperative to align business and IT interests.*

Sturgeon, W. (2006, January 27). Could your laptop be worth millions? *C-NetNews.com* http://news.com.com/Could+your+laptop+be+worth+millions/2100-1029_3-6032177.html

Sutton, W., & Linn, E. (2004). *Where the money was: The memoirs of a bank robber.* New York: Broadway.

Viscusi, W. K., & Aldy, J. (2003). The value of a statistical life: A critical review of market estimates throughout the world. *Journal of Risk and Uncertainty, 27*(1), 5–76.

von Neumann, J., & Morgenstern, O. (1944). *Theory of games and economic behavior.* Princeton: Princeton University Press.

Yukota, F., & Thompson, K. (2004). The value of information in environmental health risk management decisions: Past, present, future. *Risk Analysis, 24*(3), 635–650.

Chapter 4
Valuing the Potential Impacts of GEOSS: A Systems Dynamics Approach

Michael Obersteiner, Felicjan Rydzak, Steffen Fritz, and Ian McCallum

Abstract Global earth observations are perceived as instrumental to attaining sustainable development goals. Methods to assess the long-run socioeconomic benefits of the emerging global Earth observation system of systems (GEOSS) as an integrated multisensor infrastructure have been missing to date. This chapter presents a systems dynamics approach to assess the effect of improvements in Earth observations across the nine societal benefit areas of the Group on Earth Observation (GEO). Two types of integration are assessed with the proposed model structure: (1) measuring benefits in an integrated assessment environment (e.g., improved weather forecasting through better measurement of cloud properties could lead to benefits in the agriculture, energy and water sectors); and (2) measuring benefits of an integrated observing system (e.g., in areas with high cloud cover, improvements in the resolution of optical sensors will lead to benefits only if linked to supplementary observing systems such a radar or ground surveys). The benefits from integration relate mostly to economies of scope on both the observation and benefit system sides. Cost reduction from economies of scale are derived from a global or large scale observing system vis-à-vis the currently prevailing patchwork system of national or regional observing systems. Results indicate that the total system benefits of GEOSS are usually orders of magnitude higher than their costs. Benefits are also policy dependent and tend to increase with the degree of implementation of mainly international environmental agreements.

Keywords Dynamic modeling • Earth observation • Earth system • Global Earth Observation System of Systems • Society benefit areas • Value of information

---

M. Obersteiner (✉) • F. Rydzak • S. Fritz • I. McCallum
International Institute for Applied Systems Analysis, Ecosystems Services and Management Program, Laxenburg, Austria
e-mail: oberstei@iiasa.ac.at; rydzak@iiasa.ac.at; fritz@iiasa.ac.at; mccallum@iiasa.ac.at

R. Laxminarayan and M.K. Macauley (eds.), *The Value of Information: Methodological Frontiers and New Applications in Environment and Health*, DOI 10.1007/978-94-007-4839-2_4, © Springer Science+Business Media Dordrecht 2012

4.1 Introduction

Managing global change involves managing risk in a complex system undergoing major transitions. The Earth system in the Anthropocene (Crutzen and Stoermer 2000) is defined by its interdependencies between social, economic, and environmental subsystems constituting a complex dynamic system. Appropriate management of such a system can come only from improved monitoring and understanding of the underlying processes and their interdependence. Recent developments in the fields of information technology, data infrastructures, and Earth observation enable knowledge gains and consequently higher predictive performance, which provide the basis for improved decision making across spatial scales. The Global Earth Observation System of Systems (GEOSS), coordinated by the Group on Earth Observations (GEO),[1] aims at connecting the diverse sets of monitoring systems to support decision making of policymakers, resource managers, scientists, and average citizens.

Despite the obvious advantages that Earth observations can bring to decision making, we lack appropriate theoretical and methodological frameworks to assess the economic and wider societal benefits of a GEOSS-like infrastructure (Craglia et al. 2008). There is extensive literature on the benefits of weather forecasting (Adams et al. 1995; Katz and Murphy 1997) but relatively little assessment work in other fields of Earth observation. Furthermore, the available studies are mostly sectoral and focus on particular areas, such as biodiversity (Leyequien et al. 2007; Muchoney 2008). Case studies on the value of improvements in Earth observation systems are usually very specific and not generalizable. For example, Considine et al. (2004) analyzed the benefits of improved hurricane forecasting in oil and gas production in a confined geographic area. Bouma et al. (2009) examined the effect on water quality management in the North Sea of improved in situ observation networks or remote sensing–based observing systems. Wieand (2008) quantified the effects of an integrated ocean observation system on recreational fishing. Despite their thorough, in-depth analysis and high level of sophistication, these studies do not provide a methodological framework, and integrated assessments of the total global consequences within and across all areas remain lacking.

The need for such evaluation led to a European Commission–sponsored project, Global Earth Observation—Benefit Estimation: Now, Next and Emerging (GEOBENE)[2]—the world's first attempt to systematically and comprehensively study the benefits of a global system of system of Earth observations (European Commission 2008). GEOBENE's goal is to develop methodologies and analytical tools to assess the economic, social, and environmental effects of improved quantitative and qualitative information delivered by GEOSS, in and across nine societal benefit areas (SBAs)—disasters, health, energy, climate, water, weather, ecosystems, agriculture, and biodiversity. This chapter begins with the presentation of the systems

[1] http://earthobservations.org

[2] http://www.geo-bene.eu

dynamics model that was built to evaluate the total effects of Earth observation. The following section describes the methodology used for the systems analysis. Section 4.3 discusses a selected set of results assessing the effect of GEOSS improvements. The final section makes some closing methodological remarks.

4.2 Methods and Tools

4.2.1 Concept

The basic concept behind the work presented in this chapter is to adapt and apply methods and tools typically used in technological foresight studies for impact assessment of GEOSS scenarios. This is illustrated in Fig. 4.1. The first and principal challenge of the modeling approach is to assimilate the many heterogeneous sources of information in VOI studies carried out in the area of Earth observation into an integrated global impact study. The primary sources were direct results from GEOBENE models, impact figures from published articles and sector reports, and information obtained from expert interviews or online research. Generally, VOI studies are confined to a particular place, time and sector. Impacts are rarely reported on global aggregates or carried out using a wider economic system representation to account for the many potential feedbacks. Therefore, existing information usually needs to be adapted through aggregation to mimic effects on a global level and over long time horizons.

The next challenge is to integrate these aggregated technology storylines or economic assessment estimates in the dynamic modeling of global change. The effects across many components of the socioeconomic system are quantified using the Full of Economic-Environment Linkages and Integration dX/dt (FeliX) model. To achieve integration, a "logic model" is typically constructed first, outlining the value chain of the use of new products for Earth observation (EO). In a second step, an adapted representation of the value chain is coded in the FeliX model as a new component. The impacts of changes in the EO infrastructure are played out on a general scenario storyline that includes the major developments of global change. The basic socioeconomic and Earth system drivers are provided exogenously through socioeconomic scenarios and storylines as well as their respective Earth system projections. Assumptions on the technological development and deployment of EO technologies are harmonized with the global change storylines. Assessment of GEOSS scenarios is then carried out through the combination of the various VOI information feeds and a global change scenario.

The societal benefit areas set the boundaries of the FeliX model. For the formulation of SBA-specific model structures, literature reviews and expert consultation were carried out to identify physical properties of GEOSS improvements and how they might further propagate through the benefit system defined by the SBAs. For example, specific model structures on phenomena closely related to climate

Fig. 4.1 Basic approach aggregating and integrating VOI knowledge in GEOSS foresight studies

include atmospheric concentration of CO_2 caused by human activities and the associated carbon cycle. The basic dynamics of the climate system have been intensively researched and described in the literature (Oeschger et al. 1975; Goudriaan and Ketner 1984; Bolin 1986; Rotmans 1990; Nordhaus 1992; Fiddaman 1997), which allowed for adoption of quantitatively expressed relations of the system components in the FeliX model structure. In cases where such relations have not been quantitatively established, group model building sessions (Richardson and Andersen 1995; Vennix 1996; Andersen et al. 1997) or online research was conducted, and subject matter experts defined and quantified the relations of interest and constructed parts of the model. The outcome of this work is a system dynamics model consisting of a set of interrelated differential equations allowing for computer simulation that gives quantitative results. In our work to tease out the different relationships and links between the SBAs and the effects of GEOSS, we found the discussion around model outcomes and the creation of the model scenarios very useful. In addition to our efforts to set realistic model links and to compare the scenarios with other global projections, we found that the group model sessions provided insight into the influence of GEOSS on the SBAs and the total system.

4.2.2 FeliX Model

The FeliX model is a system dynamics type of model, following an approach originally developed by Forrester (1958, 1961), Meadows et al. (1972), Richardson and Pugh III (1981), and Sterman (2000). System dynamics models attempt to capture the interactions within a closed system. Most variables are therefore endogenous (i.e., contained within the system represented by a system dynamics model). To describe the system structure, the model focuses on the flow of feedbacks that occur throughout its parts (feedback loops): a change in one variable affects other variables over time, which in turn affects the original variable, and so on. The dynamic behavior then occurs when flows accumulate in stocks (e.g., atmospheric carbon). Special dynamic notions are also given by delays and nonlinear relations between the system elements. All these elements produce changes in the way the system has performed in the past and might evolve in the future.

The FeliX model, following the system dynamics approach, attempts a full systems perspective, where the underlying social, economic, and environmental components of the Earth system are interconnected to allow for complex dynamic behavior characterizing the Anthropocene (Schellnhuber 2009). A change in one area often results in changes in other areas. For instance, depletion of oil and gas, a source of energy, may affect population growth but also put pressure on the agriculture sector to produce more energy crops as a substitute. As a dynamic model, FeliX captures important stock changes (e.g., depletion of natural resources, accrual of carbon dioxide in the atmosphere) or consequences of certain policies (e.g., afforestation, emissions reduction) over time. The FeliX model was built to achieve congruence with the nine SBAs of GEO. The model structure of FeliX is illustrated in Fig. 4.2, and a detailed description of the FeliX model is provided in Rydzak et al. (2009).

At the core of the economy module is a neoclassical growth model. Capital is an accumulation of investments whereby in FeliX, investments in the energy and the GEOSS sector are accounted for separately. Growth in gross world product is driven by increases in the labor force, which is modeled explicitly in the population module, along with capital accumulation and technological change. The economy module contains a representation of the climate system and takes into account the effects of global average temperature change, according to the DICE model (Nordhaus 1992, 1994). In addition to the climate mitigation measures (i.e., reduction in emissions of greenhouse gases, GHGs) in the DICE model, the FeliX model accounts for climate adaptation to more intense storms, forest fires, droughts, floods, and heat waves and also incorporates prevention and adaptation activities. However, the range of effects from climate change is uncertain, the assumed model parameters were revised and some of the damages explicitly modeled. The DICE model is known to potentially underestimate climate impacts (e.g., Stern 2007).

The FeliX model accounts for CO_2 emissions with a detailed representation of emissions in the energy sector and land-use change. Energy production technologies differ in their carbon intensities. The model accounts for CO_2 emissions from oil, gas, coal, biomass, solar, and wind energy technologies for their full life-cycle. Furthermore, the FeliX model uses the carbon cycle model proposed by Fiddaman (1997): CO_2 emissions accumulate in the atmosphere and are reabsorbed through fluxes to the terrestrial biosphere and the ocean. The model also accounts for CO_2 flux between living biomass and humus and also distinguishes between the ocean's mixed layer and the deep ocean.

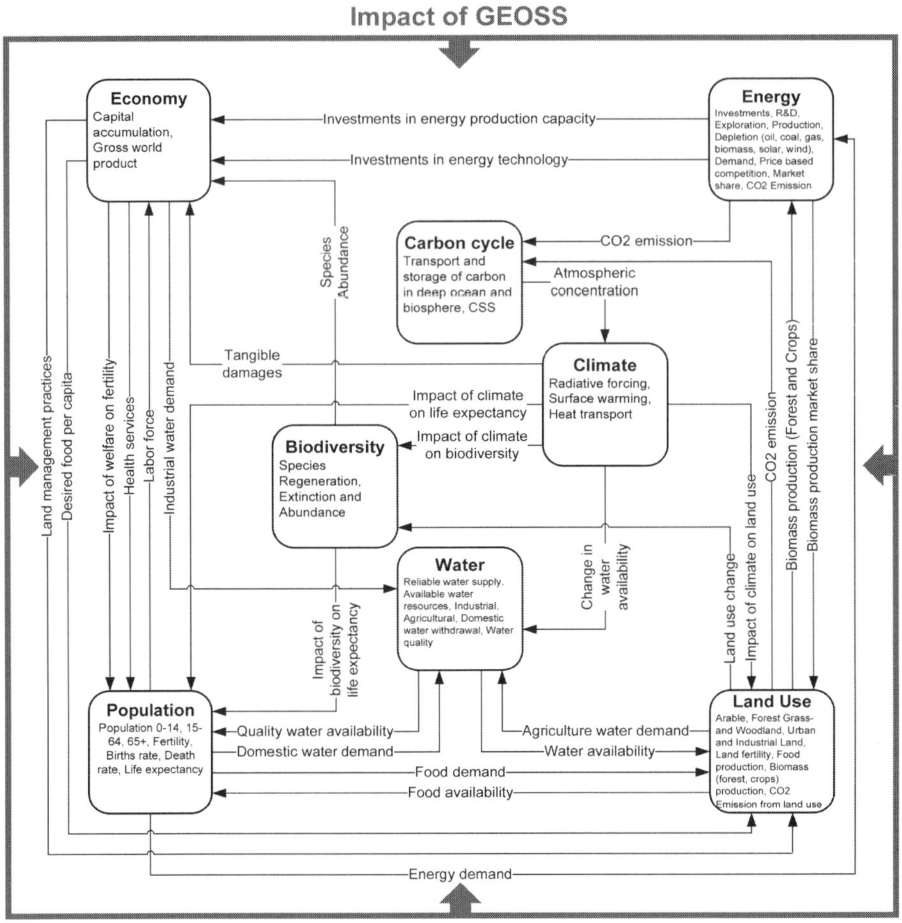

Fig. 4.2 Overview of the FeliX model structure

The FeliX model takes into account the greenhouse effect and, following Nordhaus (1994) and Fiddaman (2002), captures the additional surface warming from the accumulation of CO_2. Positive forcing increases the atmospheric and upper ocean temperatures. Additionally, heat transfer between the atmosphere and the upper ocean and deep ocean is modeled. This disturbance of the climate system, measured by changes in temperature, leads to climate change, accounted for in various sectors of the model. Thus, the consequences of climate change are spread out across the whole model, affecting land quality, population growth, and biodiversity (explicitly accounted for in a biodiversity model module).

Energy demand is driven by population development and the evolution of per capita energy demand. Exploration and production activities, investments in the deployment of energy technologies, R&D activities, and costs of energy carriers are explicitly modeled for each source of primary energy. An economic mechanism of price-based competition between energy sectors determines the market share of primary energy. Technological development is explicitly modeled in the energy and land-use sectors. R&D investments lead to increased growth of either sector- and technology-specific or economy-wide technological change. Technological change is a major driver of economic growth.

The FeliX model contains a "competition for land" module. Various social and economic activities as well as natural processes may change the characteristics of a land type and also cause transformation from one land type to another. A growing population and changing food preferences to more protein-rich diets increase the demand for food production and cause agricultural land expansion into forests and grasslands. The model accounts for more intensive agriculture due to fertilization, irrigation, and genetic improvement. Furthermore, it accounts for new demands for biomass resources for energy purposes and material use, from both forest biomass and biomass from energy crops. The intensification of competition for land between food and energy crops is explicitly modeled. Water resources are explicitly accounted for in a water module. The model allows for additional irrigation according to a water supply function reflecting marginally increasing scarcities of irrigation water.

4.2.3 Benefit Chain Definition Using FeliX

The socioeconomic benefits of GEOSS are quantified using a benefit chain approach (Fritz et al. 2008). Figure 4.3 outlines the five basic steps of benefit assessment using the FeliX model. The first step was building the basic global change model, whose components were described in the above section. The components were adapted to best address the issue of a GEOSS benefit assessment by improvements across SBAs. The basic model structure was designed based on expert consultations identifying the best model structure and feedback along with basic input data. Model components were then validated with designated SBA experts. The FeliX model maps out relations within and among the nine SBAs.

In a second step, the FeliX model was calibrated to historical data using a highly aggregated representation of the Earth system. The calibration was carried out to match multiple observations over the twentieth century. The third step was to use the calibration parameters and conjectured adjustment factors mimicking anticipated technological and societal change to construct a baseline scenario for the twenty-first century. This baseline scenario constitutes the reference for the impact analysis of GEOSS improvements. In a fourth step, the GEOSS scenarios were constructed within and across the SBAs. This step involved working with SBA experts to identify the parameter constellations that would mimic a GEOSS case and choosing the most appropriate parameter values. In the last step, the business-as-usual (BAU) scenario is compared with the GEOSS scenarios. The difference indicates the benefit that GEOSS might have across the SBAs. Multiple indicators are used, including GDP, population, ecosystem value, and the United Nation's Human Development Index.

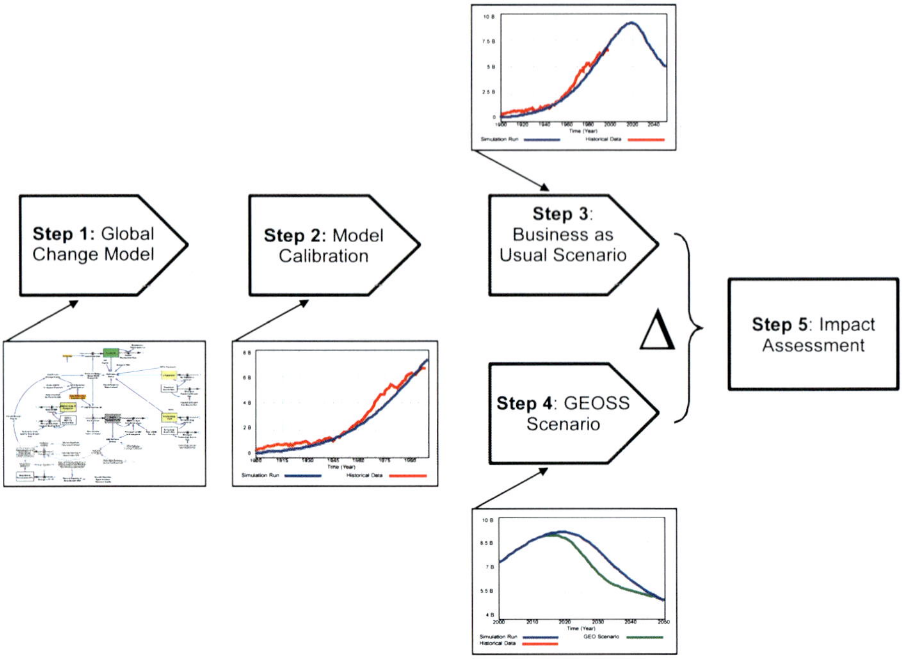

Fig. 4.3 Process of benefit assessment using FeliX

4.2.4 *Example: Population Module*

World population is modeled as an aging chain (Sterman 2000) and accounts for labor market participation by age and gender. Average reproductive lifetime and total fertility is influenced by the degree of economic development and environmental variables. Mortality is influenced by health services, food availability, pollution availability, and quality of health services.

The core sector structure of the population module is presented in Fig. 4.4. There are three population cohorts—*Population 0 to 14*, *Population 15 to 64,* and *Population 65Plus*. The population *Birth Rate* is determined by average *Reproductive Lifetime* and *Total Fertility*, which in turn is influenced by *World GDP per Capita Ratio*. In the GEOSS scenarios we assume that there is no direct EO impact on reproduction.

Each population cohort differs with regard to mortality. As is illustrated in Fig. 4.5 *Life Expectancy* is determined by the degree of health service provision, adequacy of food supply, and the level of pollution.

It is assumed that wealthy societies can invest more in health services and thus extend life expectancy. Health services can range from access to preventive vaccination programs to life-saving measures in case of incidences of cardiovascular diseases. Adequate food supply is determined by minimum calorie intake and the total amount of food supplied. Beyond basic food supplies, an impact function of

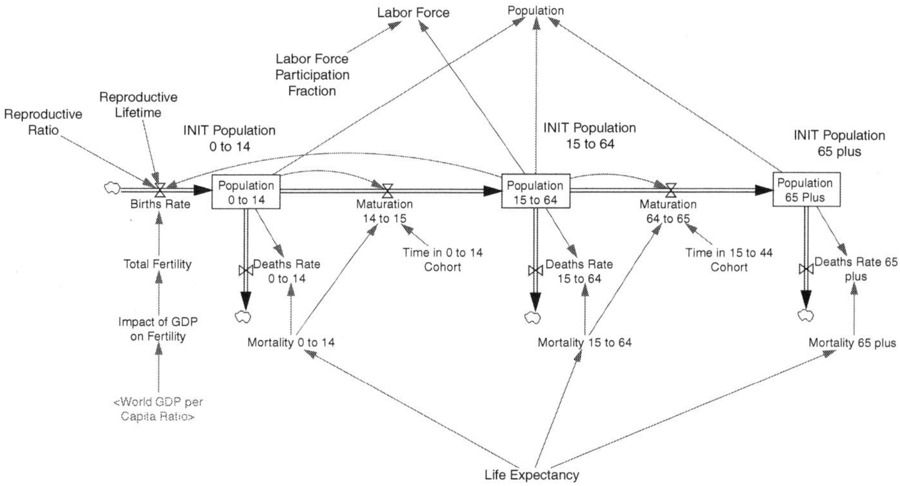

Fig. 4.4 Maturation and deaths of population cohorts

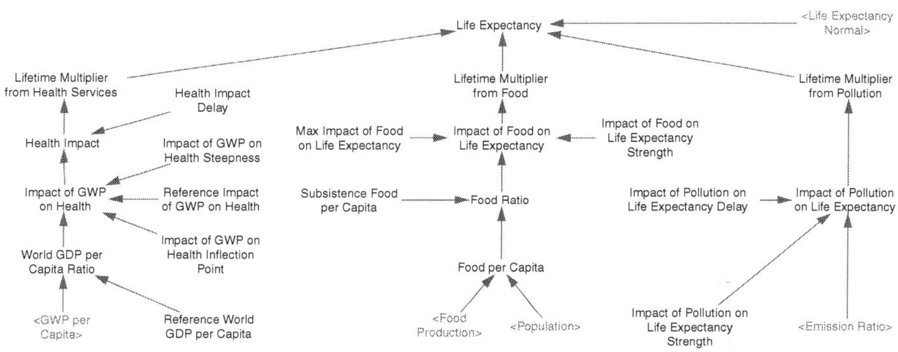

Fig. 4.5 Factors having an impact on population Life Expectancy

life expectancy mimics the degree of healthy diets. Here, indirectly, the level of technology in agriculture and competition for land are the main drivers. Pollution is a combination of air and water pollution. Air pollution is directly calculated from the energy module, where an increasing share of renewable energy is directly linked to less air pollution, and through the reduced GHG emissions, the decrease in climate change–related disaster incidences raises life expectancy. Similarly, more intensive agriculture and subsequent improved nutrition and reduced demand for industrial water consumption and associated reduced pollution levels are associated with longer life expectancy. The effects of pollution can be modeled to be active immediately or through a lag effect accounting for the delayed outbreak of chronic diseases.

4.3 Results

4.3.1 Model Calibration

The FeliX model was calibrated in an iterative process of structure formulation, parameter estimation, analysis of fit and residuals, and model reformulation. This process was conducted in two stages: (1) developing and improving sub-modules; and (2) model integration. The process was repeated until a good fit with the historical data was achieved. Calibration involved not only goodness-of-fit criteria but also the plausibility of the model per se in terms of its ability to explain the observed behavior.

Data for calibration and validation came from various sources, including IEA Key World Energy Statistics,[3] BP Statistical Review of World Energy,[4] Carbon Dioxide Information Analysis Center,[5] and FAOSTAT.[6] The calibration was conducted for a period of one century (1900 to 2000). If 100 years' worth of data was not available, the historical data for the available period were used and extrapolated.

The model went through a set of standard structure and behavior tests to build confidence in system dynamics modeling (see Sterman 1984; Oliva 1995). Figure 4.6 presents results of the calibration effort for a subset of model variables across the FeliX modules, and Table 4.1 presents historical fit summary statistics for each of the chosen variables.

4.3.2 Baseline Scenario

Once the model structure was finalized and the model calibrated to historical data constituting an acceptable representation of the Earth system, the baseline scenario was constructed by extending the model time scale to 2050. Additional policy assumptions were introduced to the model for alignment with the United Nations' Millennium Development Goals. These policies, which include investments in alternative sources of energy, improved cropping systems, and better health care, align the baseline with the spirit of the Second Earth Summit in Johannesburg, where the GEO idea was born. Thus, our baseline is more a sustainability scenario than a BAU forecast of highest likelihood. The idea is to establish a reference for GEO impact analysis. The baseline scenario was purposefully designed to assess the question of what would happen to aggregate output indicators (e.g., GHG intensity of energy production, population, ecosystem health) if particular

---

[3] http://www.iea.org/stats/index.asp

[4] http://www.bp.com/productlanding.do?categoryId=6929&contentId=7044622

[5] http://cdiac.ornl.gov/

[6] http://faostat.fao.org/

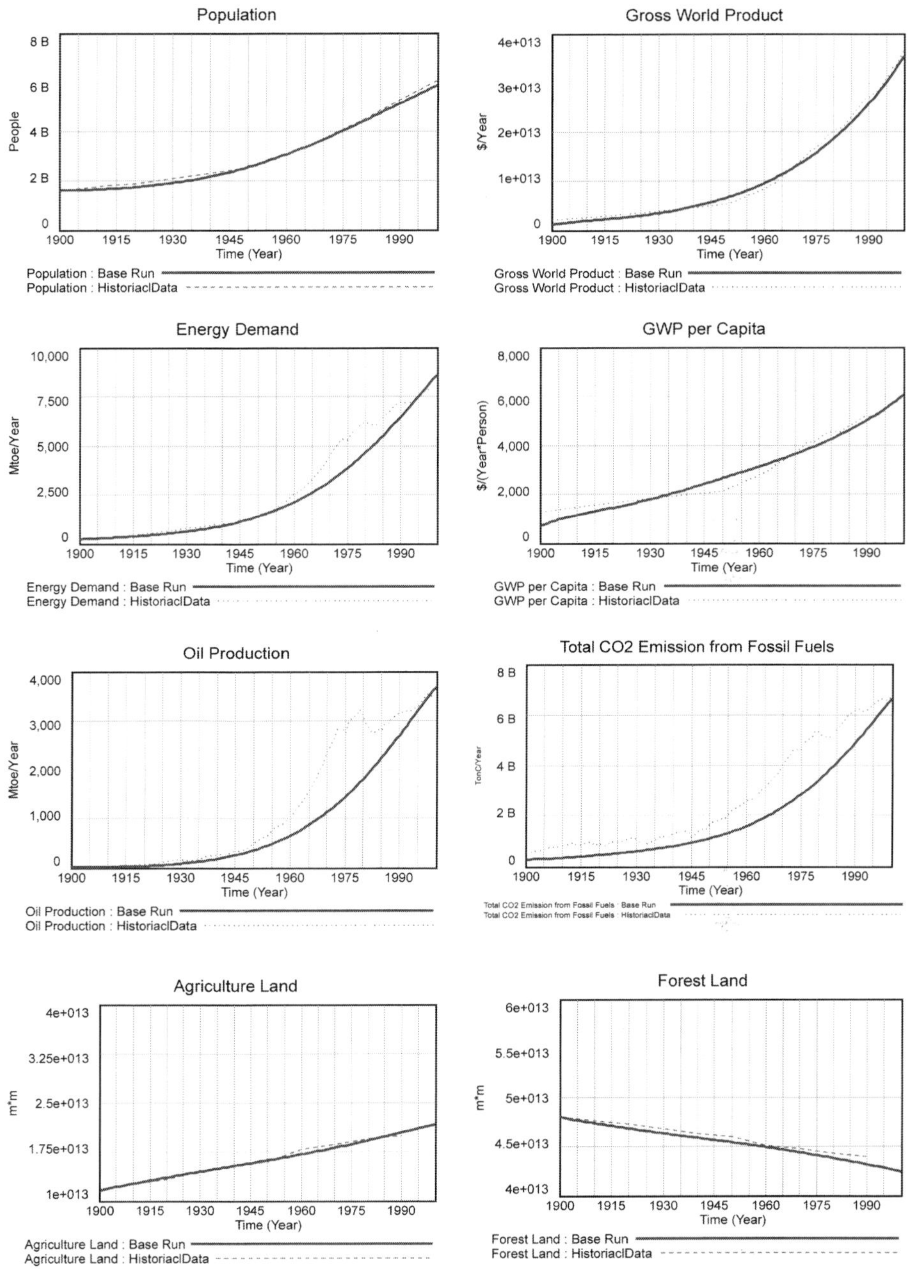

Fig. 4.6 Overview of FeliX model calibration outcome. Note: *Dashed lines* are historical data; *solid lines* are the outcomes of the calibration experiment

Table 4.1 Historical fit summary statistics (Theil inequality statistics)

	R^2	MAPE	MSE	RMSE	U^M	U^S	U^C
Population	1.000	0.01	7.84E + 15	8.86E + 07	0.41	0.56	0.03
Gross world product (GWP)	0.993	0.08	1.14E + 24	1.07E + 12	0.06	0.51	0.43
GWP per capita	0.965	0.07	6.72E + 04	2.59E + 02	0.04	0.42	0.55
Energy demand	0.964	0.14	4.27E + 05	6.53E + 02	0.34	0.14	0.52
Oil production	0.895	0.39	3.23E + 05	5.68E + 02	0.37	0.18	0.45
CO_2 emissions from fossil fuels	0.950	0.35	7.98E + 17	8.93E + 08	0.68	0.09	0.23
Agricultural land	0.986	0.02	1.25E + 23	3.53E + 11	0.02	0.08	0.90
Forestland	0.984	0.01	2.15E + 23	4.64E + 11	0.77	0.08	0.15

R^2 coefficient of determination, *MAPE* mean absolute percent error, *MSE* mean square error, *RMSE* root mean square error, U^M bias component of MSE, U^S variation component of MSE, U^C covariation component of MSE

economic, social, and environmental policies were in place but GEOSS-related improved data and data policies were not available. Figure 4.7 presents the baseline runs used for the GEOSS impact assessment.

4.3.3 GEOSS Scenarios

To assess the socioeconomic and environmental impacts of GEOSS improvement, we constructed six storylines in the energy, disaster, health, climate, agriculture, and water societal benefit areas (the other three SBAs, weather, ecosystem, and biodiversity, are considered under the six scenarios). Various storylines were expressed as incremental or more abrupt changes and new relations in the FeliX model. The range of parameter changes was either informed by particular studies or conjectured by the experts. For illustration, a few conjectured storylines that affect health outcomes are listed in Table 4.2.

Each of the six GEOSS scenarios can be considered an integrated scenario in the sense that the changes it brings to the model affect not one particular domain of interest but propagate through the whole model. For instance, changes in GHG emissions from the energy sector affect agricultural productivity. Sector-specific scenario analysis was conducted in such a way that impact assessments were performed with a sectorial view or together with the other SBA scenarios. Likewise, the effect of improved Earth observations can be analyzed from a sectorial angle or a full systems view.

Instead of considering each predefined GEOSS scenario separately, we focus here on the combined scenarios: all six GEOSS scenarios are enabled for the model simulation runs and subsequently the impact assessment. The following section presents some results of the combined scenario exercise, bringing together GEO effects on population indicators.

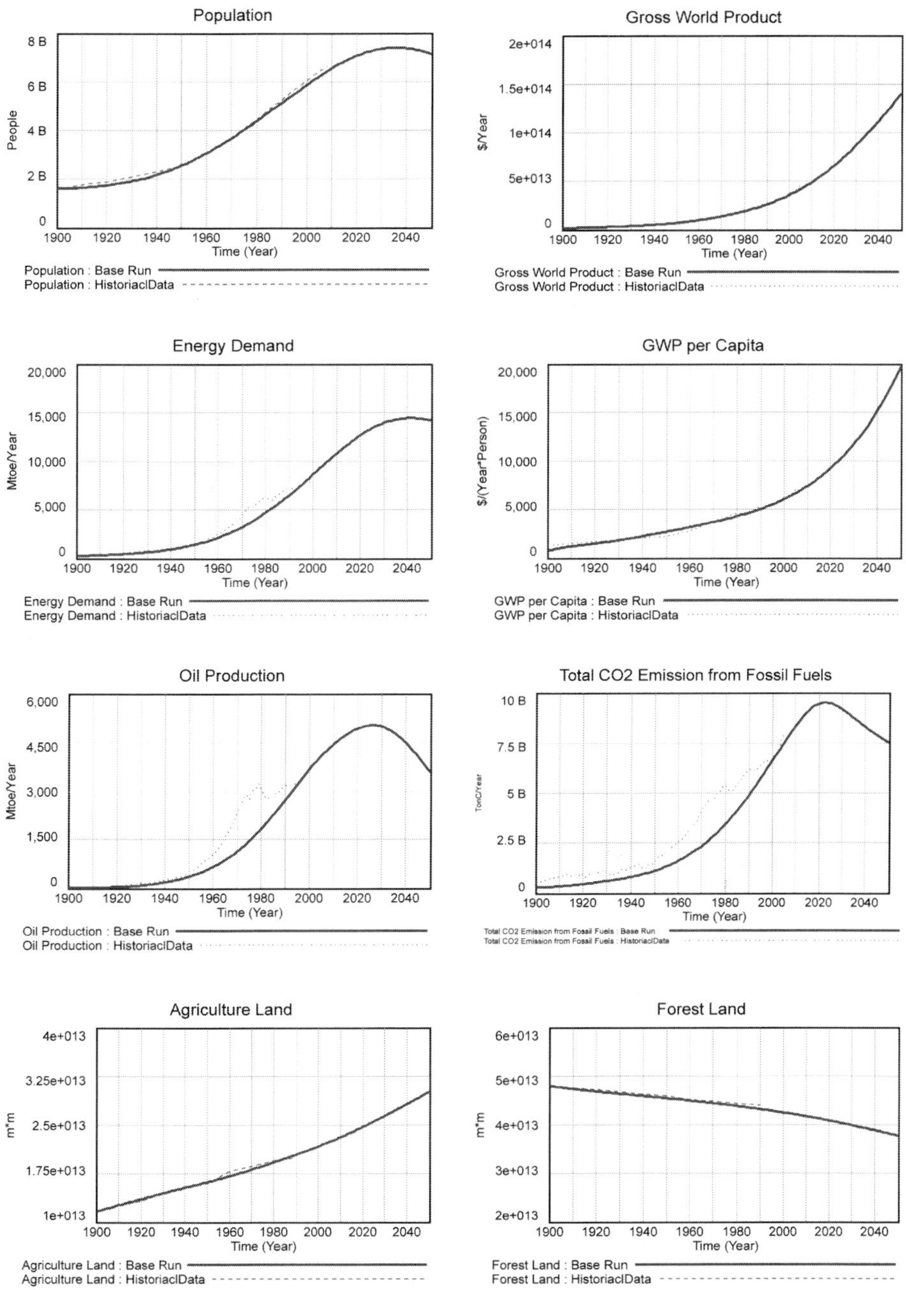

Fig. 4.7 Overview of the baseline scenario (*Dashed lines* are historical data; *solid lines* are the outcomes of the baseline scenario experiment)

Table 4.2 Example storylines influencing life expectancy

Storyline	Variable name	Base run value	Scenario value	SBA
GEOSS triggers improved prevention of cardiovascular diseases, malaria, diarrheal diseases, meningitis, and others	Life expectancy normal H	0	Ramp up to 0.14 by 2030	Health
Use of GEOSS data improves planning and commissioning of solar energy installations. Rural development is enhanced	Solar available area	5.00e + 11	Ramp up to 6e + 011 by 2020	Energy
Use of GEOSS data improves natural disaster alerts and response	Life expectancy normal D	0	Ramp up to 0.05 by 2030	Disaster
Correlation of emissions of air pollutants and GHGs reduces pollution, incidence of chronic diseases, and in long run, climate change–related hazards	Impact of CO_2 concentration on life expectancy strength	0.01	Ramp up to 0.007 by 2030	Climate
GEOSS weather forecasting enables improved crop management for consistently higher yields and global coordination of food production	Effect of GDP on agriculture management practices increment	0	Ramp up to 0.3 by 2030	Agriculture
Better water management planning and water stress monitoring reduce water pollution and increase irrigation efficiency	MAX agricultural water use	0.1	Ramp up to 0.06 by 2030	Water

4.3.4 *Benefit Assessment of GEOSS*

The approach used to measure the value of creating and improving GEOSS can be defined as deviations of the GEOSS scenarios from the baseline scenario. Since FeliX is a dynamic model, it is possible to capture the deviation of the GEOSS scenarios from the baseline scenario as it develops over time or as an accumulated value at the end of a specified period. The starting point for the GEOSS impact assessment presented here is the year 2010. An open architecture of the FeliX model (as opposed to so-called black box models) and detailed documentation of the reasoning and actual changes in model parameters let us analyze and track any differences between the GEOSS and baseline scenarios (see Rydzak et al. 2009). Model transparency is necessary when dealing with aggregated but highly interrelated complex systems.

For the purpose of illustration, only six of the storylines used for the GEOSS assessment in the GEOBENE project were chosen (see Table 4.2), all of which affect life expectancy. However, as will be illustrated, their impact spread across various SBAs.

Health, disaster, and climate storylines have a direct and, as modeled by a RAMP function, positive effect on life expectancy when compared with the baseline (dashed line in Fig. 4.8). Life expectancy starts to increase in year 2010 and causes an increase in population. To year 2050, the accumulated increase of total population in the GEOSS scenario is equal to 1.2 billion. The greatest increase (70%) is in the proportion of the population over 65 years.

Population growth has a significant effect on the global use of resources (Fig. 4.8), increasing demand for food, energy, and water. However, the GEOSS simulation scenario indicates less extensive land use compared with the baseline. Over the period of the GEOSS scenario, about 10 million km^2 of land is saved from conversion to agricultural use; of this land, 5.7 million km^2, or 57%, is forested and thus is saved from being deforested.

When tracking the reason for such outcomes in the FeliX model structure, we noticed the effect of the other storylines in the combined GEOSS scenario (Fig. 4.8). As indicated in the agriculture storyline (see Table 4.2), GEOSS enables improved crop management and thus higher yields per hectare. The yield was also increased by GEOSS-related improvements in agricultural water use management, as indicated in the water storyline (see Table 4.2): over the considered period, a total of 5,400 billion m^3 of water was saved, compared with the baseline scenario. Cumulative food production is estimated to equal ten billion vegetable-equivalent kilograms. This explains why agricultural land did not expand commensurately with the increase in population.

As a side effect of those dynamics, CO_2 emissions are lower than in the baseline scenario (Fig. 4.8). Over the considered period an accumulated difference in CO_2 emissions from land use accounts for 7.3 billion tons of carbon. However, there is also a noticeable increase in energy production. The decrease in CO_2 emissions comes not only because GEOSS enabled a more developed solar energy sector (as indicated in the energy storyline in Table 4.2) but also because of the avoided deforestation—the forest biomass that was spared from conversion to agricultural use. Both sources of energy are cleaner than fossil fuels, and that drives the decrease of CO_2 emissions even further. This climate mitigation is associated with increasing life expectancy (see the climate storyline in Table 4.2). Tracking these chains of influences, one notices important feedback loops responsible for the dynamics of the whole system. These feedback loops are able to reinforce or balance the effect of GEOSS across the SBAs.

The value of GEOSS might be assessed based on the outcomes in a particular SBA embedded into the FeliX model structure but also can be measured using such indicators as the Human Development Index and total change in ecosystem value (Fig. 4.8). For the given GEOSS scenario there is a noticeably faster human development combined with slower loss of the ecosystem value.

The outcomes of the simulation scenario described above constitute only a small portion of the GEOSS impact assessment results across all the SBAs. Although the FeliX model has an open architecture, its structure—mimicking the society-technology-environment interrelations of the Earth system—is complex, and understanding the model dynamics requires considerable time. Although this level of

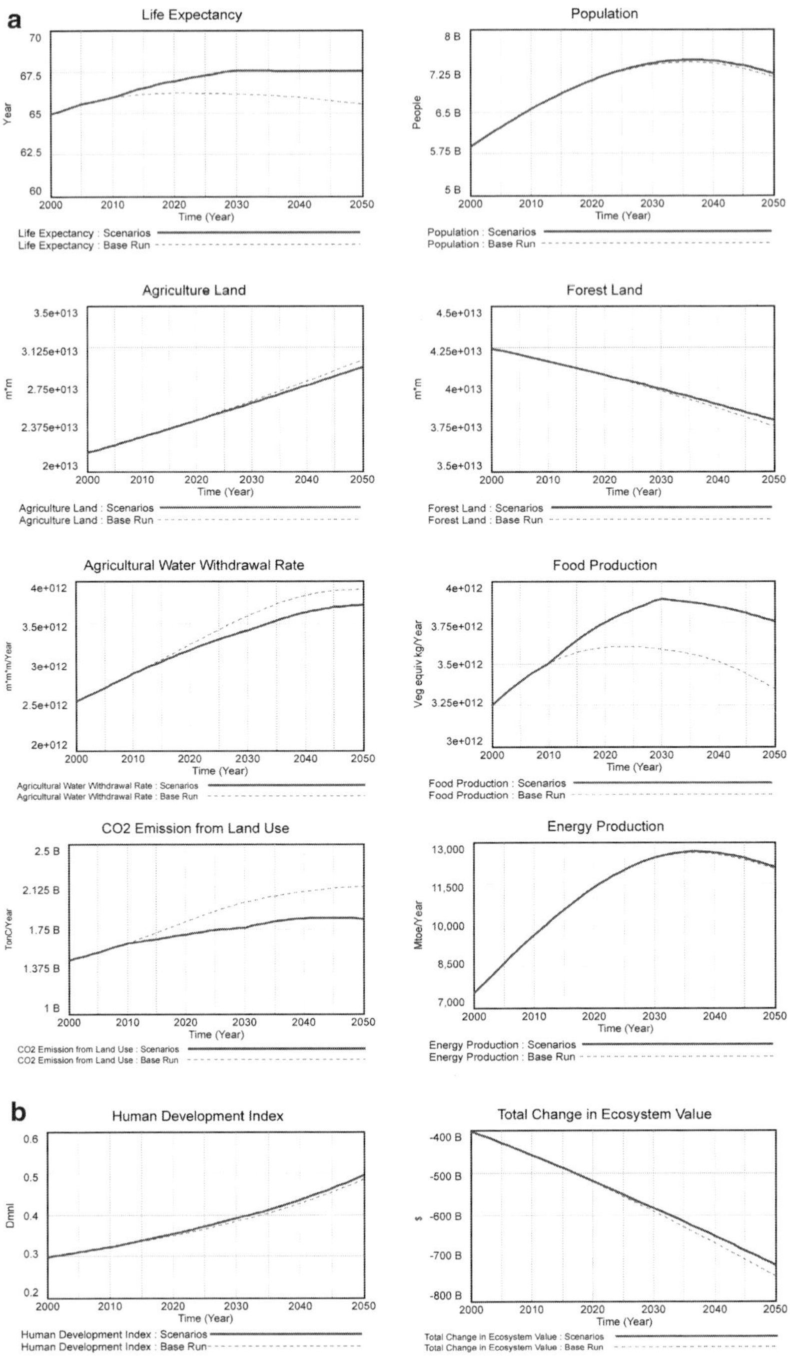

Fig. 4.8 FeliX comparison of base run and GEOSS scenarios

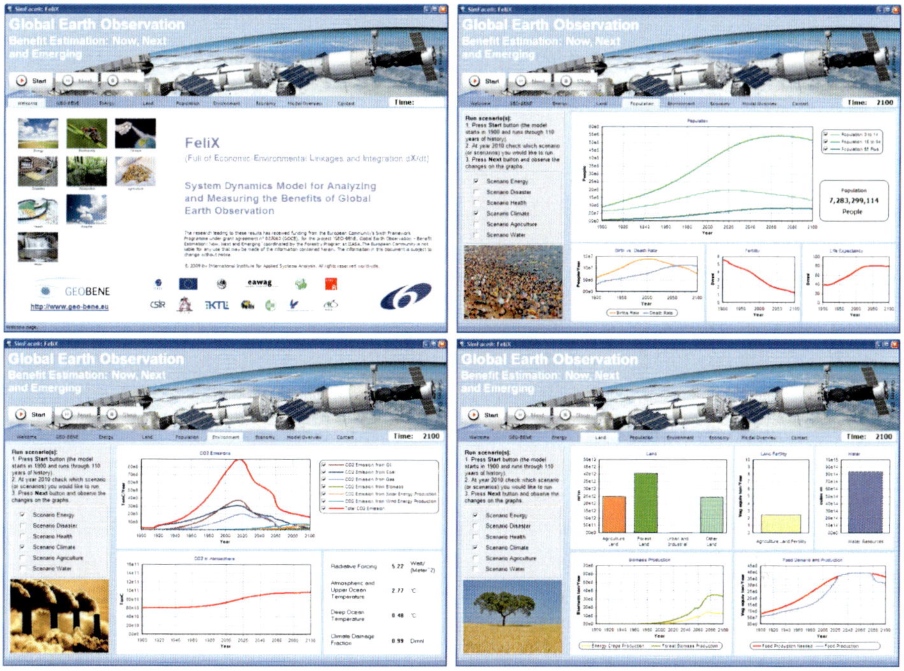

Fig. 4.9 FeliX simulator interface while running GEOSS scenarios

complexity is necessary, the model itself is too complicated to be directly presented to higher-level decisionmakers. For that reason the FeliX model-based simulator was constructed. As illustrated in Fig. 4.9, it is equipped with a user-friendly interface that allows for easy use and navigation through the simulations.

Users can run illustrative GEOSS-related scenarios and observe the potential consequences across all model sectors along several indicators. The simulator is an appropriate tool that enables decision makers to view the outcomes of various GEOSS scenario assumptions, extend their knowledge, and explore relationships in the system. The simulator is freely available from the GEOBENE project website.[7]

4.4 Conclusion

In these times of strained public budgets, any decision on how to develop a global Earth observation system of systems (GEOSS) requires international coordination of efficient and effective investments and operations. The FeliX model presented in this chapter was developed to assess the benefits from use of global Earth

[7] http://www.geo-bene.eu/

observations. FeliX's open architecture was designed to support strategic decision processes to develop GEOSS and identify areas where GEOSS initiatives might have significant economic, social, and environmental benefits.

In this chapter we have developed a methodology and analytical tool and applied it to assess the societal benefits of improving GEOSS across various benefit areas, following a benefit chain concept. The basic idea is that the costs incurred by an incremental improvement in the observing system—including data collection, interpretation, and information sharing—will deliver benefits through information cost reduction or better-informed decisions. The resulting incremental societal benefit can be judged against the incremental cost of production. Since in many cases there are large uncertainties in the estimation of costs and particularly the benefits, we expressed benefits not only in monetary terms but also by social and environmental indicators. Therefore, in most cases impacts where benefits are orders-of-magnitude larger than their production costs can be regarded as robust guidance signals to support decision making in GEOSS processes.

In particular, we have assessed two categories of benefit generation from GEOSS. The first is benefits from economies of scale of a global observation system versus the current patchwork of national or regional observation systems. We call these aggregation benefits. The second category relates to economies of scope, which emerge when changes in the observation system affect multiple benefit sectors or dimensions. These integration benefits are often considered a "public good." Quantifying them proved a significant challenge.

Because of the public good nature of the benefits, GEOSS effects are highly dependent on the type of baseline policy scenario. Apart from the choice of baseline definition, there are several other limitations to the FeliX model. Some subjects may be modeled in great detail, while others that might contribute more to the benefit are covered in less detail. This uneven coverage arises where data are very sparse or where lower anticipated benefit levels attracted less investment in data development. Like any other model, the FeliX model is a purposeful simplification of reality. There are also some questions regarding the existence or strength of a particular relation defined in the FeliX model. For instance, the functional shape and parameterization of the climate change impact function is a highly contested area of research. In addition, in many areas impact functions were not available, and we had to base our assessment on knowledge from subject matter experts. To deal with the uncertainty in the FeliX model, sensitivity analysis can be conducted, which is a subject for future work with the model.

As defined by Craglia et al. (2008), the systems approach and tools similar to the FeliX model might become part of a laboratory for learning via multidisciplinary education and science. The first step in that direction has already been made with the construction of the FeliX simulator, available for free at http://www.geo-bene.eu.

Earth observation has great potential for helping to ensure a sustainable future for the planet. According to our analysis, its value is apparent, to varying degrees, across all social, economic, and environmental indicators of the Earth system. Better climate change mitigation, increased food security, sustainable water use, available land resources, and clean energy technologies are among the many examples where improved observations of the Earth system might be beneficial from a global societal perspective.

4. Commentary: The Value of a Comprehensive Model

Molly K. Macauley

4.C.1 Introduction

In their contribution to this volume, Steffen Fritz, Ian McCallum, Michael Obersteiner, and Felicjan Rydzak use a systems engineering model of the global economy to illustrate how value could be ascribed to information obtained from Earth-observing satellites. Rydzak and coauthors constructed the model in previous research (Rydzak et al. 2010) to characterize the Earth processes and human interactions that are the focus of the Group on Earth Observations (GEO). GEO is a voluntary collaboration of 80 governments, the European Commission, and regional and other organizations. GEO seeks to coordinate the Earth-observing satellites of different countries across nine themes, called societal benefit areas: public health, climate, energy, water, agriculture, ecosystems, weather, disaster management, and biodiversity.

Rydzak and his colleagues modeled the subcomponents of their engineering model on these themes. For example, subcomponents include representation of the global carbon cycle, energy resources, and land use. With this model, Fritz and coauthors show how the model could be used to ascribe value to Earth observations. For instance, if GEO Earth observations data improve disease prevention or air quality, then the Rydzak model would show an increase in life expectancy. The value of Earth observations data in this engineering framework is expressed in changes in the physical outputs of the model (such as years of life expectancy). The examples in their chapter are hypothetical, not based on actual applications of Earth observation data.

4.C.2 Choice of the Model

An advantage of using a systems engineering model as a method to ascribe value to Earth observations information is that engineering is the language of the engineers and, although perhaps to a lesser extent, the scientists who design the satellites and their observing instruments. More challenging is the attempt to model the global economy. Discussing the Rydzak model in detail is outside the scope of

M.K. Macauley (✉)
Resources for the Future, Washington, DC, USA
e-mail: macauley@rff.org

this commentary, but as with all models of the global economy, specifying all the interrelationships and interactions of industrial sectors, natural resources, and people is difficult. The authors' example of life expectancy is a good example of the difficulty. Many factors, including existing health of the population, access to clean water and sanitation, and nutrition and diet, influence life expectancy. The "black box" in global models in which these factors combine with agricultural productivity, international trade in agriculture, peoples' behavior, technological innovation, and government policy—all of which affect life expectancy—is difficult to formulate.

Fritz and his coauthors want to use a global model because one of their objectives is to replicate the interrelationships among the GEO societal benefit areas. They argue that the value in GEO in coordinating Earth-observing systems of different countries is the complementarity of different kinds of data. To continue with their life expectancy outcome as an example, the complementarity is in data about air quality and water, which combine to influence agriculture and, in turn, life expectancy.

Such an approach is ambitious as a basis for identifying a role of Earth observations. The traceability of attribution of the role of Earth-observing information on each of these influences is difficult at best. Moreover, there are other black boxes in which actions are assumed rather than empirically accounted for: the approach doesn't permit disentangling Earth observations data from other data sources, and it assumes that the Earth observations data are in fact used by people taking action within the various subcomponents of the model.

Alternative modeling approaches are available to characterize the relationships among economic sectors, natural resources, and people. Examples of some of these alternatives include general equilibrium models and integrated assessment models.[8] These models combine physical and economic relationships of producers, consumers, and the government sector. Unlike systems engineering models and similar input-output models, these alternatives tend to emphasize the role of relative prices and the capacity of consumers and producers to make substitutions in their decisions in response to changes in prices. Depending on their purpose, the models often include international trade, assumptions about technological change, estimates of stocks and flows of natural resources, and demographic data. The models often draw some of their inputs from purely physical models. One example is integrated assessment models that use, as inputs, the outputs of global circulation models, such as centimeters of sea level rise or parts per million of atmospheric concentrations of greenhouse gases.

[8] An example of a computable general equilibrium model is the Global Trade Analysis Program (GTAP). GTAP is optimized to characterize global trade. Examples of integrated assessment climate models include the Integrated Global Systems Model (IGSM) of the Massachusetts Institute of Technology's Joint Program on the Science and Policy of Global Change, the Model for Evaluating the Regional and Global Effects (MERGE) of greenhouse gas reduction policies developed jointly at Stanford University and the Electric Power Research Institute, and the MiniCAM Model of the Joint Global Change Research Institute, a partnership between the Pacific Northwest National Laboratory and the University of Maryland.

There are many shortcomings in these alternative models, including the constraints imposed by functional forms used to characterize production and consumption decisions. The characterization of technical change and uncertainty is also problematic. An advantage of the models, however, is that they usually explicitly allow for interactions such as substitution among inputs, the effects of government policy, and as noted, changes in relative prices. Another advantage is that their outputs are usually expressed in economically relevant measures, such as changes in productivity or overall social welfare.

But even these models are subject to the same challenges as the engineering model. In all many global-scale representations, identifying the role and value of information can be a search for a needle in a haystack. In addition, changes in the quality of natural resources (air quality, water availability) or the effects of these changes (on production relationships of industry, on health and quality of life of consumers) is not typically explicit—there are no prices for these resources. This lack of explicit characterization of the role of resources further confounds the ability to identify the value of Earth observations about them.[9]

4.C.3 Other Approaches?

For the representations of the GEO societal benefit areas, a smaller-scale approach might be more tractable. Using one of the existing integrated assessment models for climate is an example. Different runs of the models under different assumptions about information would allow for a set of scenarios: "what if the Earth-observing data allow enhanced use of renewable energy" or "suppose the data show trends in allocation of land away from forests to agricultural production." Even in these models, however, the tractability of the effect of "information" as a model input is difficult, and the effect of Earth observations data, in particular as a subset of information, is also hard to identify.

Perhaps the most important contribution of Fritz and his coauthors in their assessment of benefits from GEO is to point out the desirability of accounting for the complementarity of different types of Earth observations data. The coordination of different Earth-observing systems, owned and operated by different countries, is the overall goal of GEO. The group describes this goal as GEOSS, the global Earth observation system of systems. Fritz and his coauthors seek a comprehensive model in which, for instance, the air quality observations of one country's satellite system together with the precipitation data of another country's system can be valued for their joint information content. I commend this effort.

[9] Darmstadter (2008), Banzhaf (2004), and Boyd (2008) are among the many scholars describing the desirability of including measures of natural resources, or ecological wealth, in national income accounts. This step would make it easier to identify the contribution of Earth observations information to management of natural resources.

References

Adams, R. M., Bryant, K. J., McCarl, B. A., Legler, D. M., OBrien, J., Solow, A., & Weiher, R. (1995). Value of improved long-range weather information. *Contemporary Economic Policy, 13*(3), 10–19.

Andersen, D. F., Richardson, G. P., & Vennix, J. A. M. (1997). Group model building: Adding more science to the craft. *System Dynamics Review, 13*(2), 187–201.

Banzhaf, S. (2004, Winter). Establishing a bureau of environmental statistics. *Issues in Science and Technology, 20*, 26–30.

Bolin, B. (1986). Requirements for a satisfactory model of the global carbon cycle and current status of modeling efforts. In J. R. Trabalka & D. E. Reichle (Eds.), *The changing carbon cycle: A global analysis*. New York: Springer.

Bouma, J. A., van der Woerd, H. J., & Kuik, O. J. (2009). Assessing the value of information for water quality management in the North Sea. *Journal of Environmental Management, 90*(2), 1280–1288.

Boyd, J. (2008, May). *Don't measure, don't manage: GDP and the missing economy of nature* (Resources for the Future Issue Brief 08-01). Washington, DC: Resources for the Future.

Considine, T. J., Jablonowski, C., Posner, B., & Bishop, C. H. (2004). The value of hurricane forecasts to oil and gas producers in the Gulf of Mexico. *Journal of Applied Meteorology, 43*(9), 1270–1281.

Craglia, M., Goodchild, M., Annoni, A., Camara, G., Gould, M., Kuhn, W., Mark, D., Masser, I., Maguire, D., Liang, S., & Parsons, E. (2008). Next-generation digital earth. A position paper from the vespucci initiative for the advancement of geographic information science. *International Journal of Spatial Data Infrastructures Research, 2008*(3), 146–167.

Crutzen, P. J., & Stoermer, E. F. (2000). The 'Anthropocene'. *Global Change Newsletter, 41*, 17–18.

Darmstadter, J. (2008, November). *Incorporating Resource and Environmental Change in a Nation's Economic Accounts* (Resources for the Future Issue Brief 08-04). Washington, DC: Resources for the Future.

European Commission. (2008). In: J. A. Bueno (Ed.), *Research for Europe. A selection of EU success stories.* http://bookshop.europa.eu/eubookshop/bookmarks.action?target=EUB:NOTICE:KINA22984:EN:HTML&request_locale=EN. Accessed 22 Oct 2009.

Fiddaman, T. (1997). *Feedback complexity in integrated climate-economy models.* PhD thesis, MIT Sloan School of Management, Cambridge.

Fiddaman, T. (2002). Exploring policy options with a behavioral climate-economy model. *System Dynamics Review, 18*(2), 243–267.

Forrester, J. W. (1958). Industrial dynamics: A major breakthrough for decision makers. *Harvard Business Review, 26*(4), 37–66.

Forrester, J. W. (1961). *Industrial dynamics.* Cambridge, MA: Productivity Press.

Fritz, S., Scholes, R. J., Obersteiner, M., Bouma, J., & Reyers, B. (2008). A conceptual framework for assessing the benefits of a Global Earth Observation System of Systems. *IEEE Systems Journal, 2*(3), 338–348.

Goudriaan, J., & Ketner, P. (1984). A simulation study for the global carbon cycle, including man's impact on the biosphere. *Climatic Change, 6*, 167–192.

Katz, R. W., & Murphy, A. H. (1997). *Economic value of weather and climate forecasts.* Cambridge, UK: Cambridge University Press.

Leyequien, E., Verrelst, J., Slot, M., Schaepman-Strub, G., Heitkönig, I. M. A., & Skidmore, A. (2007). Capturing the fugitive: Applying remote sensing to terrestrial animal distribution and diversity. *International Journal of Applied Earth Observation and Geoinformation, 9*(1), 1–20.

Meadows, D. H., Meadows, D. L., Randers, J., & Behrens, W. W., III. (1972). *The limits to growth.* New York: Universe Books.

Muchoney, D. M. (2008). Earth observations for terrestrial biodiversity and ecosystems. *Remote Sensing of Environment, 112*(5), 1909–1911.

Nordhaus, W. D. (1992). *The "DICE" model: Background and structure of a dynamic integrated climate-economy model of the economics of global warming* (Discussion Paper No. 1009). New Haven: Cowles Foundation for Research in Economics at Yale University.

Nordhaus, W. D. (1994). *Managing the global commons*. Cambridge, MA: MIT Press.

Oeschger, H., Siegenthaler, U., Schotterer, U., & Guglemann, A. (1975). A box diffusion model to study the carbon dioxide exchange in nature. *Tellus, XXVII*(2), 167–192.

Oliva, R. (1995). *A Vensim module to calculate summary statistics for historical fit*. D-4584, System Dynamics Group, Massachusetts Institute of Technology, Cambridge, MA.

Richardson, G. P., & Andersen, D. F. (1995). Teamwork in group model building. *System Dynamics Review, 11*(2), 113–137.

Richardson, G. P., & Pugh, A. L., III. (1981). *Introduction to system dynamics modeling with DYNAMO*. Cambridge, MA: Productivity Press.

Rotmans, J. (1990). *IMAGE: An integrated model to assess the greenhouse effect*. Boston: Kluwer Academic Publishers.

Rydzak, F., Obersteiner, M., Kraxner, F. (2009). *GEO-BENE Global earth observation—benefit estimation: Now, next and emerging* (Synthesis Report). Laxenburg: International Institute for Applied Systems Analysis (IIASA).

Rydzak, F., Obersteiner, M., & Kraxner, F. (2010). Impact of global earth observation—Systemic view across GEOSS Societal Benefit Areas. *International Journal of Spatial Data Infrastructures Research, 5*, 216–243.

Schellnhuber, H. J. (2009). From the cover: Tipping elements in earth systems special feature: Tipping elements in the earth system. *Proceedings of the National Academy of Sciences of the United States of America, 106*, 20561–20563.

Sterman, J. D. (1984). Appropriate summary statistics for evaluating the historical fit of system dynamics models. *Dynamica, 10*(Winter), 51–66.

Sterman, J. D. (2000). *Business dynamics: Systems thinking and modeling for a complex world*. Boston: Irwin/McGraw-Hill.

Stern, N. (2007). *The economics of climate change. The stern review*. Cambridge, UK: Cambridge University Press.

Vennix, J. A. M. (1996). *Group model building: Facilitating team learning using system dynamics*. Chichester: Wiley.

Wieand, K. (2008). A Bayesian methodology for estimating the impacts of improved coastal ocean information on the marine recreational fishing industry. *Coastal Management, 36*(2), 208–223.

Chapter 5
The Informative Role of Advertising and Experience in Dynamic Brand Choice: An Application to the Ready-to-Eat Cereal Market

Yan Chen and Ginger Zhe Jin

Abstract We study how consumers make brand choices when they have limited information. In a market of experience goods with frequent product entry and exit, consumers face two types of information problems: first, they have limited information about a product's existence; second, even if they know a product exists, they do not have full information about its quality until they purchase and consume it. In this chapter, we incorporate purchase experience and brand advertising as two sources of information and examine how consumers use them in a dynamic process. The model is estimated using the Nielsen Homescan data in Los Angeles, which consist of grocery shopping history for 1,402 households over 6 years. Taking ready-to-eat cereal as an example, we find that consumers learn about new products quickly and form strong habits. More specifically, advertising has a significant effect in informing consumers of a product's existence and signaling product quality. However, advertising's prestige effect is not significant. We also find that incorporating limited information about a product's existence leads to larger estimates of the price elasticity. Based on the structural estimates, we simulate consumer choices under three counterfactual experiments to evaluate brand marketing

---

This paper is based on Yan Chen's dissertation, "Information, Consumer Choice and Firm Strategy in an Experience Good Market" (August 2008). We are deeply grateful to Mark Denbaly and Ephraim Leibtag at the U.S. Department of Agriculture's Economic Research Service for providing access to the Nielsen Homescan data. We have also benefited from the insightful comments of John Rust, Roger Betancourt, Erkut Ozbay, Pallassana Kannan, and seminars participants at the University of Maryland and the Economic Research Service. All errors are ours.

Y. Chen
Bates White, Washington, DC, USA
e-mail: yan.chen@bateswhite.com

G.Z. Jin (✉)
Department of Economics, University of Maryland, College Park, MD, USA

National Bureau of Economic Research, Cambridge, MA, USA
e-mail: jin@econ.umd.edu

R. Laxminarayan and M.K. Macauley (eds.), *The Value of Information: Methodological Frontiers and New Applications in Environment and Health*, DOI 10.1007/978-94-007-4839-2_5,

strategies and a policy on banning children-oriented cereal advertising. Simulation suggests that the advertising ban encourages consumers to consume less sugar and more fiber, but their expenditures are also higher because they switch to family and adult brands, which are more expensive.

Keywords Consumer choice • Experience goods • Informative and prestige advertising • Ready-to-eat cereal market • Child-oriented advertising • Childhood obesity

5.1 Introduction

We examine how advertising and experience influence consumers' choice of products in a dynamic setting. In a market with many brands, consumers may not recognize the existence of every brand, especially when there are frequent entries and exits. Even if a consumer knows a brand exists, she may not know all its attributes until she has consumed it. Using ready-to-eat (RTE) cereal as an example, we consider both types of information problems and find that they can be partially addressed by the manufacturer's advertising and the consumer's experience.

The importance of informative and prestige advertising has long been recognized in economics and marketing. Since Ackerberg (2001), a growing literature attempts to empirically distinguish these types of advertising, based on the assumption that informative advertising targets new customers but prestige advertising increases consumption utility for both new and experienced customers (Stigler and Becker 1977; Becker and Murphy 1993). We separate two types of informative advertising: one for indicating the existence of a product (Butters 1977; Grossman and Shapiro 1984), and the other for imparting information about the product's quality (Nelson 1970, 1974; Kihlstrom and Riordan 1984; Milgrom and Roberts 1986).

Both types of informative advertising focus on new consumers but have different implications in consumer choice. By definition, the information about a product's existence brings the product into the choice set, and this effect is the same for all brands, conditional on advertising intensity. In contrast, if consumers are already aware of a product's existence, the information about product quality affects the trade-off between product quality and other observable product attributes; hence its effect on consumer choice will differ across brands. Similar identification has been used by Goeree (2008), but to our knowledge, this paper is the first to distinguish the two types of informative advertising from prestige advertising in a dynamic setting using transaction-level panel data.

In a market of experience goods, dynamic considerations can be important for two reasons: first, experience allows consumers to acquire better knowledge of product attributes. For breakfast cereal, taste and freshness are difficult to ascertain beforehand, but a single instance of consumption could yield plenty of information. Second, past experience may influence the current choice through habit formation. This is different from the informative role of experience because most information about breakfast cereal can be learned by consuming it once, but habit formation may be gradual. Our model also incorporates the potential interaction between experience and advertising: for example, if advertising makes a consumer aware of a brand, the consumer may choose it and form a habit of consuming it. Not only do these effects influence a manufacturer's pricing and advertising strategies,[1] they could also have profound implications for public policy regarding advertising.

Breakfast cereal is heavily advertised toward children, and there has been a long standing debate on whether such advertising should be prohibited. In as early as 1978, the U.S. Federal Trade Commission (FTC) issued a staff report concluding that "television advertising for any product directed to children who are too young to appreciate the selling purpose of, or otherwise comprehend or evaluate, the advertising is inherently unfair and deceptive," and that "it is hard to envision any remedy short of a ban adequate to cure this inherent unfairness and deceptiveness." Naturally, this statement generated strong opposition from broadcasters, ad agencies, and food and toy companies. In 1980 Congress passed the Federal Trade Commission Improvements Act of 1980 and barred FTC from issuing industry-wide regulations to stop unfair advertising practices.[2] However, as concern about childhood obesity grows, policymakers and consumer advocates are calling for restrictions on advertising to children about candy, sugary cereal, and other junk food.

A study by the Kaiser Family Foundation indicates that children of all age groups are exposed to many food ads every day; of all food ads that target children or teens, 28% are for sugary cereal.[3] Kid brands have significantly more sugar and less fiber per serving than adult and family brands.[4] Based on our empirical estimates, we simulate what would happen to consumer expenditures and nutritional intake if cereal TV advertising directed at children were banned. Results suggest that, following the advertising ban, consumers would consume more fiber and less sugar; this effect is more pronounced for consumers who are younger, have lower income, and have children. Consumers also increase their expenditures after the policy change because they consume more adult and family cereals, which are more expensive than kid cereals.

Although the simulation highlights various roles of advertising and experience, it is worth noting that we do not model the potentially "misleading" effects of advertising. Hence, in our model, the ban of advertising is welfare reducing from the consumer's point of view because the ban leads to a smaller choice set and a less informative choice within the choice set. Since we find little evidence in support of the prestige advertising, our findings rule out the psychological gain from

[1] For instance, if consumers are habituated to a product, then the introductory price of a new product may need to be set lower than when there is only learning to warrant a product switch.

[2] See the article "Limiting Food Marketing to Children," at www.cspinet.org/nutritionpolicy.

[3] See "Food for Thought: Television Food Advertising to Children in the United States," released by The Kaiser Family Foundation, March 28, 2007.

[4] In particular, the average sugar content of kid brands is 10.98 g per serving, compared with 5.88 in adult brands and 7.68 in family brands. The average fiber content of kid brands is 5.41 g per serving compared with 9.92 in adult brands and 7.38 in family brands.

consuming highly advertised brands. However, if advertising misleads consumers into choosing sugary cereals—either because they are unaware of the "unhealthfulness" of the advertised food or because they like the sugary taste without much health consideration into the future—limiting their choice set could be beneficial to them.

The rest of this chapter is organized as follows. Section 5.2 describes the industry background of RTE cereal and summarizes the transactional-level panel data from Nielsen Homescan, manufacturers' advertising data from TNS Media Intelligence Company, and the brand nutritional data collected from the Internet. Section 5.3 reviews the literature. Section 5.4 lays out the dynamic model of consumer choice, with an emphasis on empirical identification. Section 5.5 reports the estimation results. Section 5.6 describes three counterfactual experiments, two on manufacturers' pricing strategy and one on a ban of advertising for kids' brands. A brief conclusion is offered in Sect. 5.7.

5.2 Background and Data

Several features of the RTE cereal market make it suitable for our study.[5] First, cereal is an experience good, the attributes of which are not completely known before consumption. Second, brand entry and exit happen frequently in the RTE cereal market, and none of the national brands have a truly dominant hold on the market, which imposes a considerable informational burden on consumers.

Using Nielson Homescan data (to be described below), Table 5.1 shows the entries and exits of RTE cereal brands[6] from December 1997 to December 2003 in the Los Angeles market. In the 6-year period, a total of 62 (46) brands enter (exit)[7] the market, which accounts for about 47.3% (35.1%) of the total number of brands existing at the end of 1997. Column 2 of Table 5.2 displays sales-based market shares of major brands from December 1997 to December 2003. Because there are so many brands, we select the top 50 (which together account for about 79% of the market) and combine the rest into a composite brand, Brand 51. The biggest brand (Brand 1[8]) has a market share of 6%; most brands take up <1% of the market.

The third reason the RTE cereal market makes an interesting case is that it is heavily advertised. The advertising-to-sales ratio for RTE cereal was 13% in 2001.

---

[5] Readers can refer to Section 2 of Nevo (2001) for a more complete picture of the RTE cereal industry.

[6] Brand definition follows the classification on each manufacturer's website. Different box sizes are treated as the same brand, but extensions of a brand name are distinct brands. For example, Cheerios, Honey Nut Cheerios, and Berry Burst Cheerios are three different brands.

[7] Brand entry and exit are defined using the Nielsen Homescan data. A brand entry is observed if the first transaction of the brand occurs after June 1998. A brand exit is observed if the last transaction of the brand occurs before June 2003.

[8] Brand names are not revealed because of a confidential agreement with the data provider.

Table 5.1 Brand entry and exit

	Exit year						
Enter year	1998	1999	2000	2001	2002	Remaining	Total
1998 and before	5	3	3	10	7	103	131
1999	0	1	4	1	2	10	18
2000	0	0	0	1	3	3	7
2001	0	0	0	1	3	8	12
2002	0	0	0	0	10	30	12
2003	0	0	0	0	0	13	13
Total	5	4	7	13	17	147	193

Data source: Neilsen Homescan data, December 1997 to December 2003, Los Angeles market
A brand entry is observed if the first transaction with the brand occurred after June 1998. A brand exit is observed if the last transaction with the brand occurred before July 2003

Table 5.2 Brand summary statistics

Brand number	Sample market share[a](percentage)	Average transaction price (cents/oz.)	Average monthly advertising ($k)	Fiber content (percentage daily value per 30g)	Sugar content (percentage daily value per 30g)	Segment[b]
1	5.73	17.22	1718.89	14.00	1.00	Family
2	4.51	18.44	1977.76	6.45	4.59	Family
6	4.45	12.26	2036.62	11.25	6.23	Adult
11	4.07	14.01	1045.85	5.90	6.39	Adult
8	3.99	11.95	1667.88	2.42	9.68	Family
7	3.69	11.96	1445.58	11.49	10.37	Family
3	3.56	15.85	1701.03	7.00	11.00	Family
12	2.88	14.98	406.50	4.12	12.39	Kid
4	2.71	17.74	785.01	5.00	10.00	Kid
16	2.50	13.71	319.99	3.56	6.71	Family
20	2.49	11.43	1623.61	10.34	4.40	Adult
9	2.38	18.22	878.28	0.03	7.91	Family
10	2.36	22.81	2143.30	9.15	5.21	Adult
15	2.35	18.66	437.65	6.00	13.00	Kid
13	2.09	13.80	1377.95	13.91	1.86	Adult
14	1.92	16.38	634.15	2.92	12.46	Kid
17	1.62	14.07	1293.78	7.50	7.03	Kid
23	1.62	9.60	5.65	14.75	8.64	Family
21	1.56	16.77	604.60	0.97	14.52	Family
18	1.55	15.61	698.44	8.30	8.61	Family
19	1.55	16.80	1243.06	1.59	13.39	Kid
38	1.48	17.48	435.98	4.00	13.00	Kid
5	1.39	20.91	1611.56	7.48	6.98	Adult
24	1.29	15.77	459.00	4.00	15.00	Kid
22	1.26	21.80	56.11	1.00	6.00	Adult
42	1.09	16.88	739.14	7.94	8.02	Adult
30	1.06	18.64	379.89	3.00	14.00	Kid
26	0.76	19.15	72.82	5.00	13.00	Family

(continued)

Table 5.2 (continued)

Brand number	Sample market share[a](percentage)	Average transaction price (cents/oz.)	Average monthly advertising ($k)	Fiber content (percentage daily value per 30g)	Sugar content (percentage daily value per 30g)	Segment[b]
25	0.72	19.11	423.71	4.00	11.00	Kid
47	0.72	15.08	746.76	3.10	11.38	Kid
46	0.71	17.01	87.48	8.69	5.88	Family
39	0.66	18.01	3.26	49.00	4.33	Adult
43	0.63	18.19	108.85	27.00	5.00	Adult
50	0.59	15.38	157.45	8.57	4.82	Adult
48	0.57	18.91	303.66	6.67	12.22	Kid
49	0.56	19.91	280.83	6.32	7.89	Family
27	0.54	19.55	0.00	7.09	7.64	Adult
40	0.52	15.64	102.95	1.94	10.65	Kid
45	0.46	21.43	177.74	4.36	6.00	Family
29	0.46	20.01	0.00	6.00	9.27	Family
31	0.44	23.30	208.99	2.00	13.00	Kid
37	0.43	21.49	381.66	0.00	12.00	Family
28	0.41	24.39	1653.81	12.00	10.00	Family
33	0.41	16.35	13.43	4.00	13.00	Family
44	0.35	16.95	229.33	8.13	6.10	Family
32	0.35	25.79	6.58	58.00	0.00	Adult
34	0.33	24.19	1.68	11.00	6.00	Family
41	0.31	14.64	0.00	4.44	16.67	Kid
35	0.29	17.64	62.32	9.00	9.27	Adult
36	0.25	17.39	0.00	10.91	8.73	Adult
51[c]	21.41	14.90	47.44	8.83	8.00	Family

Data source: Columns II, III from Neilsen Homescan data, December 1997 to December 2003; column IV from TNS Media Intelligence data, January 1999 to December 2003; columns V, VI from www.nutritiondata.com

[a]Sample market is the Los Angeles market from December 1997 to December 2003

[b]Brand segment categorization is based on each brand's description on the manufacturer's website

[c]Characteristics of the 51st brand are computed as the average of the nontop 50 brands

For well-established brands, the ratio was 18%.[9] In comparison, the average ad-to-sales ratio across 200 industries was 3.2%.[10] Heavy advertising indicates that firms believe advertising is effective in promoting sales.

Our data consist of four parts. On the consumer side, we use the Nielsen Homescan data on RTE cereal products from December 1997 to December 2003. Tracking 1,402 demographically balanced households in Los Angeles, the Homescan data tie consumer purchasing behavior with demographic measures. Homescan panelists scan items at home from each shopping trip, recording price and quantity purchased as

[9] See Nevo (2001, 311).

[10] See *Advertising Age*, March 1, 2006.

Table 5.3 Summary statistics of Homescan data

Variable	Definition	NumObs	Mean	Std. Dev.	Min	Max
size	household size	1,402	3.25	1.53	1	9
inc	household income ($K)	1,402	57.11	29.58	2.5	125
age	age of female household head	1,402	48.86	12.99	20	70
nokid	=1 if no kid in the household	1,402	0.55	0.50	0	1
price	transaction price (cent/oz)	69,134	17.84	4.73	0	797.44

Data source: Neilsen Homescan data, December 1997 to December 2003, Los Angeles market

well as the age, income, and other demographic information of the shopper. When available, Nielson uses store-average price instead of the consumer's self-recorded transaction price. Einav et al. (2010) document the measurement error in this data set and conclude that the magnitude of errors in the Homescan data is comparable to that of commonly used, government-collected economic data sets.

Homescan keeps track of on-going purchasing from the same household over time and thus offers insights into households' consumption habits and dynamics. On average, a household stays in the Homescan panel for 48 months. Once a household leaves the panel, a new one that is similar in all demographic measures is selected to take its place. Table 5.3 contains definitions and summary statistics of the major variables in the Homescan data.

Using the Homescan data, we can summarize the consumption pattern in the RTE cereal market. On average, a household makes 14 shopping trips for RTE cereal per year. The households usually have two or three brands that they purchase repeatedly over time. Most brands are purchased once and never again (Fig. 5.1). After a brand is first purchased by a household, the probability of the household's repurchasing the brand is 14.1% on the next shopping trip, 12.9% on the second trip, and about 11% on the following trips (Fig. 5.2). This suggests that learning in the cereal market is mainly done after one shopping trip. Figure 5.2 also suggests that a household that repurchases a brand after the first experience then exhibits loyalty to that brand.

On the product side, we obtain advertising data from TNS Media Intelligence, which tracks advertising expenditures of cereal manufacturers from January 1999 to December 2003. The advertising data cover 278 cereal brands across 11 media types.[11] The brand advertising expenditures include both national and local advertising. On average, national advertising accounts for 98.1% of the total advertising expenditure and is mainly on network TV and cable TV, whereas local advertising accounts for 1.9% of the total advertising and is mainly on local newspapers and outdoor billboards. Average monthly advertising expenditures of the top 50 cereal brands in Los Angeles are shown in column 3 of Table 5.2, above.

[11] The media types include network TV, cable TV, sport TV, magazines, syndication, national sport radio, network radio, Sunday magazines, local newspaper, outdoor billboard, and national newspaper. In this paper advertising particularly refers to cereal manufacturers' advertising expenditures in these media types. Although retailer advertising, such as retailer deal and store featuring, is common in the RTE cereal market, it is not included in the estimation because of a lack of data on retailers.

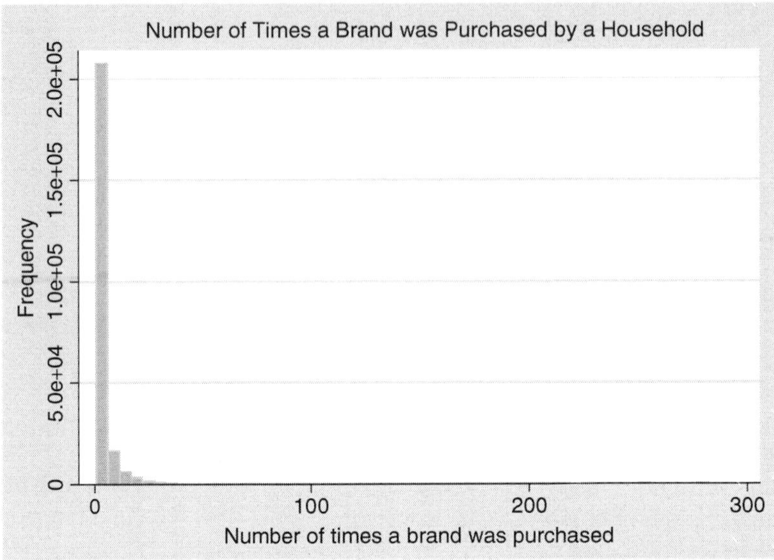

Fig. 5.1 Purchase frequency per brand

The third part of the data is nutritional information on 111 cereal brands, collected from www.nutritiondata.com[12]; it includes calories, sugar, dietary fiber, protein, and so forth. The fiber and sugar content per 30-g serving for the 50 top brands is displayed in columns 4 and 5 of Table 5.2. These two nutrients are selected because there is little variation in other nutrients across brands. In all data sets, the characteristics of brand 51, the composite brand, are calculated as the average of all non-top-50 brands.

The fourth part of the data involves cost factors that could serve as instruments to address the potential endogeneity of price and advertising. From the website of the Bureau of Labor Statistics, we collect hourly wage data for food workers (under the category Food and Tobacco Roasting, Baking, and Drying Machine Operators and Tenders) and for advertising mangers (under the category Advertising and Public Relations Managers) in the Los Angeles–Long Beach metropolitan statistical area from 1999 to 2003. Corn and wheat prices are obtained from the Farmdoc project of the University of Illinois. Gasoline and electricity prices are collected from the website of the Energy Information Administration, an office in the Department of Energy. As detailed in Sect. 5.4.4, these cost factors are likely to correlate with a firm's decisions about price and advertising but are uncorrelated with unobserved demand for RTE cereal.

[12] The nutritional information was collected on September 10, 2006, from the website. There is no variation of nutrients over time for the same brand.

5.3 Literature

Several lines of literature are relevant to our inquiry. The consumer learning literature addresses the problem of limited information about product quality. In their pioneer work, Erdem and Keane (1996) estimate how consumers learn about the cleaning power of laundry detergents. Both experience and advertising give consumers noisy signals about a detergent's quality, and consumers update their beliefs about quality in a Bayesian way. Following this research are many studies that model consumer learning in a Bayesian framework in various markets (e.g., Ackerberg 2003; Crawford and Shum 2005; Chintagunta et al. 2009). However, the consumer learning literature usually takes the consumer choice set as homogeneous. It does not account for the fact that different consumers may be exposed to different sets of products because of limited awareness, which is the central research question in the literature on heterogeneous choice set (also called consideration set in the marketing literature).

There have been very few economic studies that consider heterogeneity in the choice set. Goeree (2008) presents a model in which advertising influences the set of products from which consumers choose to purchase. Specifically, the probability that a consumer is informed of a product is a function of the effectiveness of the product advertising and the observed consumer characteristics. In the marketing literature, there are relatively more papers allowing for heterogeneity in consideration set. Brand choice is usually modeled as a two-stage process: at the first stage consumers identify a subset of brands which constitute their consideration set, and at the second stage they choose the brand with the highest utility. Roberts and Lattin (1997) review the theoretical and empirical marketing studies that develop an individual-level model of consideration set and analyze how marketing mix affects consideration set and consumer choice, including Andrews and Srinivasan (1995) and Allenby and Ginter (1995). They also point out some directions for future research, including dynamics in consideration set, which is captured in this chapter. Swait (2001) assumes that the probability a specific consideration set is formed is a function of the expected maximum utility from the alternatives in that set. Mehta et al. (2003) formulate the process of consideration set formation as a trade-off between the expected benefit from including an additional brand and the additional search cost incurred. Eliaz and Spiegler (2011) study a market model in which firms use irrelevant alternatives to influence consumers' consideration set. All these studies, however, model one-time purchases in a static setting and do not account for variation in choices and choice sets over time.

In terms of how to model advertising, this chapter learns from both theoretical (as cited in the introduction) and empirical literature on advertising (e.g., Ackerberg 2001, 2003; Anand and Shachar 2011). This chapter also benefits from the insights of the literature on RTE cereal market that involves demand estimation (Hausman 1996; Nevo 2001; Shum 2004; Hitsch 2006) and simulation of counterfactual pricing and advertising strategies (Dubé et al. 2005). Compared with the previous studies, the richness of the data allows us to include more dynamics in consumer choice, identify consumer learning from habit formation based on the difference in choice dependence structure of new and old consumers (as in Osborne 2006), and distinguish different effects of advertising.

Last but not the least, this chapter is an extension of the literature on analyzing demand systems in differentiated product markets (Berry 1994; Berry et al. 1995, 2004a, b). With household-level data, the parameters that vary with individual households can be identified without any constraints on the distribution of unobserved brand characteristics. The parameters that do not vary with individuals, such as the mean price coefficient, need to be estimated with the market share data and instrumental variables. This chapter applies the estimation method to a limited information environment.

5.4 Model

5.4.1 Setup

Consider a number of consumers (index by i) choosing from a set of brands (indexed by j) on different shopping trips (indexed by t). The brand choice is a two-stage process. At the first stage, based on previous purchase experience and brand advertising, the consumer is informed of a subset of brands that constitute her choice set on that shopping trip. At the second stage, the consumer chooses a brand from her choice set that maximizes her expected utility. On a specific shopping trip, the consumer's information set includes the quality and characteristics of brands he has purchased before, and prices and advertising intensities of all brands in her current choice set. Note that brands are differentiated both horizontally and vertically. The horizontal differentiation is on brand characteristics, such as taste, fiber, and sugar content. The vertical differentiation is on brand quality, including the quality of ingredients such as types of grains and rice, the processing techniques, and freshness.

Before going into the details of the model, we mention two simplifications implicit in the above framework. First, we focus on consumer brand choice conditional on purchasing RTE cereal. There are two reasons for not including nonpurchase of RTE cereal as the outside good. Consumers may choose not to purchase because they have cereal at home, not because the utility of nonpurchase is higher than all cereal brands. Treating nonpurchase as the outside good, therefore, would bias the parameter estimates downward in the utility function. In addition, consumers choose not to purchase RTE cereals on about two-thirds of all shopping trips. Including those shopping trips will further add to the already large computation burden.

The second simplification of the model is the absence of quantity choice. Taking quantity into consideration requires tracking consumer's stockpiling and inventory, which will greatly complicate the model. About 52 % of the purchases are associated with only one brand. Multiple-brand transactions are treated as independent transactions, following Shum (2004).[13] For example, if on a shopping trip a consumer purchased brands A, B, and C, it is estimated as if he had made three separate transactions with A, B, and C within the same day. Suppose on the previous shopping trip the consumer purchased brand A. Then in the transaction with brand A, the past choice dummy would be set to 1. In the other two transactions, the past choice dummy would be set to 0. On the next shopping trip, if the consumer purchased any one of brands A, B, and C, her last-time choice dummy would be set to 1. Apart from not estimating the quantity choice, the model also does not consider the store choice or the brand choice conditional on visiting a store, since store-level data are not available.

Now we return to the first stage of the model. Two assumptions are made about the choice set formulation. First, a brand purchased before would stay in the choice set. In other words, once a consumer tries a brand, he never forgets about it, even though he may dislike it and choose not to purchase it again. Second, the probability of consumers' being informed of a previously untried brand is a function of the brand's advertising stock. Formally, at time t, the probability that consumer i has choice set C_{it} is

$$P(C_{it}) = \prod_{j \in C_{it}} q_{ijt} \prod_{k \notin C_{it}} (1 - q_{ikt}) \tag{5.1}$$

where q_{ijt} is the probability of consumer i being informed of brand j at time t, and

$$q_{ijt} = \begin{cases} \dfrac{\exp(\varphi_0 + \varphi_1 adv_{jt} + \varphi_2 adv_{jt} inc_i + \varphi_3 adv_{jt} nokid_i + \varphi_4 adv_{it}^2)}{1 + \exp(\varphi_0 + \varphi_1 adv_{jt} + \varphi_2 adv_{jt} inc_i + \varphi_3 adv_{jt} nokid_i + \varphi_4 adv_{it}^2)}, \forall j \notin E_{it} \\ = 1, \forall j \notin E_{it} \end{cases} \tag{5.2}$$

where E_{it} is consumer i's experience set as of time t—that is, the set of brands previously purchased by consumer i up to time t. In the estimation, transactions in the first year of a consumer's purchase history are used to initialize her experience set. The variable adv_{jt} is a depreciated stock of advertising expenditures for brand j at time t. Specifically,

$$adv_{jt} = \sum_{\tau=0}^{T} \delta^{\tau} a_{jt-\tau} \tag{5.3}$$

[13] Shum (2004) fails to find across-brand synergies in demand patterns of RTE cereals that would require modeling the multiple-brand purchase decision. See Hendel (1999) and Dubé (2004) for examples of a multiple-discrete choice model that allows multiple-unit and multiple-brand purchases on one shopping trip; and see Hendel and Nevo (2006) for an example of a consumer inventory model. Multiple brand purchases on one shopping trip are treated as independent events.

where a_{jt} denotes brand j's advertising expenditure at time t,[14] and δ is the discount factor. Using stock instead of current flow of advertising allows advertising to have a lagged effect on consumer choice in the form of goodwill stock. Specifically, if a brand entered a consumer's choice set on the previous shopping trip but was not purchased, the probability of its reentering the consumer's current choice set may still be high even if the brand is not advertised in the current period, because of the lagged effects of previous advertising. The termadv_{jt}^2 is included to account for the potential increasing or decreasing returns to scale of advertising. In Eq. (5.2), adv_{jt} is also interacted with household income and whether there are any children in the household, to reflect the heterogeneity in exposure to advertising for different types of households.

At the second stage, consumer i chooses brand j to maximize expected utility conditional on her choice set. As is now standard in the discrete choice literature, the expected utility consumer i obtains from brand j is a function of brand j's characteristics.

$$U_{ijt} = E(X_j)\beta_i + \alpha_i price_{ijt} + \rho_i adv_{jt} + \kappa \cdot unused_{ijt} + \lambda_i \cdot unused_{ijt} \cdot adv_{jt} + pastchoice_{ijt} \bullet \gamma + \eta_{jt} + \varepsilon_{ijt} \quad (5.4)$$

where $X_j = [\textit{fiber sugar}]_j$, $\beta_i = [\beta_{1i}\ \beta_{2i}]'$, $price_{ijt}$ is the price of brand j when consumer i it at time t. In the Nielsen Homescan data, the price of a brand is recorded as the weekly average price of that brand in the store where the brand was sold. In the estimation, we subtract the manufacturer's coupon value and the retailer's deal value from the price if a coupon or a deal is used.[15]

Note that β_i, α_i, ρ_i, and λ_i are individual coefficients. Specifically,

$$\begin{bmatrix} \beta_{1i} \\ \beta_{2i} \\ \alpha_i \\ \rho_i \\ \lambda_i \end{bmatrix} = \begin{bmatrix} \beta_1 \\ \beta_2 \\ \alpha \\ \rho \\ \lambda \end{bmatrix} + \Pi \bullet D_i + \Sigma \bullet v_i \quad (5.5)$$

where D_i is a vector of observed household characteristics, including household income, age of female household head, and presence of children; v_i represents a vector of unobserved household characteristics with standard normal distribution.

[14] Advertising data are monthly; purchase data are daily. Therefore advertising expenditure at time t means advertising expenditure in the month that day t belongs to. In the empirical results, reported in Sect. 5.5, $\delta = 0.95$ and $T = 6$. We also estimate the model with δ varying from 0.8 to 0.99 and T from 3 to 12. The robustness checks do not yield significant qualitative differences.

[15] We are not able to control for coupons and deals systematically, as in Nevo and Hendel (2006), because we do not have store-level data and do not observe the availability of coupons and deals to consumers.

The variable $unused_{ijt}$ is a dummy equal to 1 if brand j was never purchased by consumer i before time t. It interacts with adv_{jt}, implying that advertising may provide information about the quality of unused brands. For example, the fact that the cereal manufacturer is able to spend a huge amount on promoting a brand may signal to consumers that the manufacturer is in a good financial condition and can therefore produce cereals with better ingredients and better technology. The vector $pastchoice_{ijt} = [chosen_{ijt-1}\ chosen_{ijt-2}, \ldots, chosen_{ijt-\tau}]$, where $chosen_{ijt-\tau}$ equals 1 if brand j was chosen τ shopping trips before t.[16] The term η_{jt} represents brand j's characteristics that are observable to the consumer but not to the researcher at time t. In the case of RTE cereals, η_{jt} encapsulates packaging, shelf space, etc. Lastly, ε_{ijt} is a mean-zero stochastic term independent across time, brands and consumers.

If brand j has not been purchased before, the consumer holds expectations of its fiber and sugar content according to the following rule: $E(fiber_j) = mean(fiber_k)$, and $E(sugar_j) = mean(sugar_k)$ $\forall$ brand k tried by consumer i before and belonging to the same segment as brand j . Following Hausman (1996) and Shum (2004), we divide the brands into family, adult, and kid segments. The segment categorization is shown in column 7 of Table 5.2. If the brand has been purchased before, then the consumer knows its characteristics.

The utility maximization stage generates $P(j|C_{it})$, the conditional probability that brand j is chosen by consumer i at time t. By the law of conditional probability, multiplying $P(j|C_{it})$ and $P(C_{it})$ yields P_{ijt}, the unconditional probability of consumer i choosing brand j at time t.

$$P_{ijt} = \sum_{C_{it} \in S} \prod_{j \in C_{it}} q_{ijt} \prod_{k \notin C_{it}} (1 - q_{ikt}) P(j|C_{it}), \tag{5.6}$$

where S is the set of all choice sets that include brand j. Matching the choice probabilities predicted by the model with the observed choices by maximum likelihood yields the parameter estimates.

5.4.2 Discussion

Several features of the demand model merit additional discussion. First, the choice set formation process addresses the informational problem about a product's existence. Even though the choice set is aggregated to contain the 50 biggest national brands and a composite brand, it is still unlikely that consumers would know and compare the utility of all 51 brands on each purchase occasion. Allowing the choice set to depend on consumption experience and brand advertising brings the model closer to real

[16] In the empirical results I use T = 6. Compared with previous studies, where T is often equal to 1, my results show a more complete picture of time dependence of consumer choices. I also estimate the model with T = 12, and the results are similar.

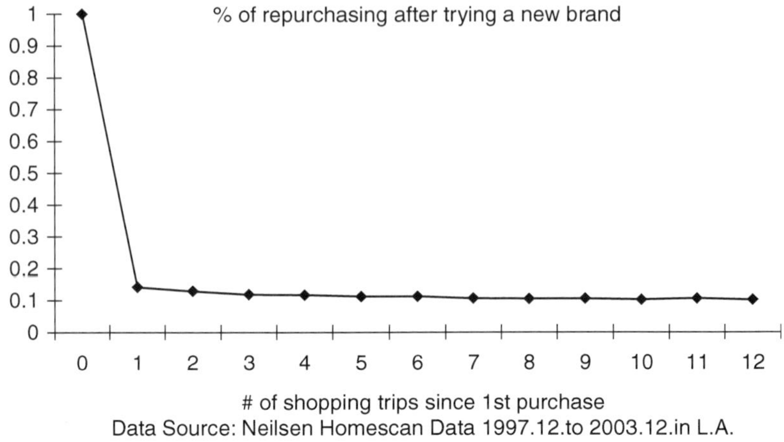

Fig. 5.2 Repurchase pattern

consumer behavior. Since the choice set is not observable in the data, we simulate them in the estimation. The details of simulation will be discussed in Sect. 5.4.4.

Second, consumers learn about brand quality after their first experience with the brand, which captures learning in the RTE cereal market reasonably well, as shown in Fig. 5.2. Unlike some complicated products, consumers usually attain precise knowledge about a cereal after consuming one box of it.

Third, compared with most previous choice models, where only the past choice is included, choices on the previous six shopping trips are included in the utility function. The coefficients on the set of past choice variables provide a better description of the temporal dependence of brand choices than when there is only the most recent choice. For example, if a consumer's brand choice history consists of A, B, A, B, . . ., A, B, and only the last-time choice dummy is included, then we would wrongly infer that she only seeks variety and is not subject to habituation. If we extend the model to have additional past choices, then it is possible to better capture the potential for habit formation. The distinction is important because if variety seeking is dominant, then temporary promotions' effect on demand would be short-lived. On the other hand, if consumers are susceptible to habit formation, temporary promotions may affect sales well into the future. Thus, adding more past choice variables not only better describes time dependence but also helps managers optimize decisions on marketing strategies.

Fourth, advertising has three roles in the model: (1) affecting consumer choice set, which represents advertising's informative effect on brand existence and is captured by the φ parameters; (2) signaling quality of an unused brand, which represents advertising's informative effect on brand quality and is captured by the parameter λ; (3) directly providing utility, which represents advertising's prestige effect and is captured by the parameter ρ. Identification of the different effects will be discussed below.

In the demand setup, we assume that the consumer is myopic and maximizes her current utility. When state dependence (habit formation) is present, a forward-looking consumer considers the future effects of her current choice. Forward-looking behavior is important in many cases, especially in situations where the experimentation cost is high, as in the choice of durable goods (computer, digital camera) and decisions about whether to accept a job offer or continue searching. It is less critical in this situation, where consumers choose a frequently purchased product and the cost of trying a new product is low because they can easily switch back to previous brands. Marketing research shows that consumers spend an average of 13 s in selecting a brand out of the shelf[17]—a very short time for a consumer to make choices. Therefore, we tend to believe that the myopic assumption is reasonable in this application and in the choice process of many other nondurable goods, such as beverages and cosmetics.

5.4.3 Identification

The parameters to be estimated (denoted as θ) include φ_0, φ_1, φ_2, φ_3, φ_4. β, α, ρ, $\prod$, $\sum$, κ, λ, and γ. Variation of brand choices corresponding to observed brand characteristics, price, and advertising for all consumers is used to identify β, α, and ρ. A cereal may also have attributes that are favored by a subgroup of consumers. For example, older consumers may prefer higher fiber content while kids may prefer higher sugar content. Substitution pattern of consumers with different demographics when brand characteristics vary helps identify $\prod$. And heterogeneity in substitution pattern of consumers with the same demographics helps identify $\sum$. Comparing the average probability of choosing a used brand with the average probability of choosing an unused brand on each purchase occasion identifies κ. Comparing the repurchase probability after purchase of a new brand with the repurchase probability of a previously purchased brand identifies learning from habit formation, and variation in brand choices over time pins down γ.

The main identification assumption of the prestige effect is that it does not vary by consumption experience. As in Ackerberg (2001, 2003), the prestige effect affects both experienced and inexperienced consumers in the same way, but the informative effects works only on consumers who have never tried the brand before. Therefore, variation in the ratio of the choice probability between experienced and inexperienced consumers as advertising intensity changes can be used to distinguish the informative effect from the prestige effect. The two types of informative effect (coefficient φ versus coefficient λ) both affect the choice probability of inexperienced consumers. An inexperienced consumer may choose to try a brand because advertising alerts him to the existence of the brand or because advertising raises the expected quality of this brand. Ignoring advertising's prestige effect for the moment, if advertising provides information only about brand existence, consumers will include the brand in their

[17] See Cesar Costantino, Ph.D. dissertation, Chapter 4, "Gone in Thirteen Seconds: Advertising and Search in the Supermarket," 2004.

choice set with a higher probability if the brand's advertising increases. In this case, advertising does not enter the consumer utility function, and hence the marginal effect of advertising on brand choice probability is independent of the observed brand characteristics. If two brands with different characteristics increase advertising by the same percentage, their choice probability will go up by the same percentage. If, furthermore, advertising provides a signal about brand quality, then consumers have two information channels to evaluate a brand—the advertising signal and the other brand characteristics. They would trade off the information inferred from advertising with the information observed from the brand characteristics. If the quality perception of the brand is already high based on the brand characteristics, the marginal effect of advertising on brand choice probability would be small: there are fewer consumers on the margin who would switch to the brand because of more exposure to advertising. If, on the other hand, the quality perception of the brand is relatively low from the brand characteristics, then the marginal effect of a surge in advertising would be big because more consumers would be persuaded to switch. Therefore, the two types of informative effect can be distinguished by whether the marginal effect of advertising on brand choice probability depends on the brand characteristics, since advertising enters the utility function and interacts with the brand characteristics only if the informative effect about brand quality exists.

To see this mathematically, let us consider a simple example with two brands in the market. Brand 1 has been established for a long time, and Brand 2 was newly introduced. Consumers all know about Brand 1, and Brand 2 launches an advertising campaign. Ignoring in this example the returns to scale of advertising in choice set formation and heterogeneity in coefficients across households, if advertising's only effect is informing consumers of the existence of Brand 2, then the probability that consumers choose it is

$$P = \frac{\exp(\varphi_0 + \varphi_1 adv_2)}{1 + \exp(\varphi_0 + \varphi_1 adv_2)} * \frac{\exp(E(X_2)\beta + \alpha^* price_2 + \Psi)}{1 + \exp(E(X_2)\beta + \alpha^* price_2 + \Psi)} \tag{5.7}$$

where ψ denotes the sum of variables in utility function other than price and observed brand characteristics. The marginal effect of advertising on the change in choice probability is

$$\frac{\partial \ln(P)}{\partial(adv_2)} = \frac{\varphi_1}{1 + \exp(\varphi_0 + \varphi_1 adv_2)} \tag{5.8}$$

Note that Eq. (5.8) is independent of Brand 2's characteristics. If advertising also provides information about quality, the choice probability of Brand 2 is

$$P = \frac{\exp(\varphi_0 + \varphi_1 adv_2)}{1 + \exp(\varphi_0 + \varphi_1 adv_2)} * \frac{\exp(E(X_2)\beta + \alpha^* price_2 + \rho^* adv_2 + \Psi)}{1 + \exp(E(X_2)\beta + \alpha^* price_2 + \rho^* adv_2 + \Psi)} \tag{5.9}$$

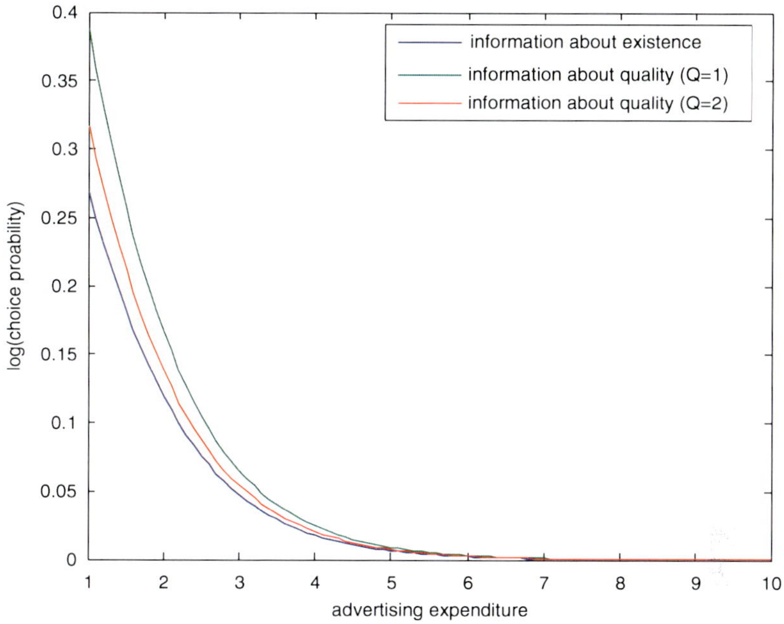

Fig. 5.3 Effect of advertising on marginal change in choice probability

The marginal effect of advertising on the change in choice probability is

$$\frac{\partial \ln(P)}{\partial(adv_2)} = \frac{\varphi_1}{1 + \exp(\varphi_0 + \varphi_1 adv_2)} + \frac{\rho}{1 + \exp(E(X_2)\beta + \alpha^* price_2 + \rho^* adv_2 + \Psi)} \tag{5.10}$$

The higher the utility consumers infer from the brand characteristics, the less the need to rely on the information in advertising. Comparing Eqs. (5.8) and (5.10), we can see that whether the marginal effect of advertising on choice probability depends on the brand characteristics distinguishes the informative effect about brand quality from the informative effect about brand existence. To illustrate this point, Fig. 5.3 depicts the marginal effect of advertising on choice probability. The nonstochastic part of the utility function other than advertising is denoted by Q. When only the informative effect about existence exists, the marginal change in choice probability is a declining function of advertising expenditure and is independent of Q. When advertising also signals quality, the marginal change in choice probability is not only a declining function of advertising but also a function of Q. As Q increases, the marginal change in choice probability decreases.

5.4.4 Estimation Issues

Here we discuss four estimation issues: unobserved consumer heterogeneity, choice set simulation, property of simulators, and the potential endogeneity of price and advertising.

5.4.4.1 Unobserved Consumer Heterogeneity

In a model with lagged dependent variables, state dependence (habit formation) is observationally equivalent to consumer heterogeneity because individual specific effects can lead to persistence in choices. State dependence can be exaggerated if unobserved consumer preferences are mistakenly assumed to be homogeneous. For example, an overweight consumer can have a high preference for a low-sugar cereal and repeatedly purchase it. If the consumer's specific preference is not controlled for, repeated purchases will be captured by the past choice variables and regarded as strong habit. Therefore, it is important to disentangle the true state dependence from consumer heterogeneity. In the estimation we use consumer-brand random effects to control for unobserved consumer heterogeneity. The details of the implementation are provided in Appendix 5.1.

5.4.4.2 Choice Set Simulation

To address the informational problem about brand existence, we allow for heterogeneity in consumer choice sets. The underlying choice sets over which consumers make utility comparisons are unobservable to researchers. Moreover, the number of potential choice sets can be very large—with 51 brands in the market, the number of possible choice sets is 2^{51}. Hence, instead of attempting to exhaust all possibilities, we simulate the choice sets. In the simulation, the probability of a brand's being included in a consumer's choice set is a function of brand advertising and purchase experience according to Eq. (5.2). The details of the choice set simulation process are provided in Appendix 5.2.

5.4.4.3 Estimation Procedure Without Instruments

After simulating the choice sets, we can calculate $\hat{P}_{ijt}$, the simulated choice probability of each brand for each household on every purchase occasion, and conduct a maximum simulated likelihood (MSL) estimation. The joint simulated likelihood function is

$$SL(\theta) = \prod_i \prod_t \hat{p}_{ijt}(\theta)^{Y_{ijt}} \tag{5.11}$$

where $Y_{ijt} = 1$ if consumer i purchases brand j at time t, and $Y_{ijt} = 0$ otherwise. The joint simulated log likelihood is

$$SLL(\theta) = \sum_i \sum_t Y_{ijt} \log(\hat{p}_{ijt}(\theta)) \tag{5.12}$$

The MSL estimator $\widehat{\theta}$ is a vector of parameters that maximize Eq. (5.12). Train (2003) shows that if the number of simulation draws rises faster than the square root of sample size, then the MSL estimator is not only consistent but also asymptotically equivalent to the maximum likelihood estimator.[18] Specifically, the MSL estimator is distributed

$$\hat{\theta} \overset{a}{\sim} N(\theta^*, -H^{-1}/N) \tag{5.13}$$

where $\theta^*$is the true parameter value, N is sample size, and $-H = -E(\frac{\partial^2 LL(\theta^*)}{\partial\theta\partial\theta'})$ is the information matrix. In practice, we use $\hat{H} = \frac{\partial^2 SLL(\hat{\theta})}{\partial\theta\partial\theta'}$ to approximate the value of H and calculate the estimated variance.

5.4.4.4 Endogenous Price and Advertising

If the manufacturer sets up prices and advertising levels according to consumers' willingness to pay, then an endogeneity problem may arise, since price and advertising levels could be correlated with unobserved brand characteristics in the utility function. For example, if the brand manager coordinates media advertising and store promotion activities, then the unobserved brand characteristics, such as shelf space or store featuring, can be correlated to the price and advertising expenditures of the brand. As a result, the coefficients on price and advertising can be overestimated. It is worth noting that we include brand fixed effects to control for unobserved brand characteristics invariant over time. For example, if government dietary policies promote the health effects of whole-grain foods, then the price and advertising levels of the whole-grain cereals may be increased. Whether a cereal is made with whole grains is invariant over time and is absorbed by brand dummies. However, unobserved time-varying brand characteristics, such as shelf space, are not absorbed by brand dummies and could create endogeneity.

One way to deal with the endogeneity problem is using instrumental variables (IV). Competition among differentiated products suggests that the optimal price and advertising levels depend on the characteristics, prices, and advertising levels of all brands offered. Brands facing more competition (due to existence of close substitutes in the characteristic space) will tend to have lower markups relative to brands facing less competition. If brand characteristics are exogenous, then the

[18] Monte-Carlo studies done by Keane (1994) and Geweke et al. (1994) also suggest that MSL has excellent small sample properties if reasonably good simulators are used.

characteristics of other brands are valid instruments for price and advertising. In the RTE cereal market, characteristics of a brand will not change once the brand is introduced into the market. Therefore, the exogeneity of brand characteristics is a reasonable assumption. However, the price and advertising levels of other brands are not valid instruments, since they are correlated with unobserved brand characteristics through consumer utility maximization. On the other hand, variables that shift production costs (ingredient prices, wages of food workers) are candidates for instruments, too.

In the nonlinear discrete choice model, IV estimation cannot be directly implemented on the consumer-level data. Following Berry et al. (1995, 2004), we first aggregate individual consumer choices into market shares and then match predicted and observed brand market shares to recover the component of utility that does not vary with individuals. This component is a linear function of price, advertising, and other brand characteristics, and one can estimate this function with IV for price and advertising.

Formally, let $\chi_{jt} = [fiber_{jt} sugar_{jt} price_{jt} adv_{jt}]$,[19]

$$\beta_i = \begin{bmatrix} \beta_{1i} \\ \beta_{2i} \\ \alpha_i \\ \rho_i \end{bmatrix} = \bar{\beta} + \Pi \bullet D_i + \Sigma \bullet v_i, \quad \text{where} \quad \bar{\beta} = \begin{bmatrix} \beta_1 \\ \beta_2 \\ \alpha \\ \rho \end{bmatrix},$$

$D_i = [income_i \; age_i \; nokid_i]'$, and $v_i = [v_{1i} \; v_{2i} \; v_{3i} \; v_{4i}]'$.

Then we can write the utility as

$$U_{ijt} = \chi_{jt}\beta_i + \kappa \cdot unused_{ijt} + \lambda_i \cdot unused_{ijt} \cdot adv_{jt} + pastchoice_{ijt} \bullet \gamma + \eta_{jt} + \varepsilon_{ijt}. \quad (5.20)$$

Let

$$\delta_{jt} = x_{jt}\bar{\beta} + \eta_{jt}. \quad (5.21)$$

Note that although Π, Σ, κ, λ, and γ can be estimated with micro data, we cannot estimate $\bar{\beta}$ without a further assumption to separate the effect of η from the effect of χ on δ. To provide consistent estimates of $\bar{\beta}$, we need IV for price and advertising.

We use two sets of instruments. The first set includes the fiber and sugar content of all other brands. The second set of instruments comprises the cost shifters, including wage of food workers, wage of advertising managers, corn price, wheat

[19] Although the true fiber and sugar content of brands do not vary over time, the expected fiber and sugar content do.

price, gasoline price, and electricity price. The data sources of these cost factors are described in Sect. 5.2.

The IV estimation involves three sets of moment conditions: (1) the model's predicted brand choice probabilities are matched to observed individual brand choices; (2) the model's prediction for brand j's market share in year t is matched to its observed market share in year t; and (3) the unobserved time-varying brand characteristics are assumed to be orthogonal to all the observed brand attributes and the instruments.

More specifically, the estimation algorithm consists of four steps.

Step 1. Given an initial guess of Π, Σ, κ, λ, and γ, we first find the values of δ_{jt} that equate the predicted market shares ($\sigma_{jt}(\delta, \Pi, \Sigma, \kappa, \lambda, \gamma)$) and the observed market shares (S_j) using the iteration $\delta_{jt}^{h+1} = \delta_{jt}^{h} + \ln(S_{jt}) - \ln(\sigma_{jt}(\delta^h))$. The details of calculating $\sigma_{jt}(\delta, \Pi, \Sigma, \kappa, \lambda, \gamma)$ and the proof that the above iteration is a contraction mapping are provided in Appendix 5.3.

Step 2. Given δ_{jt}, we provide random draws for unobserved consumer heterogeneity and for choice set formation, then use maximum simulated likelihood to obtain estimates of Π, Σ, κ, λ, and γby matching the observed choices with the predicted choice probabilities. Note that these estimates do not depend on any distributional assumptions of η. The probability that a household with observed characteristics D_i will choose brand j given δ,Π,Σ,κ, λ, and γb is given by

$$\Pr(j|D_i, \delta, \Pi, \Sigma, \kappa, \lambda, \gamma) = \int_v \frac{\begin{array}{l}\exp(\delta_{jt} + \chi_{jt} g \Pi g D_i + \chi_{jt} g \Sigma g v + \kappa \cdot unused_{ijt} \\ \quad + \lambda_i \cdot unused_{ijt} \cdot adv_{jt} + pastchoice_{ijt} \bullet \gamma)\end{array}}{\begin{array}{l}\sum_{k=1}^{51} \exp(\delta_{kt} + \chi_{kt} g \Pi g D_i + \chi_{kt} g \Sigma g v \kappa \cdot unused_{ijt} \\ \quad + \lambda_i \cdot unused_{ijt} \cdot adv_j + pastchoice_{ijt} \bullet \gamma)\end{array}} f(v) d(v) \tag{5.22}$$

The integrals are computed by simulation.

Step 3. Given the new values of Π, Σ, κ, λ, and γ, repeat the first two steps until δ,Π, Σ, κ, λ, and γ converge.

Step 4. Using the δ_{jt} obtained in step 3,construct the moment condition $E(\eta_{jt}|z) = 0$, where $\eta_{jt} = \delta_{jt} - \chi_{jt}\bar{\beta}$ and z represents instrument variables, and estimate $\bar{\beta}$ by minimizing the sample moments $G(\bar{\beta}) = \sum_j \eta_j Z_j$[20]

[20] Both Berry, Levinsohn and Pakes (2004) and Berry, Linton and Pakes (2004) show that in this type of BLP model with two sources of errors, the sampling error and the simulation error, both the number of observations and the number of random draws for simulation need to grow at rate J^2 for the parameter vector to have an asymptotically normal distribution.

5.5 Results

This section presents the demand estimates with and without instrumental variables. We carry out the demand estimation based on the panel data in the Los Angeles RTE cereal market.[21] There are 1,402 households with 69,134 cereal purchases in the LA market from December 1997 to December 2003. The first 12 months of each household's purchase history is used to construct its experience set; households staying in the Homescan panel for <12 months are dropped. The unit of observation in the estimation is a transaction—that is, a household-purchase date-brand combination. Observations with missing values on key estimation variables are dropped.[22] The regressions begin with July 1999, since the earliest advertising data are available in January 1999, and to calculate the advertising stock, we need advertising data for the previous 6 months. The estimation sample consists of 844 households and 37,858 transactions and remains unchanged in all specifications. Values of the key variables in the estimation sample are summarized in Table 5.4. In all specifications 50 brand dummies are used.

To guide the choice of variables, we first run a preliminary regression, a conditional logit regression with full information. Consumers are assumed to know all brands for sale and also their quality. The interaction terms of household demographics and brand characteristics that are not significant are excluded in later regressions. Since the logit model is subject to independence of irrelevant alternatives and does not capture the realistic substation patterns, the random coefficient logit model is used instead where a random component is added in the coefficients of price, advertising, fiber content, and sugar content.

5.5.1 Estimation Without Instrumental Variables

After the variable selection guided by the conditional logit, we run three random coefficient logit models with different informational assumptions. First, we assume that consumers have full information about both brand quality and brand existence. The choice set is the same over time and across consumers. The second specification is a regression with learning about quality information, where consumers are assumed to know all brands for sale in the market but not the quality of untried brands. In the third specification, consumers are assumed to have limited information about both the quality of untried brands and the brand existence. A random coefficient logit with quality learning and heterogeneous choice sets is estimated.

[21] The modeling technique and estimation method in this paper are not specific to a particular geographical market or a particular experience good. We can apply the model to environments where consumers face the two types of informational problems—for example, consumer choice of cosmetics, credit cards, and health care plans.

[22] The missing values do no happen systematically, so we are not concerned with a selection bias.

Table 5.4 Summary of variables in estimation sample[a]

Variable	Definition	Mean	Std. Dev.	Min	Max
chosen	1{brand is chosen in current transaction}	0.02	0.14	0	1
price	transaction price (cent/oz)	17.84	4.75	0	797.44
price*inc	transcation price(cent/oz)*household income($K)	1031.84	629.38	0	13,000
price*nokid	transcation price(cent/oz)*1{household has nokid}	892.50	348.03	0	7,400
adv	stock of advertising expenditure ($M)	3.22	4.02	0	22.30
adv*inc	advertising stock($M)*household income ($K)	188.47	281.56	0	2787.46
adv*nokid	advertising stock($M)*1{household has nokid}	1.95	3.51	0	22.30
unused	1{brand not purchased previously}	0.11	0.32	0	1
unused*adv	1{brand not purchased previously}*adv	1.82	3.29	0	22.30
unused*adv*inc	1{brand not purchased previously} *adv*household income($K)	106.57	223.03	0	2787.46
unused*adv*nokid	1{brand not purchased previously}*adv*1 {household has nokid}	1.21	2.79	0	22.30
chosen_1	1{brand chosen on last shopping trip}	0.03	0.17	0	1
chosen_2	1{brand chosen 2 shopping trips ago}	0.03	0.17	0	1
chosen_3	1{brand chosen 3 shopping trips ago}	0.03	0.17	0	1
chosen_4	1{brand chosen 4 shopping trips ago}	0.03	0.17	0	1
chosen_5	1{brand chosen 5 shopping trips ago}	0.03	0.17	0	1
chosen_6	1{brand chosen 6 shopping trips ago}	0.03	0.17	0	1
sugar	sugar content(% daily value per 30g)	8.81	3.73	0	16.67
sugar*age	sugar content(% daily value per 30g)*age of female head	439.51	227.30	0	1166.67
sugar*nokid	sugar content(% daily value per 30g)* 1 {household has no kid}	5.22	5.19	0	16.67
fiber	fiber content(% daily value per 30g)	8.58	10.28	0	58
fiber*age	fiber content(% daily value per 30g)*age of female head	428.18	544.41	0	4,060
fiber*nokid	fiber content(% daily value per 30g)* 1 {household has no kid}	5.08	8.96	0	58
adult*size	1{brand is adult brand}*household size	0.98	1.72	0	9
adult*inc	1{brand is adult brand}*household income ($K)	17.25	30.98	0	125
adult*age	1{brand is adult brand}*age of female head	14.69	23.51	0	70
adult*nokid	1{brand is adult brand}*1{household has no kid}	0.16	0.37	0	1
kid*size	1{brand is kid brand}*household size	0.98	1.72	0	9
kid*inc	1{brand is kid brand}*household income ($K)	17.22	30.95	0	125
kid*age	1{brand is kid brand}*age of female head	14.67	23.50	0	70
kid*nokid	1{brand is kid brand}*1{household has no kid}	0.16	0.37	0	1

[a]Estimation sample consists of 890 households with 42,396 transactions from January 1999 to December 2003 in Los Angeles market

Note that the three specifications are nonnested. To compare them, ideally we would like to construct a test statistic with a limiting distribution. However, our panel data do not satisfy the distributional assumptions of tests for nonnested models (e.g., Vuong 1989 and Chen and Kuan 2002). Therefore, to assess the goodness of fit, we use two methods. First we compare the different specifications using the Akaike information criterion (AIC) and using a measure of predictive performance developed by Betancourt and Clague (1981). Then we construct a variable that measures market share prediction errors of the three specifications to see how well they predict consumer choices.

5.5.1.1 Estimation with Full Information

The benchmark specification is a random coefficient logit regression where consumer choice sets include all 51 brands and the characteristics of all brands are known. The benchmark model allows us to examine in a simple way how price and advertising affect demand and have a sense of the temporal dependence of consumer choices.

The parameter estimates of the benchmark specification are reported in column I of Table 5.5. Price is negative and significant. The price sensitivity decreases as household income increases and if the household has no children. On average, advertising's prestige effect is negative and marginally significant. But the prestige effect increases as income grows and when there are no children in the household. The *unused* (untried) variable is negative and significant. If we calculate the odds ratio, we can see that the fact that a brand was never purchased before decreases the brand choice probability by 75%. However, *unused*adv* is positive and significant, suggesting that more advertising signals better quality to inexperienced consumers. The signaling effect diminishes with income and when the household has no children. All six past choice variables are positive and significant. The coefficient of *chosen_2* is slightly higher than that of *chosen_1*, consistent with the fact that consumers usually switch away from the brand last purchased if they were trying the brand for the first time. Both *fiber* and *sugar* are negative and significant. Older consumers without children prefer more fiber and less sugar.

5.5.1.2 Estimation with Limited Information About Brand Quality

In the second specification, we run a random coefficient logit regression where all consumers face the same choice set of 51 brands but do not know the quality of brands not bought before. Consumers form expectations of brand characteristics based on their previous experience with brands in the same segment. They also infer brand quality from advertising, and brand quality can be ascertained after one purchase. The signs of many coefficients (column II of Table 5.5) are the same as those of the benchmark regression, and for most coefficients the magnitudes are

Table 5.5 Estimation results

	I	II	III	IV
	RCL	RCL + Learning	RCL + Learning + HCS	IV Estimation
price	−0.164***	−0.187***	−0.233***	−0.368***
	(0.003)	(0.054)	(0.009)	(0.051)
price*inc	0.002***	0.001**	0.001***	0.001***
	0.000	(0.001)	0.000	0.000
price*nokid	0.127***	0.129***	0.118***	0.014***
	(0.003)	(0.007)	(0.007)	(0.005)
adv	−0.007*	−0.018	−0.023	0.274
	(0.004)	(0.065)	(0.050)	(1.731)
adv*inc	0.001**	0	0.001***	0.000***
	0.000	(0.001)	0.000	0.000
adv*nokid	0.073***	0.009	0.073***	−0.003
	(0.003)	(0.006)	(0.003)	(0.003)
unused	−1.874***	−2.071***	−1.800**	−1.801***
	(0.021)	(0.017)	(0.043)	(0.091)
unused*adv	0.335***	0.083***	0.286***	0.696***
	(0.006)	(0.010)	(0.023)	(0.041)
unused*adv*inc	−0.004**	0	0.027***	0.028***
	0.000	(0.001)	(0.001)	(0.001)
unused*adv*nokid	−0.155***	−0.041	−0.019**	−0.035
	(0.005)	(0.030)	(0.009)	(0.083)
chosen1	0.614***	0.649***	0.578***	0.563***
	(0.017)	(0.018)	(0.018)	(0.021)
chosen2	0.638***	0.629***	0.603***	0.590***
	(0.017)	(0.018)	(0.018)	(0.021)
chosen3	0.612***	0.574***	0.577***	0.565***
	(0.017)	(0.018)	(0.019)	(0.019)
chosen4	0.548***	0.504***	0.514***	0.503***
	(0.017)	(0.018)	(0.019)	(0.019)
chosen5	0.531***	0.466***	0.497***	0.488***
	(0.017)	(0.019)	(0.019)	(0.019)
chosen6	0.534***	0.468***	0.501***	0.491***
	(0.017)	(0.019)	(0.019)	(0.019)
fiber	−0.112***	0.014	−0.068***	4.466***
	(0.005)	(0.046)	(0.011)	(0.853)
fiber*age	0.001***	0.001	0	0.001***
	0.000	(0.001)	0.000	0.000
fiber*nokid	0.018***	0.037***	0.047***	0.026***
	(0.003)	(0.011)	(0.005)	(0.004)
sugar	−0.083***	−0.06	−0.099***	1.996
	(0.009)	(0.059)	(0.020)	(2.293)
sugar*age	0.001	−0.001	0	−0.001***
	(0.001)	(0.003)	0.000	0.000
sugar*nokid	−0.062**	−0.007	−0.066***	−0.032***
	(0.005)	(0.006)	(0.007)	(0.005)
φ0			−7.350***	−7.350***

(continued)

Table 5.5 (continued)

	I	II	III	IV
	RCL	RCL + Learning	RCL + Learning + HCS	IV Estimation
			(0.005)	(0.001)
φ1			2.996***	2.996***
			(0.001)	(0.001)
φ2			−0.002***	−0.002***
			0.000	0.000
φ3			0.001	0.001***
			(0.001)	0.000
φ4			−0.001***	−0.001***
			0.000	0.000
Σ11	0.187***	0.081***	0.512***	0.512***
	(0.001)	(0.023)	(0.007)	(0.002)
Σ22	0	0	0.005	0.005
	(0.003)	(0.094)	(0.014)	(0.004)
Σ33	0.036***	0.018	0.004	0.006**
	(0.001)	(0.033)	(0.005)	(0.003)
Σ44	0	0	0.002	0.010***
	(0.004)	(0.090)	(0.009)	(0.003)
log likelihood	−106,389	−100,617	−82,177	

RCL represents random-coefficient logit model. Learning uses expected product attributes for untried ("unused") brands in the utility function. *HCS* stands for heterogeneous choice set. ***p<0.01, **p<0.05, *p<0.1

comparable. The coefficient on *adv* is still negative but no longer significant. The only coefficient that changes sign is the *fiber* coefficient, but it is not significant.

The similarity of the coefficients (and the log likelihood) to the benchmark suggests that limited information about brand quality does not significantly affect consumer behavior. This is probably due to the nature of the RTE cereal market: the cost of experimenting with an untried brand is low, thus uncertainty about brand quality may not be an important factor when consumers decide which brand to buy.

5.5.1.3 Estimation with Limited Information About Both Brand Quality and Brand Existence

In the third specification consumers have limited information on both brand quality and brand existence. They still infer quality of untried brands from experience and advertising, but their choice sets are now heterogeneous and vary over time. The probability of having a particular choice set for each consumer on each purchase occasion follows Eqs. (5.1) and (5.2), and the choice set is simulated as described in Appendix 5.2.

The price coefficients (column III of Table 5.5) suggest that allowing for heterogeneous choice sets increases price sensitivity. The coefficient on price is significantly bigger than in the first two scenarios. To get a sense of how the price coefficient translates into price elasticity, we increase each brand's price by 1%

separately and simulate the consumer choices based on the parameter estimates. Consumer choices are then aggregated to calculate the percentage change in brand market shares resulting from the 1% price change. The values of own price elasticity for the top 10 brands are reported in Table 5.6. Compared with the previous two specifications, the price elasticity in the current one is much larger. The estimated price elasticities in the third specification are more plausible, since their absolute values are all bigger than 1, which is consistent with the fact that profit-maximizing firms should be operating at the elastic part of the demand curve.

When consumers have limited information about brand existence, they are not aware of brands outside their choice set and therefore cannot respond to the price changes of those brands. If we estimate the model as if consumers had full information about brand existence, we are in essence imposing the idea that consumers know the price changes of all brands but choose not to respond to some of them. As a result, the price elasticity is lower in the case of full information. The price estimate in the third specification suggests that consumers are actually much more sensitive to price changes of the brands that they are aware of. Should the consumers have lower information search costs and know more brands for sale, they would switch more frequently when the price is reduced. Therefore, if the information problem about a product's existence is alleviated, the market should be more competitive because consumers would be more responsive to price variations.

In the utility function, the coefficient on *adv* is negative but not significant, implying that advertising's prestige effect is not important. The coefficient on *unused*adv* is positive and insignificant, suggesting that advertising's informative effect on brand quality is not significant. In the choice set formation, φ_1 (coefficients on *adv* in Eq. (5.2)) is positive and significant, whereas φ_4 (coefficients on adv^2 in Eq. (5.2)) is negative and significant. Advertising raises the probability that consumers are informed of the brand, but this effect exhibits decreasing returns to scale. The coefficient on *adv*inc*, φ_2, is negative and significant, suggesting that the informative effect of advertising on brand existence decreases with household income. In contrast, the coefficient on *unused*adv*inc* in the utility function is positive and significant, suggesting the informative effect of advertising on brand quality increases with household income. This makes sense if richer consumers have higher opportunity cost of time and watch fewer TV commercials, but once they are alerted to the availability of an untried brand, they rely more on advertising to obtain the quality information than other methods of searching. The coefficient on adv*nokid in choice set formation, φ_3, is positive but not significant, implying that the effect of advertising does not vary with the presence of children. Figure 5.4 plots the probability of a brand's entering a consumer's choice set against the brand's advertising expenditure evaluated at the mean level of household income and presence of children. At the mean of advertising stock ($3.22 million), the probability of that a brand is included in the choice set is 88%. Increasing advertising stock by $1 million from the mean will result in a 99% probability that the brand is included in the choice set. What is consistent over the three specifications is that advertising plays a significant role in providing information to consumers, but it does not have a significant prestige effect.

Table 5.6 Predicted market shares

Brand number	Sample market share (percentage)	RCL	RCL + QualityLearning	RCL + QualityLearning + HCS
1	6.03	8.04	7.42	6.64
2	5.07	3.96	3.28	3.72
3	3.63	1.62	1.94	2.38
4	2.84	2.04	2.29	2.57
5	1.56	0.76	0.75	0.92
6	4.67	3.36	4.8	4.18
7	4.04	2.19	3.53	3.44
8	4.11	6.21	3.42	3.52
9	2.32	3.85	1	1.13
10	2.75	3.35	2.18	2.25
11	4.56	7.25	5.2	4.26
12	2.61	1.47	2.12	2.91
13	2.12	0.97	0.94	1.64
14	1.82	1.03	1.27	2.42
15	2.47	1.27	2.34	2.97
16	2.32	3.28	1.42	1.98
17	1.49	0.42	0.39	1.62
18	1.78	1.02	1.1	1.12
19	1.47	0.87	1.18	1.43
20	2.84	1.25	1.62	2.03
21	1.5	0.31	0.43	0.56
22	1.19	0.84	0.19	0.54
23	1.72	0.23	0.48	0.65
24	1.25	0.18	0.6	0.99
25	0.76	0.14	0.15	0.25
26	0.93	0.2	0.31	0.42
27	0.53	0.18	0.17	0.18
28	0.51	0.23	0.71	0.41
29	0.46	0.08	0.09	0.13
30	0.97	0.14	0.2	0.83
31	0.44	0.04	0.03	0.14
32	0.36	0	0.27	0.29
33	0.37	0	0.02	0
34	0.28	0.01	0.11	0.01
35	0.26	0	0	0.02
36	0.23	0	0.01	0
37	0.44	0.08	0.06	0.05
38	1.44	0.31	0.45	0.98
39	0.83	0	0.59	0.62
40	0.38	0.02	0.02	0.04
41	0.3	0.01	0.07	0.08
42	1.09	0.21	0.35	0.88
43	0.62	0.07	0.27	0.34
44	0.26	0.01	0.01	0.01

(continued)

Table 5.6 (continued)

Brand number	Sample market share (percentage)	RCL	RCL + QualityLearning	RCL + QualityLearning + HCS
45	0.5	0.23	0.09	0.24
46	0.75	0.18	0.23	0.36
47	0.61	0.1	0.08	0.19
48	0.53	0.04	0.14	0.17
49	0.49	0.12	0.1	0.24
50	0.66	0.04	0.03	0.05
Prediction error	0	7.26	5.29	3.81

Data: estimation sample for all regressions
Prediction error square root of sum of squared deviations of predicted market share to sample market share, *RCL* random-coefficient logit model, *Learning* using expected product attributes for untried ("unused") brands in the utility function, *HCS* heterogeneous choice set

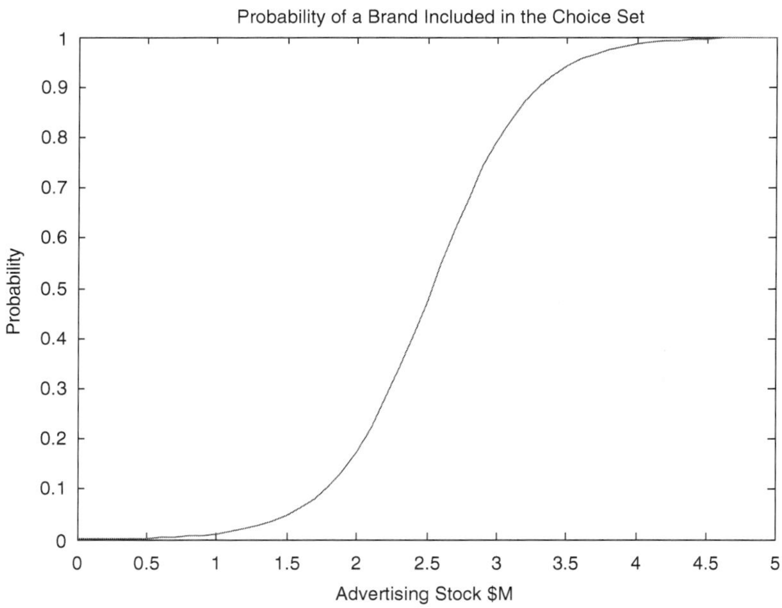

Fig. 5.4 Relationship between advertising stock and the probability of being included in the choice set

The past choice variables are still positive and significant, suggesting that consumers form persistent habits in cereal purchases. Compared with the results obtained without heterogeneous choice sets, the dependence on the past choice variables falls. The smaller coefficients on past choices are consistent with the larger (in absolute value) coefficient on price: consumers are more likely to switch brands in response to price changes when they rely less on previous experience.

5.5.1.4 Goodness of Fit

To compare the goodness of fit of the three specifications, two measures are computed. The first measure is the Akaike information criterion, which equals 2*k−2lnL, where k is the number of parameters and lnL is the log likelihood. The AIC imposes a penalty on more parameters, and the smaller the value of AIC, the better the model fit. The AIC for the first specification is 212930, for the second one, 201386, and for the third, 164516. Hence, according to the AIC, the third specification fits the data best.

Second, we compute a measure of predictive performance for discrete choice models developed by Betancourt and Clague (1981). The measure is based on the idea of information entropy. It rewards correct predictions when predicted choices are the same as observed choices and penalizes wrong predictions when predicted choices are different from observed choices. Moreover, the summary measure scores each choice prediction by giving it points not only in accordance with whether the prediction is correct but also in a way that reflects the degree of certainty of the prediction.[23] To obtain the measure, we first need to calculate the entropy for an observation in terms of predicted probabilities, $E_{it} = -(\sum_{j=1}^{51} P_{ijt} \log P_{ijt})$. Then the amount of information contained in the predicted probabilities P_{ijt} is defined as $I_{it} = 1 - E_{it}/E_{\max}$, where $E_{\max} = -\frac{1}{51}\log(\frac{1}{51})$ [24] and represents the maximum amount of uncertainty associated with the data distribution. Defining a correct prediction as $P_{ijt} > 1/51$ when brand j is chosen at time t and $P_{ijt} < 1/51$ when it is not chosen, we can calculate the amount of information contained in the sample set of predictions as $\bar{I} = (I_1 - I_2)/N$, where I_1 is the sum of information for all correct predictions, I_2 is the sum of misinformation for all incorrect predictions, and N is the number of observations. The specification with the highest value of $\bar{I}$ predicts the data best.[25] Applying the formula to our data, we find that the $\bar{I}$for the first specification is −11.5, for the second one, −13.2, and for the third, 0.8.[26] Again, the third specification represents the best fit.

[23] For a more detailed discussion of the measure, refer to Betancourt and Clague (1981, Section 4.6). The original measure is defined for cross-section data but can be easily extended to panel data. When choice sets are simulated, the probabilities used in the calculation are the mean of simulated probabilities.

[24] The formula is $E_{\max} = -\frac{1}{J}\log(\frac{1}{J})$, where J is the number of alternatives. In our case J = 51.

[25] Betancourt and Clague (1981) continue to develop several measures that capture the amount of information provided by the introduction of the theoretical model relative to the information contained in the sample. Since our goal is to compare only the three specifications, we do not calculate the other measures. Interested readers should refer to Betancourt and Clague's book for more information.

[26] A negative value of $\bar{I}$ suggests that the misinformation contained in wrong predictions exceeds the information contained in correct predictions. It can arise for two reasons: (1) there more wrong predictions than correct predictions; and (2) the wrong predictions generate probabilities farther away from 1/51 relative to the correct predictions.

Table 5.7 Own price elasticity for top 10 brands

Brand	RCL	RCL&Learning	RCL&Learning&HCS
1	−1.01	−1.27	−2.32
2	−1.27	−0.61	−2.63
6	−0.82	−0.68	−1.71
11	−1.27	−0.65	−1.39
8	−0.98	−0.74	−2.23
7	−0.68	−0.79	−1.66
3	−1.24	−1.04	−1.46
12	−0.98	−1.48	−2.82
4	−1.13	−1.63	−2.42
16	−1.42	−1.59	−2.57

RCL random−coefficient logit model, *Learning* using expected product attributes for untried ("unused") brands in the utility function, *HCS* heterogeneous choice set

Next we construct a variable to check how well the three specifications predict aggregate consumer behavior. Using the parameter estimates, we first predict consumer brand choice on each shopping occasion, which is the brand that generates the highest utility for the consumer on that shopping trip. Assuming that the consumer would purchase the same quantity of cereal as in the data, we can then calculate the consumer expenditure on that shopping trip. Summing up the consumer expenditures for each brand in the sample period, we get the predicted brand sales and brand market shares. Then we square the difference of predicted market share and observed market share for each brand, sum up the squared differences for all brands, and take the squared root of it to obtain the measure of market share prediction error. As shown in Table 5.7, the third specification generates a smaller market share prediction error than the first two.

In summary, introducing limited information about brand existence into the model improves the data fit and better captures consumer behavior. Therefore, we will base the following estimation on the limited information specification where consumer choice sets are heterogeneous.

The estimated parameters have important implications for brand pricing and advertising strategies. The pricing decision for a brand depends on the price elasticity of demand. Advertising provides product information and affects the composition of consumer choice sets, which can also affect consumer substitution. Therefore, a brand's advertising level also depends on the consumers' sensitivity to changes in advertising.

Given the parameter estimates in Column III of Table 5.5, above, we calculate the own and cross price elasticities for the top 25 brands,[27] which are reported in Table 5.8. The formula for computing the price elasticities is in Appendix 5.4. The price elasticities are evaluated at the median of each brand's price and the sample market shares.

[27] The remaining 25 brands have market shares of less than 1 % and relatively few observations, and therefore the simulation errors might be big.

Table 5.8 Estimated price elasticities for top 25 brands based on IV estimation

Brand	1	2	3	4	5	6	7	8	9	10	11	12
1	−2.428	0.367	0.146	0.01	0.099	0.338	0.22	0.315	0.003	0.673	0.034	0.002
2	0.136	−2.768	0.265	0.018	0.178	0.612	0.398	0.569	0.005	1.217	0.062	0.003
3	0.12	0.594	−1.545	0.015	0.157	0.54	0.352	0.503	0.005	1.075	0.054	0.002
4	0.092	0.455	0.179	−3.703	0.12	0.414	0.269	0.385	0.004	0.823	0.042	0.002
5	0.07	0.349	0.137	0.009	−2.762	0.317	0.206	0.295	0.003	0.63	0.032	0.001
6	0.292	1.446	0.569	0.038	0.383	−1.994	0.856	1.223	0.012	0.815	0.133	0.006
7	0.212	1.051	0.414	0.027	0.278	0.955	−1.561	0.889	0.008	1.901	0.096	0.004
8	0.249	1.234	0.485	0.032	0.326	1.121	0.73	−2.209	0.01	1.331	0.113	0.005
9	0.047	0.232	0.091	0.006	0.061	0.211	0.137	0.196	−1.679	0.42	0.021	0.001
10	0.141	0.698	0.275	0.018	0.185	0.635	0.413	0.591	0.006	−4.991	0.064	0.003
11	0.045	0.222	0.087	0.006	0.059	0.202	0.131	0.188	0.002	0.401	−1.561	0.001
12	0.022	0.109	0.043	0.003	0.029	0.099	0.064	0.092	0.001	0.197	0.01	−2.798
13	0.153	0.756	0.298	0.02	0.2	0.688	0.448	0.64	0.006	1.368	0.069	0.003
14	0.022	0.108	0.043	0.003	0.029	0.099	0.064	0.092	0.001	0.196	0.01	0
15	0.033	0.165	0.065	0.004	0.044	0.15	0.098	0.14	0.001	0.299	0.015	0.001
16	0.044	0.217	0.085	0.006	0.057	0.197	0.128	0.183	0.002	0.392	0.02	0.001
17	0.116	0.573	0.225	0.015	0.152	0.52	0.339	0.484	0.005	1.035	0.052	0.002
18	0.034	0.168	0.066	0.004	0.045	0.153	0.1	0.142	0.001	0.305	0.015	0.001
19	0.078	0.385	0.152	0.01	0.102	0.35	0.228	0.326	0.003	0.697	0.035	0.002
20	0.168	0.831	0.327	0.022	0.22	0.755	0.492	0.703	0.007	1.503	0.076	0.003
21	0.027	0.136	0.053	0.004	0.036	0.123	0.08	0.115	0.001	0.245	0.012	0.001
22	0.014	0.07	0.028	0.002	0.019	0.064	0.042	0.059	0.001	0.127	0.006	0
23	0.116	0.573	0.225	0.015	0.152	0.521	0.339	0.485	0.005	1.036	0.053	0.002
24	0.025	0.126	0.049	0.003	0.033	0.114	0.074	0.106	0.001	0.227	0.012	0.001
25	0.031	0.155	0.061	0.004	0.041	0.141	0.092	0.131	0.001	0.28	0.014	0.001

13	14	15	16	17	18	19	20	21	22	23	24	25
0.1	0.001	0.001	0.065	0.261	0.027	0.115	0.469	0.004	0.004	0.011	0.003	0.001
0.18	0.003	0.001	0.077	0.306	0.032	0.135	0.551	0.005	0.004	0.013	0.003	0.003
0.159	0.002	0.001	0.014	0.058	0.006	0.025	0.104	0.001	0.001	0.002	0.001	0.004
0.122	0.002	0.001	0.043	0.173	0.018	0.076	0.312	0.003	0.002	0.008	0.002	0.004
0.093	0.001	0.001	0.014	0.055	0.006	0.024	0.099	0.001	0.001	0.002	0.001	0.001
0.387	0.005	0.003	0.007	0.027	0.003	0.012	0.049	0	0	0.001	0	0.002
0.281	0.004	0.002	0.047	0.188	0.02	0.083	0.338	0.003	0.003	0.008	0.002	0.006
0.33	0.005	0.003	0.007	0.027	0.003	0.012	0.048	0	0	0.001	0	0.008
0.062	0.001	0.001	0.01	0.041	0.004	0.018	0.074	0.001	0.001	0.002	0	0.001
0.187	0.003	0.002	0.013	0.054	0.006	0.024	0.097	0.001	0.001	0.002	0.001	0.004
0.059	0.001	0	0.036	0.142	0.015	0.063	0.256	0.002	0.002	0.006	0.002	0.001
0.029	0	0	0.01	0.042	0.004	0.018	0.075	0.001	0.001	0.002	0	0.001
−3.629	0.003	0.002	0.024	0.096	0.01	0.042	0.172	0.002	0.001	0.004	0.001	0.005
0.029	−1.094	0	0.052	0.206	0.022	0.091	0.371	0.003	0.003	0.009	0.002	0.001
0.044	0.001	−1.351	0.008	0.034	0.004	0.015	0.061	0.001	0	0.001	0	0.001
0.058	0.001	0	−2.409	0.017	0.002	0.008	0.031	0	0	0.001	0	0.001
0.153	0.002	0.001	0.036	−3.788	0.015	0.063	0.256	0.002	0.002	0.006	0.002	0.003
0.045	0.001	0	0.008	0.031	−1.305	0.014	0.056	0	0	0.001	0	0.001
0.103	0.001	0.001	0.01	0.038	0.004	−2.384	0.069	0.001	0.001	0.002	0	0.002
0.222	0.003	0.002	0.092	0.01	0.041	0.166	−3.958	0.001	0.004	0.001	0.002	0.005
0.036	0.001	0	0.167	0.017	0.074	0.3	0.003	−1.434	0.007	0.002	0.004	0.001
0.019	0	0	0.148	0.015	0.065	0.265	0.002	0.002	−0.944	0.002	0.004	0
0.153	0.002	0.001	0.113	0.012	0.05	0.203	0.002	0.002	0.005	−3.178	0.003	0.003
0.034	0	0	0.087	0.009	0.038	0.156	0.001	0.001	0.004	0.001	−1.171	0.001
0.041	0.001	0	0.359	0.037	0.158	0.645	0.006	0.005	0.016	0.004	0.009	−1.409

5.5.2 Estimation with Instrumental Variables

Using both sets of instruments (nutrition of competing brands and cost factors), we report the estimates of $\bar{\beta}$ in column IV of Table 5.5, above. To test the endogeneity of price and advertising, we run an ordinary least squares (OLS) regression of Eq. (5.21) after we obtain δ_{jt} in step (3) and compare the coefficients with the IV estimates. The Hausman test of the two sets of estimates yields a P value of 0.55; therefore the OLS estimates are not significantly different from the estimates with IV. Hence the endogeneity of price and advertising does not affect the coefficient estimates much in this application. Since the price and advertising coefficient estimates without IV are much more precise than the IV estimates—in the IV estimation only 255 observations (δ by brand and by year) can be used whereas in the estimations without IV, 37,858 transactions are used—we will conduct policy experiments using the estimates without IV.

5.6 Counterfactual Experiments

We conduct three counterfactual experiments to evaluate some of the brand marketing strategies and a hypothetical food policy change. In the first two experiments, we choose Brand 28 as an example because it was newly introduced into the market in January 2003. Figure 5.5 summarizes Brand 28's average monthly prices, sales, and advertising in the estimation sample. Marketing managers are usually

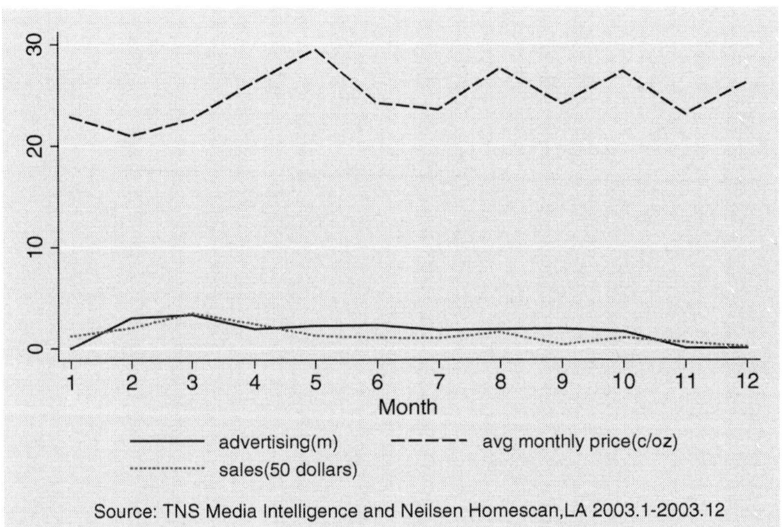

Fig. 5.5 Average monthly advertising, price and sales for Brand 28

Table 5.9 Change in sales under alternative pricing strategies

Δprice (%)	1	5	−1	−5
brand 28				
Δmarket share(%)	−1.23	−7.09	0.14	6.48
Δsales (%)	−0.78	−2.41	0.01	2.32

Δmarket share market share in the experiment – market share observed in data, *Δsales* sales in the experiment – sales observed in data

concerned with what price to charge and how to schedule advertising expenditures when a new product is launched. Therefore, looking into the data of Brand 28 offers us an opportunity to evaluate the marketing strategies of a product at the beginning of its life cycle. In the third experiment, we explore the effect of a hypothetical policy change—banning cereal advertising targeted to children—on consumer choices. A caveat should be borne in mind when we interpret the results of the experiments: the strategies of other firms are kept unchanged when we simulate the results, and thus the optimal responses of rival firms are not taken into account.[28]

5.6.1 *Pricing Strategy for Brand 28*

We first vary Brand 28's price from its observed price by +1, +5, −1, and −5 %, separately. Each time under the new pricing scheme, we calculate every household's simulated choices and aggregate them to get brand market shares and sales. The resulting changes in market share and sales of Brand 28 are reported in Table 5.9. We can see that if the price is reduced by 5%, the sales improve by 2.3%, compared with the sales figure before the price cut. The market share expands by 6.5%, which more than compensates for the reduction in price. Therefore, Brand 28's price was too high in general.

To see how the price cut affects different types of consumers, we calculate the changes in expenditures for different demographic groups after the price drops by 5%. We divide consumers by household income (high if household income >= $55,000, low otherwise), by age of female household head (old if age >= 32, young otherwise), and by the presence of children in the household. The results by demographic groups are shown in Table 5.10. Consumers with children and lower income respond more to the price cut than their counterparts, but the response does not vary with age groups.

Next we look at the average (weighted by volume) daily transaction prices of Brand 28 at its introductory stage (the first 3 months of 2003) and see whether its sales can be increased by altering the depth and frequency of the price discounts. The observed daily transaction price series for Brand 28 from January to March

[28] To derive the optimal responses, we need to solve a competitive equilibrium. However, the static Bertrand equilibrium is not realistic and the dynamic equilibrium is very hard to solve.

Table 5.10 Change in expenditure by demographic group under 5% price cut

	Δ in expenditure (%)
Highinc	1.21
Lowinc	3.59
Old	2.33
Young	2.32
Nokid	3.02
WithKid	1.09

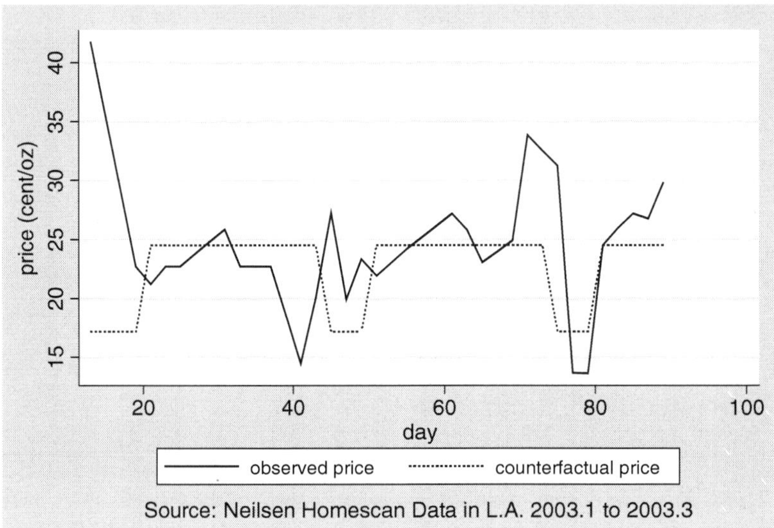

Fig. 5.6 Average daily transaction price for Brand 28

2003 is shown in Fig. 5.6. The initial price was very high, followed by a period of medium price level. Deep discounts happened twice when the price was about 60% of the average level. We consider an alternative pricing strategy, whereby price is set to be 70% of the average price in this period for the first week of each of the 3 months and 100% of the average price in the remaining weeks. The observed prices and the counterfactual prices in this period are plotted in Fig. 5.4, above. With the new pricing strategy, we find that Brand 28's market share goes up by 1.5% and sales go up by 1.2%. High introductory price is not desirable in this case because consumers are loyal to brands they are already using. To warrant a switch, the utility associated with the new brand needs to be sufficiently high, which could be achieved by a lower introductory price. Consumers who are lured into purchase by the low introductory prices will then form brand loyalty, and thus the brand manager can profit by setting the price low initially and increasing it later.

There may be two reasons why the brand manager would set a high initial price, as observed in the data. On the one hand, higher prices may be used by the brand manager as a signal for better quality in a market with limited information and hence attract consumers with higher willingness to pay. However, in the cereal market, many private label products have been introduced at low prices, and many consumers have come to realize that lower price does not necessarily affect the quality or taste.[29] Therefore, a high initial price would limit the consumer demand. On the other hand, the brand manager may have underestimated the price elasticity. As shown in Sect. 5.5, above, if demand is estimated while ignoring that consumers have limited information about product brand existence, price elasticities would be understated, which could lead the manager to set a higher than optimal price.

5.6.2 Advertising Strategy for Brand 28

A major consideration of a brand manager is to determine the best schedule of advertising expenditures for a certain budget. Conceptually, the manager could choose to do continual advertising (i.e., schedule ad expenditure smoothly over all times) or follow a strategy of pulsing (i.e., advertise in some weeks of the year and not at other times). We observe in Fig. 5.5, above, that Brand 28's advertising was relatively smooth over time. In contrast, many advertisers of consumer packaged goods use pulsing strategies. For example, Dubé et al. (2005) find that pulsing is the optimal advertising strategy in the frozen entrée market. Naik et al. (1998) develop a model of dynamic advertising that shows that pulsing strategies can generate greater total awareness than the continual advertising when the effectiveness of advertising varies over time. Specifically, ad effectiveness declines during periods of continual advertising and is restored during periods of no advertising. Such dynamics make it worthwhile for advertisers to stop advertising when ad effectiveness becomes very low and wait for ad quality to restore before starting the next campaign. They also show that the best advertising strategy for a major cereal brand is pulsing.

To mimic the pulsing strategy, we reschedule Brand 28's advertising by equally dividing the 2003 total ad expenditure into the 6 odd months and setting the budget to zero in the 6 even months (Fig. 5.7 plots the observed advertising versus the counterfactual pulsing advertising). Then we recalculate consumer choices under the new advertising strategy. The results show that Brand 28's market share and sales both increase by 1.9%. The pulsing strategy works better because it can increase the probability of Brand 28's entering the consumer choice set in the first 2 months after its introduction. In the observed data, the advertising expenditure for Brand 28 in January is zero, but in the pulsing strategy it is $3.3 million. The increase in the advertising expenditure in January raises the probability that an average consumer (with mean income, mean age, and mean presence of children)

[29] See "Eating Well," *New York Times*, September 22, 1993.

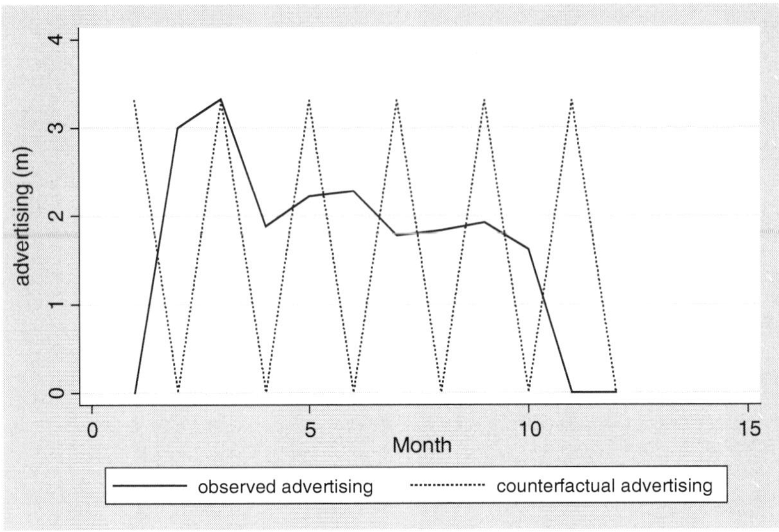

Fig. 5.7 Monthly advertising for Brand 28

Table 5.11 Change in expenditure by demographic group under pulsing strategy

	Δ in expenditure (%)
Highinc	2.64
Lowinc	1.59
Old	1.91
Young	1.91
Nokid	1.81
WithKid	2.15

will be aware of Brand 28 from almost zero to 89.7%. In February the pulsing strategy increases the advertising expenditure of Brand 28 from $2.99 million to $3.14 million, and it raises the probability of an average consumer's being aware of Brand 28 from 78.3 to 84.5%. In the following months an average consumer will be aware of the brand with probability close to 1 in both strategies. Therefore, under the pulsing strategy, more consumers are aware of the brand from the beginning and have a higher probability of choosing it. Some of these consumers become habituated to the brand, and hence the pulsing strategy can increase its overall market share. We also examine how different consumer groups respond to the pulsing strategy. The results, in Table 5.11, suggest that consumers with higher income and with children are more sensitive to the change in advertising strategy, but age does not matter.

In the advertising data, 98.9% of the advertising expenditure is spent on national media, such as network TV, national sports radio, and national newspapers. The pulsing strategy could also increase sales in other local markets without changing the advertising budget and could potentially be very profitable.

Table 5.12 Change in segment share after ban on advertising for kid brands

	Δ in mktshare (%)
Kid	−5.98
Adult	2.01
Family	3.96

Table 5.13 Effects of ban across consumer groups

	Δ in sugar (%)	Δ in fiber (%)	Δ in expenditure (%)
Highinc	−3.41	0.46	6.43
Lowinc	−5.27	2.67	4.36
Old	−4.22	0.95	4.87
Young	−5.91	4.24	8.67
Nokid	−2.69	−1.24	5.15
Withkid	−6.92	7.10	5.37

5.6.3 Effects of Banning Child-Oriented Cereal Advertising

We do not directly observe the value of ad dollars used for marketing toward children. To measure the effects of an advertising ban, we approximate the ban of child-oriented cereal advertising by eliminating the advertising expenditures for kids' cereal brands while holding other factors unchanged. In the experiment, we replace the ad stock of these brands with zero and calculate how the brand market shares change. The total changes for each brand segment (family, adult, kid) are summarized in Table 5.12. After the hypothetical policy change, the total market share of kids' brands goes down by about 6%, of which 2% goes to the adult brands and 4% goes to the family brands.

Then we look at how the policy change affects the nutritional intake and the expenditures of different consumer groups. The results are summarized in Table 5.13. Overall, after the ban of child-oriented cereal advertising, consumers consume more fiber and less sugar, which is better for their health. Consumers who are younger, with lower income, and with children reduce their sugar intake and increase their fiber intake more than their counterparts. Therefore, the policy change seems to have more effect on the "right" group of consumers. However, after the ban, consumers of all demographic groups have to increase their expenditures because they consume more adult and family cereals, which are more expensive than kids' cereals.

5.7 Conclusion

Using ready-to-eat cereal as an example of experience goods, we consider limited information on both product existence and product quality in a dynamic model. On each purchase occasion, a consumer first forms a choice set depending on her purchase experience and brand advertising. Conditional on the choice set, she then chooses the brand that maximizes her expected utility.

We have two main findings pertaining to the value of information. First, failure to account for limited information about a product's existence may significantly underestimate price elasticity. In our data, consumers are indeed sensitive to the price of the brands they know, but by assumption they cannot respond to price cuts in the brands they do not know. This finding implies that informative advertising that expands consumer choice set promotes competition because it allows price-sensitive consumers to choose among more brands. Second, advertising is much more effective on new consumers than on old consumers, which is consistent with the argument that advertising is mainly informative and not persuasive (at least in the RTE cereal market). The strong habit formation found in our data emphasizes the importance of the first-time experience and the information generated from it.

Both findings have useful implications for public policy. Since manufacturers' advertising is driven by private gains, informative advertising may be under-provided if part of the value of informative advertising is public (e.g., the value of condoms in reducing public health risk), if new entrants cannot afford informative advertising, or if manufacturers anticipate the procompetitive effect of informative advertising and collude to keep consumers uninformed of all choices. In these cases, public policies may play an active role in presenting available choices to consumers and encouraging competition among firms. By helping consumers make a smarter choice of first-time experience, these public policies can have a long-lasting effect on consumer welfare, thanks to habit formation.

On the other hand, manufacturers' advertising can be overprovided if advertising signals high quality in a dimension that is easy to tell by experience (say, the taste of the cereal) but remains silent on dimensions that are hard to know (say, the health consequences of eating sugary cereals). Since we do not model this complication, the ban of advertising on sugary cereals appears welfare reducing *from the consumer's point of view* because it leads to a smaller choice set and a less informative choice within the choice set. However, if advertising misleads consumers to choose sugary cereals—either because consumers are unaware of the unhealthfulness of the advertised food or because they like the sugary taste and do not consider their future health—limiting consumer's choice set could be beneficial to consumers.

There are other reasons why the counterfactual predictions on the ban of advertising should be taken with caution. In all the counterfactual experiments, we do not consider the competitive responses of other firms to the change in brand strategies. Nor do we account for the fact that firms may change the way they promote kids' brands once the government regulation comes into play. To control for these responses, we would need to solve the firm's profit maximization problem. In a model with brand loyalty on the consumer side, the firm's problem should involve dynamic optimization: the firm considers not only the effect of pricing and advertising on current consumer choices, but also the effect on future demand and future profits. However, the dynamic optimization problem with multiple firms, each with multiple brands, is extremely hard to solve and thus left for future research. In addition, many brand marketing strategies are decided by manufacturers and retailers together. This chapter focuses only on the role of manufacturers. A vertical competition model is needed to analyze the role of retailers.

5. Appendixes

5.1. Controlling for Unobserved Consumer Heterogeneity

We introduce consumer-brand random effects to capture the unobserved consumer heterogeneity in brand preferences. Specifically, the utility function can be written as

$$U_{ijt} = Z_{ijt} \bullet \Phi + v_{ij} + \varepsilon_{ijt}$$

where Z_{ijt} represents the vector of explanatory variables, Φ represents the vector of coefficients corresponding to Z_{ijt}, and v_{ij} represents consumer i's unobserved preference for brand j, which is independent from Z_{ijt} and ε_{ijt}.

Let $v_{ij} = \mu_{ij} + \omega_j$. $\mu_{ij} : N(0, \varsigma_{ij}^2)$, and $\omega_j = E(v_{ij})$ is a constant. Assuming ε_{ijt} has a generalized extreme value distribution, then we can write the probability that consumer i will choose j conditional on $\mu_{i1}, \mu_{i2}, \ldots \mu_{i51}$, and choice set C_{it} as

$$P(j|\mu_{i1}, \mu_{i2}, ..., \mu_{i51}, C_{it}) = \frac{\exp((Z_{ijt} - Z_{i51t}) \bullet \Phi + \mu_{ij} + \omega_j - \omega_{51})}{\sum_{l=1}^{51} \exp((Z_{ilt} - Z_{i51t}) \bullet \Phi + \mu_{il} + \omega_l - \omega_{51})} = \frac{\exp(z_{ijt} \bullet \Phi + \mu_{ij} + \xi_j)}{\sum_{l=1}^{51} \exp(z_{ilt} \bullet \Phi + \mu_{il} + \xi_l)}$$

where for the second equal sign we use $z_{ijt} = Z_{ijt} - Z_{i51t}$ and $\xi_j = \omega_j - \omega_{51}$.

$p(j|C_{it})$ is equal to $P(j|\mu_{i1}, \mu_{i2}, \ldots, \mu_{i51}, C_{it})$ integrated over the marginal distribution of the μ_{ij}'s. Specifically, it is equal to

$$\int_{-\infty}^{\infty} \int_{-\infty}^{\infty} \cdots \int_{-\infty}^{\infty} \frac{\exp(z_{ijt} \bullet \Phi + \mu_{ij} + \xi_j)}{\sum_{l=1}^{51} \exp(z_{ilt} \bullet \Phi + \mu_{il} + \xi_l)} f(\mu_{i1}) f(\mu_{i2}) \ldots f(\mu_{i51}) d\mu_{i1} d\mu_{i2} \ldots d\mu_{i51}$$

It is hard to compute $p(j|C_{it})$ analytically, and we simulate it by taking S draws from the distribution of μ_{ij}, for all j. The simulator for $p(j|C_{it})$ is

$$\hat{p}(j|C_{it}) = \frac{1}{S} \sum_{s=1}^{S} \frac{\exp(z_{ijt} \bullet \Phi + \mu_{ij}^s + \xi_j)}{\sum_{l=1}^{51} \exp(z_{ilt} \bullet \Phi + \mu_{il}^s + \xi_l)}$$

To reduce the number of parameters to be estimated, we allow ω_j to vary across brand segment, and ς_{ij}^2 to vary across both brand segment and whether the household has children. There are a total of eight parameters to estimate for unobserved consumer-brand preferences, of which six are scale parameters: $\varsigma_{FK}^2, \varsigma_{FN}^2, \varsigma_{AK}^2, \varsigma_{AN}^2$, $\varsigma_{KK}^2, \varsigma_{KN}^2$, where the first subscript denotes whether the brand belongs to the family, adult, or kid segment, and the second subscript denotes whether there are any children in the household; two are location parameters: ω_A and ω_K, where the subscript denotes whether the brand belongs to the adult or kid segment. ω_F is normalized to zero.

5.2. Choice Set Simulation Details

In the simulation, we assume that choice set is a function of brand advertising and purchase experience, as shown in Eqs. (5.1) and (5.2). The specific choice set simulation process is outlined as follows.

Step 1. Calculate $q_{ijt}(\varphi)$ for each consumer, each brand, and each time, where $\varphi = (\varphi_0, \varphi_1, \varphi_2)$.
Step 2. For each consumer-time-brand combination, draw a random number u_{ijt}^r from the uniform distribution between 0 and 1.
Step 3. If $u_{ijt}^r < q_{ijt}$, then brand j is included in consumer i's choice set at time t; otherwise it is not. This defines the choice set in the rth simulation C_{it}^r. After simulating the choice set, we can calculate simulated brand choice probabilities for each consumer.
Step 4. Calculate $P^r(j|C_{it})$, consumer i's probability of choosing brand j conditional on C_{it}^r. (The formula for calculating $P^r(j|C_{it})$ depends on the distributional assumption on the error term in the utility function).
Step 5. Calculate $p_{ijt}^r = \prod_{j \in C_{it}^r} q_{ijt} \prod_{k \notin C_{it}^r} (1 - q_{ikt}) \times P^r(j|C_{it})$, consumer i's unconditional probability of choosing brand j at time t in the rth simulation.
Step 6. Draw the random numbers u_{ijt}^r repeatedly for R times, and each time repeat steps 2–5.
Step 7. Calculate the simulated choice probability $\hat{p}_{ijt} = \frac{1}{R}\sum_{r=1}^{R} p_{ijt}^r$.

5.3. Contraction Mapping Details

In the instrumental variable estimation, we need to find the δ that makes predicted market shares based on the model equal to the observed market shares. Given an initial guess of δ, Π, and Σ, the predicted market share for brand j, $\sigma_j(\delta^h, \Pi, \Sigma, \kappa, \lambda, \gamma)$, is calculated as follows.

First, based on advertising data and household characteristics, simulate choice sets for each consumer on each shopping occasion.

Second, given δ, Π,Σ, κ, λ, and γ, a consumer compares the utility levels of all brands in his choice set on the shopping occasion and chooses the one that yields the highest utility.

Third, sum the consumer brand choices in a year to get predicted brand market shares.

To obtain the values of δ that solve $\sigma_j(\delta^h, \Pi, \Sigma, \kappa, \lambda, \gamma) = S_j$, we use the iteration $\delta_j^{h+1} = \delta_j^h + \ln(S_j) - \ln(\sigma_j(\delta^h, \Pi, \Sigma, \kappa, \lambda, \gamma))$. The proof that the iteration is a contraction mapping follows Goeree (2008).

Define $f(\delta_j) = \delta_j + \ln(S_j) - \ln(\sigma_j(\delta^h, \Pi, \Sigma, \kappa, \lambda, \gamma))$. To show that f is a contraction mapping, we need to show that ∀j and m, $\partial f(\delta_j)/\partial \delta_m \geq 0$, and $\sum_{m=1}^{J} \partial f(\delta_j)/\partial \delta_m < 1$.

We can write $\sigma_j = \int \sum_{C_i \in \Omega_j} \prod_{l \in C_i} q_{ilt} \prod_{k \notin C_i} (1 - q_{ikt}) P(j|C_i) f(v) dv$,where, and Ω_j denotes the set of choice sets that include j.

$$\partial f(\delta_j)/\partial \delta_m = \frac{1}{\sigma_j} \int \sum_{C_i \in \Omega_j} \prod_{l \in C_i} q_{ilt} \prod_{k \notin C_i} (1 - q_{ikt}) P(j|C_i) Q_j^m f(v) dv,$$

where

$$p(j|C_i) = \int_v \frac{\exp(\delta_j + \chi_j g \Pi g D_i + \chi_j g \Sigma g v + \kappa \cdot unused_{ij} + \lambda_i \cdot unused_{ij} \cdot adv_j + pastchoice_{ij} \bullet \gamma)}{\sum_{k=1}^{51} \exp(\delta_k + \chi_k g \Pi g D_i + \chi_k g \Sigma g v \kappa \cdot unused_{ij} + \lambda_i \cdot unused_{ij} \cdot adv_j + pastchoice_{ij} \bullet \gamma)} f(v) d(v)$$

$$Q_j^m = \frac{\exp(\delta_m + \chi_m g \Pi g D_i + \chi_m g \Sigma g v + \kappa \cdot unused_{im} + \lambda_i \cdot unused_{im} \cdot adv_m + pastchoice_{im} \bullet \gamma)}{\sum_{l \in C_i} \exp(\delta_l + \chi_l g \Pi g D_i + \chi_l g \Sigma g v + \kappa \cdot unused_{il} + \lambda_i \cdot unused_{il} \cdot adv_l + pastchoice_{il} \bullet \gamma)}, \; if\ m \in \Omega_j$$
$$= 0, \; if\ m \notin \Omega_j$$

Note that for m = j, $Q_j^m = P(j|C_i)$
Since all elements in the integral are nonnegative, we have $\partial f(\delta_j)/\partial \delta_m \geq 0$.
Moreover, $\sum_{m \in \Omega_j, m \neq 51} Q_j^m < 1$, therefore $\sum_{m \in \Omega_j, m \neq 51} \partial f(\delta_j)/\partial \delta_m < 1$ is satisfied.

5.4. *Price Elasticity Calculation*

Suppressing the time subscript, we can write the consumer utility function as

$$U_{ij} = \alpha_i p_j + \Upsilon_j g \beta_{\Upsilon i} + \varepsilon_{ij}$$

where $\alpha_i = \alpha + \Pi_3 g D_i + \Sigma_{33} \cdot v_3$, Υ_j represents the vector of variables other than price, and $\beta_{\Upsilon i}$ the vector of coefficients for Υ_j.

The formula for price elasticity is given by

$$\rho_{jk} = \frac{\partial s_j}{\partial p_k} \cdot \frac{p_k}{s_j} = \begin{cases} \frac{p_j}{s_j} \frac{1}{N} \sum_{i=1}^{N} \int \alpha_i \hat{p}_{ij}(1 - \hat{p}_{ij}) f(v) dv, \ j = k \\ -\frac{p_k}{s_j} \frac{1}{N} \sum_{i=1}^{N} \int \alpha_i \hat{p}_{ij} \hat{p}_{ik} f(v) dv, \ j \neq k \end{cases}$$

where p_{ij} represents the probability that consumer i will choose brand j.

In the estimation, we take NR random draws of v from f(v) to get α_i and compute ρ_{jk} using the formula

$$\hat{\rho}_{jk} = \begin{cases} \frac{p_j}{s_j} \frac{1}{N*NR} \sum_{i=1}^{N} \sum_{nr=1}^{NR} \alpha_i^{nr} \hat{p}_{ij}(1 - \hat{p}_{ij}), j = k \\ -\frac{p_k}{s_j} \frac{1}{N*NR} \sum_{i=1}^{N} \sum_{nr=1}^{NR} \alpha_i^{nr} \hat{p}_{ij} \hat{p}_{ik}, j \neq k \end{cases}$$

5. Commentary: Explaining Market Dynamics: Information Versus Prestige

Mead Over

Information is valuable to cereal manufacturers, who pay for advertising. Information is valuable to consumers, who reveal by their expenditure response that they attend to advertising. Information is valuable to nutrition activists, as a policy instrument to manipulate in the paternalistic hope that consumers deprived of advertising for sugary cereals will feed their children less sugar. And finally, information is valuable to the authors of the chapter, because using more of it enables them to explain more of the variation in market shares across the cereal brands and to predict more plausibly the reaction of consumers to price or advertising interventions for an individual brand or by a government consumer protection agency

Advertising is one of the industries whose business model involves the packaging and delivery of information. In contrast to the commercial publishing industry, wherein the author and originator of the information profits when the consumer values the information enough to buy the book, profits of the advertising industry derive from the advertiser's willingness to pay to subsidize information provision to the consumer. The distinction is due to the fact that consumers of books value them for their own sake, whereas consumers of information about advertised products use that information to inform their expenditures on those products. In an imperfectly competitive market for ready-to-eat cereals, cereal manufacturers are willing to subsidize consumers' information acquisition in order to differentiate brands from one another and reduce consumers' price elasticity of demand for their own brands.

The chapter deploys a variety of interesting microeconomic modeling and computationally intense econometric techniques to exploit a large data set on consumer purchases of ready-to-eat cereals and estimate the potential effect of a specific type of government intervention in this market: a ban on the advertising of children's cereal. The authors conclude that such a ban would indeed be effective in reallocating consumer expenditure away from the least healthful types of cereals and toward more healthful, more expensive brands, but it would induce consumers to spend more on cereal than they would without the ban. But one wonders whether the extraordinarily complex econometric paraphernalia the authors would really be required to show these impacts of advertising.

Since the authors generously provide the market shares of the top 50 brands as well as their average prices, brand-specific monthly advertising expenses, and market

M. Over
Center for Global Development, Washington, DC, USA
e-mail: mover@cgdev.org

Table 5.C.1 Ordinary least squares regression of logit of average market share on log price and advertising expenditures, by market segment

Number of obs = 50 F(8, 41) = 10.73 Prob > F = 0.0000 R-squared = 0.6768 Adj R-squared = 0.6137 Root MSE = .55338

Logit of (marketshare)	Coef.			$P > \lvert t \rvert$	[95%]	
Log(price):						
Adult	−.88	.69	−1.27	0.212	−2.28	.52
Family	−2.17	.54	−4.05	0.000	−3.26	−1.09
Kid	−.07	1.16	−0.06	0.951	−2.42	2.28
Advertising:						
Adult	.00086	.0002	4.44	0.000	.00047	.0013
Family	.00092	.0002	5.14	0.000	.00056	.0013
Kid	.00105	.0004	2.50	0.016	.00020	.0019
Constants:						
Adult	−3.78	2.56	−1.47	0.148	−8.96	1.40
Family	1.09	1.55	0.70	0.485	−2.04	4.22
Kid	−5.94	3.71	−1.60	0.117	−13.43	1.55

Source: This reviewer's estimates using the grouped data from Table 5.2

segment (in Table 5.2), one can calculate a descriptive ordinary least squares regression of (the logit of) market share on this grouped data. The results of this "naïve" regression are presented here in Table 5.C.1.

Although requiring very little effort beyond the tabulation of the average market shares, prices, and advertising expenditures for the 50 top brands, these results seem somewhat informative. The point estimates of the three estimated price coefficients, one for each of the three market segments, are all negative, as expected, with the one for family cereals being large (>2 in absolute magnitude) and statistically significant. Furthermore, all three advertising coefficients are highly statistically significant, suggesting that an extra million dollars of advertising increases market share by 0.86% for adult cereals, 0.92% for family cereals, and 1.05% for kid cereals. The category of kid cereals seems to respond more to advertising expenditures than the other two.

So why do more? What have the authors' prodigious efforts added to our knowledge of the ready-to-eat cereal market?

This chapter supports the proposition that "information is valuable to economic researchers" in three ways. First, by exploiting detailed information on the thousands of individual consumer transactions summarized in Table 5.2, the authors are able to relax several of the assumptions that are maintained by the above naïve analysis. In so doing, they demonstrate the value of that detailed information to the understanding of this complex market. Second, by bringing to bear an economic theory of decision making, the authors demonstrate that this theory itself has information content—because it helps explain the market data. Third, by combining the unusually detailed and granular data with this powerful theory, the authors are able to distinguish the two channels by which advertising hypothetically affects consumer behavior, the "information" channel and the "prestige" channel, and to demonstrate that it's the information that influences the consumer's behavior—not

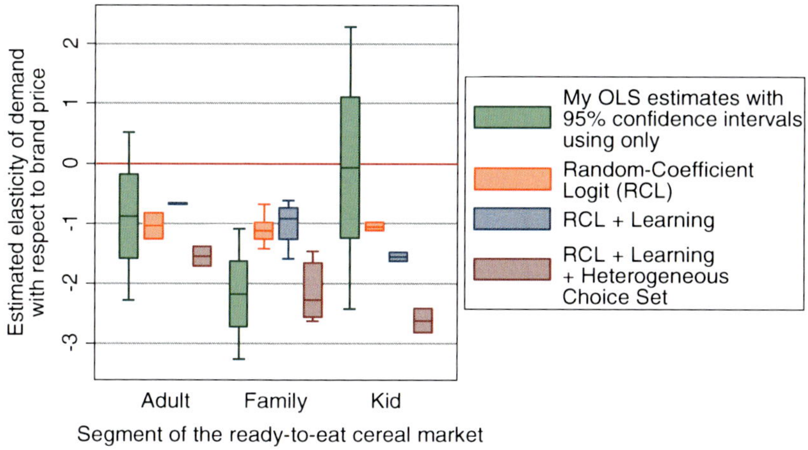

Fig. 5.C.1 Adding information either with more granular data or more theory-constrained economic structure increases both the precision and heterogeneity of estimated price elasticities across brands

the prestige. Fourth, by using information from the supply side of the cereal market, the authors are able to reject some types of endogeneity that would cast doubt not only on my naïve model, but also on their three principal models.

Consider the estimated price elasticities. Figure 5.C.1 displays for each of the three market segments the confidence intervals for my naively estimated price elasticities from Table 5.C.1 and the range of estimated elasticities for the top 10 cereal brands presented by the authors in their Table 5.8. There are two adult cereal brands in the top 10, six family brands and two kid brands. Note the extremely wide confidence intervals from my naïve estimates. Next to those confidence intervals (in green), my Fig. 5.C.1 displays the range of estimated price elasticities for each of the authors' three estimated models. Although the authors do not report confidence intervals, the point estimates of the brand-specific coefficient estimates from which these elasticities are derived (the first row of Table 5.5) are from 3 to 50 times larger than their estimated standard errors, suggesting tight confidence intervals for the elasticities. And the range of these reported estimates is also relatively tight within each market segment. Thus, one benefit of the information in the granular data appears to be tighter estimates of the brand-specific price elasticities.

The authors' basic model is a random-coefficients logit model (RCL) structured to assume that the choice sets for all consumers include all 50 brands (plus a 51st composite of all other brands) and characteristics of all brands are known. Figure 5.C.1 shows that the estimated elasticities for this model are roughly the same across the three market segments. (See the orange boxes in Fig. 5.C.1.) The authors' second model, whose elasticity estimates are represented by the blue boxes labeled "RCL + Learning," relaxes the assumption that all consumers know the characteristics of all brands. In this model the consumers again choose among all brands but only know the qualities of brands previously purchased. Advertising directly influences a brands market share. Thus, the impact of the economic theory on the estimated elasticities is to differentiate the three theoretically distinct markets, information that is useful to students of this ready-to-eat cereal market. Finally when the authors use an elaborate simulation model to require advertising to inform consumers of an unused brand's existence before it can affect their purchases (the assumption of heterogeneous choice sets), the estimated elasticities diverge even more across the three market segments (pink boxes) and also increase substantially in absolute magnitude. In the words of the authors, "[t]he estimated price elasticities in the ... specification [allowing a heterogeneous choice set] are more plausible, since their absolute values are all bigger than 1, which is consistent with the fact that profit-maximizing firms should be operating at the elastic part of the demand curve." Once more, economic theory has improved the fit of the model and contributed insight on the cereal market.

Variation in observed market shares, the naïve model contains substantial information. Its prediction error (defined by the authors as the square root of the sum of squared differences between the actual market share of Table 5.6 and the predicted share) equals 6.5, which is actually less than the 7.26 scored by the authors' random-coefficients model (bottom row of Table 5.6). However, both of the authors' more sophisticated models do better than my naïve model, scoring 5.28 and 3.81 respectively, and thus can be said to contain more valuable information.

Because they are able to simulate the consumers' choice sets each time on each visit to the grocery store, the authors can distinguish the two possible channels by which advertising might induce people to spend more on cereal—the information channel and the prestige channel. It's interesting that for this market, the authors find no support for the hypothesis that advertising persuades consumers to increase their consumption of ready-to-eat cereals that are familiar to them—which would be a prestige effect of advertising. Instead, advertising's role seems to be to induce consumers to try cereals that are unfamiliar. When they model this effect, the authors estimate much larger price elasticities (the pink boxes in Fig. 5.C.1). Since consumers have many choices in the cereal market, evidence that price elasticities are large in the children's cereal market and small in the adult cereal market suggests that the adults who purchase cereal for children see them as highly substitutable for one another, whereas they are loath to substitute one adult cereal for another. Adult cereal brands thus have more market power than children's brands.[30]

The authors' simulations of a ban on advertising for children's cereal and of a "pulsed" advertising strategy both raise the issue of the potential value to the public of government use of advertising. Using their third model, which incorporates

[30] The authors' finding of the highest price elasticity for children's cereal contrasts with the naïve model's failure to find any price effect on the market shares of children's cereals. A simple experimental manipulation of the price of a children's cereal would thus quickly demonstrate which of these two models is a more realistic portrayal of this market.

consumer learning and heterogeneous choice sets, and assuming that affected cereal manufacturers hold constant the prices of their brands, the authors simulate a ban on advertising and conclude that "the total market share of kid brands goes down by about 6%, of which 2% goes to the adult brands and 4% goes to the family brands." It's possible to perform this same experiment with the naïve model, by first computing the fitted market shares from the OLS regression in the children's market and then computing them a second time after the value of advertising has been set to zero. The result from the naïve model is that the total market share of children's brands would decline from 17.7 to 9.5% of the market, a reduction of about 8.2%. Under the assumption of the independence of irrelevant alternatives (the well-known IIA assumption typically maintained in multinomial logit models), about 2.2 percentage points of this decline would be reflected by an increase in the adult segment and about 5.8% age points would go to the family segment. Despite the simplicity of the naïve model, these results are remarkably similar to those obtained by the authors.

In contrast to the ban on advertising of children's cereals, the possible effects on the market of pulsed advertising could not be analyzed with the naïve model. The authors have used their heterogeneous choice set model to show that spreading the same advertising dollars smoothly is less effective at increasing market share than would be a strategy of bunching the advertising in specific months. The superior effectiveness of pulsing seems to be due to the lack of a prestige effect of advertising in this market. The implication is that government public awareness campaigns that intend to improve people's awareness of alternatives—and subsequently depend on their good experience with these alternatives to motivate behavior—could also benefit from pulse advertising. Whether the reverse is true for public awareness campaigns that intend to enhance the prestige of certain behavior remains to be determined.

The authors allude in passing to monopolistic pricing strategies when they point out that a monopolist operates in the elastic portion of its demand curve. Under certain conditions one could go further and assert that a profit-maximizing firm in a monopolistic or monopolistically competitive market will set its price-cost margin equal to the inverse of the elasticity of demand. According to the authors' heterogeneous choice set model, the median elasticities in the adult, family, and children's market segments are about −1.5, −2.3, and −2.8, respectively. This suggests that typical markups of price over marginal cost in these three segments are 65, 49, and 38 % of marginal costs, respectively. Furthermore, markups on individual brands vary from 34 to 72% of marginal costs. This information is of only academic interest in the market for ready-to-eat cereals, imagine if a similar analysis of the pharmaceutical market revealed such information about the prices of pharmaceutical brands. Views on pharmaceutical pricing range from the idea that monopoly profits in the pharmaceutical market are unproductive "rent" gained from branding products that largely result from government-subsidized research to the position that these profits are a just return on pharmaceutical firms' own research investments and motivate their future research. An objective observer would grant that both views have some legitimacy in various parts of the market. But policy intervention on the prices of individual drugs is hampered by the secrecy with which pharmaceutical firms guard their cost information. To the extent that the techniques employed in this chapter could be used to reveal the apparent markups of pharmaceutical prices over costs, regulators would value this information as an input to the regulation of the monopoly prices of individual pharmaceutical products.

References

Ackerberg, D. (2001). Empirically distinguishing informative and prestige effects of advertising. *The RAND Journal of Economics, 32*(2), 316–333.

Ackerberg, D. (2003). Advertising, learning, and consumer choice in experience good markets: A structural empirical examination. *International Economic Review, 44*(3), 1007–1040.

Allenby, G., & Ginter, J. (1995). The effects of in-store display and feature advertising on consideration sets. *International Journal of Research in Marketing, 12*, 67–80.

Anand, BN., & Shachar, R. (2011). *Advertising, the Matchmaker. The RAND Journal of Economics*, 42(2), 205–245.

Andrews, R., & Srinivasan, T. (1995, Feburary). Studying consideration effects in empirical choice models using scanner panel data. *Journal of Marketing Research, 32*, 30–41.

Becker, G., & Murphy, K. (1993). A simple theory of advertising as a good or bad. *Quarterly Journal of Economics, 108*, 942–964.

Berry, S. (1994). Estimating discrete choice models of product differentiation. *The RAND Journal of Economics, 25*(2), 242–262.

Berry, S., Levinsohn, J., & Pakes, A. (1995). Automobile prices in market equilibrium. *Econometrica, 63*(4), 841–890.

Berry, S., Levinsohn, J., & Pakes, A. (2004). Differentiated products demand systems from a combination of micro and macro data: The new car market. *Journal of Political Economy, 112* (1), 68–105.

Berry, S., Linton, O., & Pakes, A. (2004). Limit theorems for estimating the parameters of differentiated product demand systems. *Review of Economics Studies, 71*(3), 613 –654.

Betancourt, R., & Clague, C. (1981). *Capital utilization: A theoretical and empirical analysis*. Cambridge: Cambridge University Press.

Butters, G. (1977). Equilibrium distribution of prices and advertising. *The Review of Economic Studies, 44*, 465–492.

Chen, Y., & Kuan, C. (2002). The pseudo-true score encompassing test for non-nested hypotheses. *Journal of Econometrics, 106*, 271–295.

Chintagunta, P., Jiang, R., & Jin, G. (2009). Information, learning, and drug diffusion: The case of Cox-2 inhibitors. *Quantitative Marketing and Economics, 7*(4), 399–443.

Costantino, C. (2004). *Three essays on vertical product differentiation: Exclusivity, non-exclusivity and advertising*. Ph.D. dissertation, University of Maryland, College Park.

Crawford, G., & Shum, M. (2005). Uncertainty and learning in pharmaceutical demand. *Econometrica, 73*, 1137–1174.

Dubé, J. (2004). Multiple discreteness and product differentiation: Demand of carbonated soft drinks. *Marketing Science, 23*(1), 66–81.

Dubé, J., Hitsch, G., & Manchanda, P. (2005). An empirical model of advertising dynamics. *Quantitative Marketing and Economics, 3*(2), 107–144.

Einav, L., Leibtag, E., & Nevo, A. (2010, June). Recording discrepancies in Nielsen Homescan data: Are they present and do they matter? *Quantitative Marketing and Economics, 8*(2), 207–239.

Eliaz, K., & Spiegler, R. (2011). Consideration sets and competitive marketing, *Review of Economic Studies, 78*(1), 235–262.

Erdem, T., & Keane, M. (1996). Decision-making under uncertainty: Capturing dynamic brand choices in turbulent consumer goods markets. *Marketing Science, 15*, 1–20.

Geweke, J., Keane, M., & Runkle, D. (1994). Alternative computational approaches to inferences in the multinomial probit model. *The Review of Economics and Statistics, 76*(4), 609–632.

Goeree, M. S. (2008). Limited information and advertising in the US personal computer industry. *Econometrica, 76*(5), 1017–1074.

Grossman, G., & Shapiro, C. (1984). Informative advertising with differentiated products. *The Review of Economic Studies, 51*, 63–81.

Hausman, J. (1996). Valuation of new goods under perfect and imperfect competition. In *The economics of new goods* (pp. 209–237). Chicago: University of Chicago Press.

Hendel, I. (1999). Estimating multiple-discrete choice models. *The Review of Economic Studies, 66*(2), 423–446.

Hendel, I., & Nevo, A. (2006). Measuring the implications of sales and consumer inventory behavior. *Econometrica, 74*(6), 1637–1673.

Hitsch, G. (2006). An empirical model of optimal dynamic product launch and exit under demand uncertainty. *Marketing Science, 25*(1), 25–50.

Keane, M. (1994). A computationally practical simulation estimator for panel data. *Econometrica, 61*(1), 95–116.

Kihlstrom, R., & Riordan, M. (1984). Advertising as a signal. *Journal of Political Economy, 92*, 427–450.

Mehta, N., Rajiv, S., & Srinivasan, K. (2003). Price uncertainty and consumer search: A structural model of consideration set formation. *Marketing Science, 22*(1), 58–84.

Milgrom, P., & Roberts, J. (1986). Price and advertising signals of product quality. *Journal of Political Economy, 94*, 796–821.

Naik, P., Mantrala, M., & Sawyer, A. (1998). Planning media schedules in the presence of dynamic advertising quality. *Marketing Science, 17*(3), 214–235.

Nelson, P. (1970). Information and consumer behavior. *Journal of Political Economy, 78*(2), 311–329.

Nelson, P. (1974). Advertising as information. *Journal of Political Economy, 82*, 729–753.

Nevo, A. (2001). Measuring market power in the ready-to-eat cereal industry. *Econometrica, 69* (2), 307–342.

Nevo, A., & Hendel, I. (2006). Sales and consumer inventory. *The RAND Journal of Economics, 37*(3), 543–561.

Osborne, M. (2006). Consumer learning and habit formation in packaged goods markets, Ph.D. thesis. Stanford University.

Roberts, J., & Lattin, J. (1997). Consideration: Review of research and prospects for future insights. *Journal of Marketing Research, 34*(3), 406–410.

Shum, M. (2004). Does advertising overcome brand loyalty? Evidence from the breakfast-cereals market. *Journal of Economics and Management Strategy, 13*(2), 241–272.

Stigler, G., & Becker, G. (1977). De Gustibus Non Est Disputandum. *The American Economic Review, 67*(2), 76–90.

Swait, J. (2001). Choice set generation within the generalized extreme value family of discrete choice models. *Transportation Research Part B: Methodological, 35*, 643–666.

Train, K. (2003). *Discrete choice methods with simulation*. Cambridge: Cambridge University Press.

Vuong, Q. (1989). Likelihood ratio tests for model selection and non-nested hypotheses. *Econometrica, 57*, 307–333.

Chapter 6
The Effect of Public and Private Quality Information on Consumer Choice in Health Care Markets

Jonathan T. Kolstad

Abstract Information-based policy interventions have become increasingly common in health care markets. The rationale for such interventions is to correct a market failure in which consumers are asymmetrically informed about relevant attributes of a health care provider (e.g., quality). The magnitude of this market failure and the effect of public intervention on welfare depend on whether there exists market-based information on quality that alters consumer choice. To better understand such effects, I study consumer response to information provided by *U.S. News and World Report* hospital rankings and hospital reputation before and after the release of report cards on surgeon quality in Pennsylvania's market for cardiac bypass surgery. I estimate a model of consumer demand for surgeon quality (mortality) that integrates market-based information and quality reporting while controlling for the role of insurers and referring physicians in consumers' choice. The role of public versus market-based learning is identified using the interaction of the intertemporal change in information induced by the release of report cards with differences across providers in market-based information on those providers' quality. I find that market-based mechanisms impact patient response to quality prior to the release of report cards. After public release of information, the response to surgeon quality increases significantly. However, existing *U.S. News and World Report* rankings reduce consumer response to surgeon quality.

Keywords Consumer choice • Health care • Matching • Quality reporting • Hospital competition

---

J.T. Kolstad (✉)
The Wharton School, University of Pennsylvania, Philadelphia, PA, USA
e-mail: jkolstad@wharton.upenn.edu

R. Laxminarayan and M.K. Macauley (eds.), *The Value of Information: Methodological Frontiers and New Applications in Environment and Health*,
DOI 10.1007/978-94-007-4839-2_6, © Springer Science+Business Media Dordrecht 2012

6.1 Introduction

Market structure and function in the health care industry depend critically on the availability of information. Arrow (1963) demonstrated that information structure alone can explain many of the unique institutions that set health care delivery apart from other markets (e.g., the dominant role of not-for-profit providers, physician agency, insurance). Predicated, at least in part, on this idea, policymakers have sought to improve efficiency in health care markets by changing the way in which consumers use information by gathering, analyzing and providing health care "report cards." Whether such efforts improve welfare depends on the available mechanisms for consumers to learn about quality (and other asymmetrically held attributes of providers) in the absence of government intervention; this is called market-based learning (Dafny and Dranove 2008).

For example, U.S. News and World Report ranks hospitals in a number of specialties without any policy intervention. If such existing information sources have already informed consumers about quality, then public provision of quality data will be of little value. Alternatively, patients may rely on the advice of a physician agent (e.g., cardiologist) in choosing a specialist or may be constrained by their insurance in the set of available choices. The effect of both will be to alter the effect of report cards on observed consumer choice.

Motivated by these observations, I seek to answer two questions. First, how much do market-based learning and private information alter consumer choices in the absence of public reporting? Second, are privately provided information sources complements or substitutes for information-based public policy initiatives?

To answer those questions and to better understand the role of public and private information provision, I estimate a model of demand for cardiac bypass surgeons in Pennsylvania. Using detailed individual and provider observables, I first explore what factors alter consumers' choice of surgeons in the absence of public reporting. I estimate patients' response to a latent measure of a surgeon's quality, his or her risk-adjusted mortality rate (RAMR), and to cross-sectional variation in surgeons' attributes as reported in a privately provided information source, U.S. News and World Report.

I then turn to the effect of public reporting efforts. Changes in response to RAMR after intertemporal changes in information availability due to the release of report cards identify the effect of quality reporting. Differences in response across surgeons with differing privately provided quality estimates (measured by U.S. News and World Report rankings) allow difference-in-difference estimates for whether private and public reporting are complements or substitutes for each other.

I also include demand shifters observable to agents (but not patients)—the quality of the match between patients and surgeons given the prior types of patient treated by the surgeon. Incorporating this measure into demand allows me to characterize the role of agency in choice overall and on the effect of public reporting.

The model also allows the choice set to vary based on the breadth of the network offered to each patient based on the type of insurance they have, such as Medicare fee for service (FFS), Medicare health maintenance organization (HMO), or private HMO. In this way, the demand model accounts for two of the major agents in patient choice- referring physicians and health insurance- and isolates the impact of information with and without these effects.

The chapter proceeds as follows. Section 6.2 provides background on CABG surgery and the Pennsylvania setting. Section 6.3 introduces the data used. Section 6.4 develops an empirical model of patient choice with multiple information sources. Section 6.5 presents results and discussion, and Sect. 6.6 concludes and suggests avenues for future research.

6.2 Background and Setting

6.2.1 CABG Surgery

When a patient's blood flow to the heart is compromised by narrowing of the coronary arteries, coronary artery bypass graft (CABG) surgery is one of a range of available treatment options. Diagnosis and treatment of a patient with coronary disease are an integrated process requiring effort from both a primary-care physician and a cardiologist to diagnose the problem and to select a treatment regime. If a decision for surgical intervention is made, the patient must then choose between angioplasty and CABG and select a cardiac surgeon.

To perform CABG surgery, the surgeon opens the chest wall and creates a bypass around the blocked coronary artery, using either internal mammary arteries or arteries from the leg. The process is highly invasive, typically requiring a heart-and-lung bypass machine to support the patient during the procedure and a stay of several days in the hospital intensive care unit (ICU).

It has been well documented that production in cardiac surgery exhibits a volume-outcome relationship—that is, quality rises with the number of surgeries performed by a surgeon (Ramanarayanan 2007; Gowrisankaran et al. 2006; Gaynor et al. 2005; Huckman and Pisano 2005; Arrow 1963). This is generally attributed to learning-by-doing, though the endogeneity of volume raises the possibility of an alternative mechanism, selective referral.

6.2.2 Public Reporting in Health Care Provider Markets

Currently, 37 states in the United States mandate some form of mandatory quality reporting for providers (Steinbrook 2006). The earliest and most studied provider-reporting initiatives are New York's and Pennsylvania's CABG quality report cards.[1] Reporting of RAMR for CABG began in 1989 with New York State's release of risk-adjusted performance measures for hospitals and cardiac surgeons.[2] Beginning in 1990, the Pennsylvania Health Care Cost Containment Council (PHC4), a public-private partnership, began collecting discharge data on outcomes and patient comorbidities. The first widely available report card was released in May 1998 and included data for 1994–95.[3]

Studies generally find evidence for consumer response to the release of public information, though the economic magnitude varies substantially (Kolstad and Chernew 2008). There is evidence that higher-quality (lower-RAMR) hospitals and surgeons in New York saw increased demand after the release of quality information (Mukamel and Mushlin 1998; Cutler et al. 2004). Decomposing this effect, Cutler et al. (2004) find a statistically significant reduction of five surgeries per month (10% of the average hospital's volume) following a low quality indication but little effect of being flagged as a high-quality hospital. There is also evidence that quality reporting may lead the supply side of the market (e.g. surgeons) to improve quality solely due to their intrinsic incentives to do so (Kolstad 2010).

The only paper to date that explicitly considers market-based learning in the context of provider report cards is Dranove and Sfekas (2008). The authors estimate a discrete choice model that accounts for consumers' beliefs about provider quality prior to the release of report cards. They find a significant effect of *new* information on hospital market share. A one-standard-deviation improvement in reported RAMR results is approximately a 5% increase in market share. Dranove and Sfekas (2008) do not, however, decompose the way in which prior beliefs are established. This study extends their work by estimating a model that decomposes prior learning among private information sources, agency, and insurance.

6.3 Data

The data include 89,406 observations, consisting of every isolated CABG surgery performed in Pennsylvania in 1994–1995, 2000 and 2002–2003 (PHC4 1994, 1995, 1999, 2002, 2003). Table 6.1 presents summary statistics for the pre- and post-report

---

[1] Similar report card programs for cardiac surgery are now in use in many states, including California, Massachusetts, Florida, and New Jersey, as well as at the country level in the United Kingdom (Steinbrook 2006).

[2] Initially the project was undertaken by the state Department of Health to gather and measure outcomes only at the hospital level. However, *Newsday* sued the State under the Freedom of Information Act, leading to the public release of the data in the form of surgeon- and hospital-level quality reports.

[3] Reports based on 1990–1993 data were constructed and released between 1992 and 1995. However, these reports are no longer available, and discussions with experts suggest that these data and the reports were not widely observed. Schneider and Epstein (1996) present survey evidence consistent with very low exposure for the early paper versions of the report cards.

Table 6.1 Summary statistics

Total number	1994–1995	2000 and 2002 (Q1,2)
Hospitals	43	63
Surgeons	201	208
Teaching hospitals	19	19
U.S. News hospitals	9	9
Average		
Surgeon RAMR	3.52	2.36
Hospital RAMR	3.18	2.41

card periods. Every observation includes surgeon and hospital identifiers, patient demographics, a set of patient comorbidities, the patient's home zip code, data on the payer type, and a set of outcome variables.[4] The outcome of interest in this chapter is inpatient mortality.

In addition to the data from PHC4, I introduce data from the American Hospital Association (AHA) annual survey of hospitals in 2000. Hospitals are matched based on the name reported to PHC4 and AHA. For a small number of hospitals whose reported names could not be found in the AHA data, I match street address reported in the AHA survey with the hospital address based on each hospital's website. The AHA data include detailed information on hospital size, service offerings, teaching status, and insurance contracts.

I also merge data from the *U.S. News and World Report* rankings of hospitals. The magazine began providing ratings in 1993 and issues ratings across a range of specialties; it ranks the top 25–50 hospitals in United States in a given specialty (the number ranked varies by year). Based on the name of the hospital in the rankings and that reported in the PHC4 data, I merge data on the Pennsylvania hospital rankings in each year in either cardiology or cardiac surgery. Of the 63 total hospitals with bypass programs in Pennsylvania, nine receive a ranking in the top 50 hospitals between 1994 and 2002. These rankings range from the 22nd hospital to the 50th hospital in the country. Hospitals that received a ranking in 1994 tend to continue to be included in the list. For example, of those ranked in 1994, 82% were also ranked in 2000. Given the small number of hospitals being ranked, the variation in the actual number of the rank over the time period, and the stability of hospital inclusion, I collapse the ranking to a dummy variable equal to 1 if a hospital receives a *U.S. News and World Report* ranking at any point during the sample period. The model also includes a control for the numerical ranking of the hospital. The number of ranked hospitals as well as the number of teaching hospitals is consistent over the pre- and post-report card period (see Table 6.1).

To compute a measure of surgeon quality, I use a measure of risk-adjusted performance. Each observation includes a dummy variable equal to 1 if a patient

[4] Patient characteristics include age, indicators for cardiogenic shock, concurrent angioplasty, complicated hypertension, dialysis, female sex, heart failure, and prior CABG or valve surgery.

died in the hospital during or immediately following surgery. The log probability of death is computed as follows:

$$\ln\left(\frac{\Pr(MORT_{i,s,h} = 1 \mid x_i)}{1 - \Pr(MORT_{i,s,h} = 1 \mid x_i)}\right) = \beta_0 + \beta_1 \cdot X_i + \varepsilon_{i,s,h} \tag{6.1}$$

where i indexes patient, s surgeon, and h hospital. MORT is the indicator variable that equals 1 if the patient died in the hospital. This model is estimated for each report card period (1994–95, 2000, 2002, and 2003) (Pennsylvania Health Care Cost Containment Council 1998, 2002, 2004, 2005). The fitted values are obtained for each patient to form a predicted probability of death: the expected mortality rate (EMR). For each surgeon I then compute the risk-adjusted mortality rate:

$$RAMR_{s,h} = \left(\frac{OMR_{s,h}}{EMR_{s,h}}\right) OMR_{PA} \tag{6.2}$$

where the risk-adjusted, expected, and observed mortality rates for each surgeon s or hospital h are RAMR, EMR, and OMR respectively. These measures are computed as:

$OMR_{s,h} = \sum_{i|s,h} MORT_i$ and $EMR_{s,h} = \sum_{i|s,h} EMR_i$, where the summation is over patients i conditional on choosing surgeon s and hospital h, MORT is measured as above and EMR_i is equal to the fitted value for probability of death for patient i. Risk adjustment is accomplished by dividing the actual number of fatalities by the expected number of deaths conditioning on the actual patients selecting surgeon s or hospital h. This ratio is then normalized by multiplying this ratio by the statewide average mortality rate.

6.4 A Model of Patient Choice

6.4.1 *Patient Utility*

Each patient selects from the set of surgeons, $j \in J$, defined by the available set of surgeons in the hospital referral region (Wennberg et al. 1999). The utility for patient i from choosing a given surgeon s is a function of cost (both monetary and time costs), expected health improvement (capturing all components of quality and the ability of the patient to observe them), and an error term.

Indirect utility to consumer i who selects surgeon s is

$$u_{i,s,h} = g(X_i, \eta_i, \mu_i, Z_{s,h}, \theta_{s,h}; \rho) + \varepsilon_{i,s,h} \tag{6.3}$$

where X_i and η_i are vectors of observed and unobserved patient characteristics and μ_i is a vector of physician agent characteristics, all of which lead to differences in taste.

$Z_{s,h}$ is a K-dimensional a vector of hospital and surgeon characteristics not directly related to expected health. $\theta_{s,h,t}$ is the expected quality (beliefs about the gains in health) of surgeon s at hospital h. Finally, $\varepsilon_{i,s,h}$ is an iid error term distributed type-1 extreme value and ρ is a vector of parameters.

Learning about quality from different sources enters in the way in which a patient determines $\theta_{s,h,t}$. Each patient is assumed to infer quality from all available information sources in each period *t*. The model of expected quality for surgeon s at hospital h in period t is

$$\theta_{s,h,t} = \beta_1 RAMR_{s,t-1} + \beta_2 RAMR_{s,t-1} * Post_t + \beta_3 \varphi_{s,h,t} + \beta_4 \varphi_{s,h,t} * RAMR_{s,h,t-1} \\ + \beta_5 \varphi_{s,h,t} * Post_t + \beta_6 \varphi_{s,h,t} * RAMR_{s,t-1} * Post_t \quad (6.4)$$

The effect of formal reporting is identified by changes in patient choices between the pre- and post-report card period (1994–95 compared with 2000 and the first two quarters of 2002), captured by the dummy variable$Post_t$that takes a value of 1 if an observation in period t is after 1995.

Equation (6.4) models two main effects. The first is the average response to quality, given available information. The coefficient β_1 is a measure of the average response to surgeon RAMR with only market-based learning. β_2 captures the average differential response to quality following the release of report cards. The second is the effect of market-based learning on choice. The coefficient β_3 is a measure of patient response to privately supplied information on quality. The term on the interaction of $\varphi_{s,h,t}$ with surgeon RAMR,β_4, is an estimate for the differential effect of each component of $\varphi_{s,h,t}$ on an individual's response to latent surgeon quality. Interacting $Post_t$ with $\varphi_{s,h,t}$ captures the effect of report cards on patient response to privately supplied quality information. Finally, the coefficient β_6 captures the effect of quality reporting on the response to quality, given market-based signals about quality.

β_6 is a triple differences estimate for the role of each component of market-based information on individuals' response to RAMR after it is made public following quality reporting. If public information is a substitute for private information (and vice versa), estimates will be positive and significant. That is, given prior information from private sources that a provider is of high quality, patient response to surgeon quality (RAMR) after public release is smaller. On the other hand, if market-based learning complements public reporting, I expect that the interaction of RAMR with market-based information will have a (weakly) negative coefficient.

6.4.2 *Prices and Insurance*

Indirect utility in (6.3) is derived directly from a quasilinear utility function without wealth effects or prices. Typically, demand models include price in indirect utility. In this case, however, we do not observe the out-of-pocket price facing a patient undergoing CABG. For a procedure as expensive as CABG, the out-of-pocket cost for an insured patient is unlikely to vary in any meaningful way between surgeons. On the other hand, for patients not covered by traditional Medicare or private fee-for-service plans, network constraints can limit their choice of surgeon. To deal with this issue, I model the patient's specific network constraints for each surgeon.

Implementing this empirically is hampered by a common difficulty in estimating patient choice models in health care: data on patients' specific plans and the hospital and surgeon networks available within those plans in each period are not available. Because I do not directly observe network participation by hospitals or surgeons or a patient's specific plan brand (e.g., Blue Cross Blue Shield preferred provider organization, Aetna HMO), I infer the network constraints facing individual i using data on the general type of insurance for each patient (Medicare, Medicare HMO, private FFS, private HMO, Medicaid, and uninsured). A patient's choice probability is computed as a function of the likelihood that a given surgeon is available to that patient in period t based on observed market outcomes for that surgeon in the prior period.

Consider a specific plan (e.g., an Aetna HMO) that is included in insurance type z (Private HMO). The probability that a surgeon s is included in individual i's choice set is the joint probability that surgeon s is included in any type z network and, conditional on inclusion in type z network, the probability that the plan is included in the type z plans in market h. I assume that surgeons and hospitals prefer to serve the most profitable patients and that prior-period profitability is correlated with current-period profitability for a patient with insurance type z as follows:

$$\pi_{i,z,t} = v\pi_{i,z,t-1} + u_{i,z,t} \tag{6.5}$$

where v is a coefficient capturing the intertemporal correlation in insurer type profitability and $u_{i,z,t}$ is mean zero error term. When $v>0$, we expect the more profitable a given insurance type was in period t-1, the more likely a surgeon is to contract with that insurance type again in period t. Conditional on a surgeon's being included in any plan of type z, the probability that the plan is included in the set of type z plans in market h is $\tau_{z,h,t} \sim N(b, W)$. Under these assumptions, the share of patients in plan type z treated by surgeon s in the prior period is a measure of the relative profitability of those patients. Thus the probability that patient i will have the option of selecting a surgeon s is a function of the lagged share of patients of insurance type z seen by surgeon s weighted by the (unobserved) probability that the plan is included in the set z for surgeon s. Assuming that the error in (6.5) is distributed normally, we subsume the error into the distribution of $\tau_{z,h,t} \sim N(b, W)$.[5] Thus the probability that a surgeon s is included in the network is

$$\tau_{i,s}\left(v_z PayorShare_{z,t-1}\right) \tag{6.6}$$

[5] Distributing τ and taking the plim of the error term $\tau * u_{i,z,t} = 0$.

In this way, network constraints are captured by reducing the probability that a patient can substitute an alternative surgeon who served a smaller share of patients of the same insurance type in the prior quarter. In estimation I also interact the prior-quarter payer share with the number of HMO contracts at hospital h to allow a more flexible response.

6.4.3 Agency

No studies to date have been able to estimate a model of consumer choice that separates the role of a physician agent from patient choice in responding to information (Pope 2009). To identify the role of agency in patient choice, I rely on a demand shifter that is unlikely to be observable to an individual patient but should be known by a cardiologist: the match between a patient with a given severity and the types of patients each surgeon usually treats. That is, a referring physician is better informed regarding different surgeons' performance in treating patients with differing types of disease and severity. By conditioning demand on such a measure, I can control for the role of the physician agent in demand.

I model this match value as a function of the absolute value of the deviation between a patient's severity (measured by predicted mortality) and the lagged mean patient severity for surgeon i in period t-1 (the prior quarter). This measure is used on the assumption that patients with more comorbidities are likely to be (both observably and unobservably) more difficult cases and be better suited to surgeons who have more experience in treating complex cases. Thus agency enters demand as a (potentially nonlinear) function:

$$\mu_i = \gamma_1 f(|EMR_i - \overline{EMR}_{s,t-1}|) \tag{6.7}$$

The subscript i remains because realized behavior is the manifestation of the joint decision process of patient and agent, and I do not observe any information on the identity of the referring physician. In some specifications I also allow agency to depend on the unobserved term η_i. This accounts for unobserved patient willingness to take the advice of her agent as well as unobserved variation in the degree to which patient-surgeon matching alters the agent's referral patterns.

6.4.4 Likelihood and Estimation

A patient selects surgeon s at hospital h if and only if

$$u_{i,s,h} > u_{i,j,h} \forall j \neq s \tag{6.8}$$

Incorporating Eqs. (6.4), (6.6), and (6.7) and substituting back into (6.3), the patient's utility function is now

$$\begin{aligned} u_{i,s,h} = {} & X_i + \gamma_1 f(|EMR_i - \overline{EMR}_{s,t-1}|) + \lambda_1 PayorShare_{i,z,t-1} + \beta_1 RAMR_{s,h,t-1} \\ & + \beta_2 RAMR_{s,h,t-1} * Post_t + \beta_3 \varphi_{s,h,t} + \beta_4 \varphi_{s,h,t} * RAMR_{s,h,t-1} + \beta_5 \varphi_{s,h,t} * Post_t \\ & + \beta_6 \varphi_{s,h,t} * RAMR_{s,h,t-1} * Post_t + \varepsilon_{i.s.h} \end{aligned} \quad (6.9)$$

Note that in (6.9) I do not include unobserved taste components. I begin by estimating (6.9) as a standard multinomial logit model (McFadden 1974). I then turn to a model that incorporates unobservable patient responses to quality, market-based information, and agency as well as unobserved network constraints. Individual utility in this model is

$$\begin{aligned} u_{i,s,h} = {} & X_i + [\gamma_1 f(|EMR_i - \overline{EMR}_{s,t-1}|) + \beta_1 RAMR_{s,h,t-1} \\ & + \beta_2 RAMR_{s,h,t-1} * Post_t + \beta_3 \varphi_{s,h,t} + \beta_4 \varphi_{s,h,t} * RAMR_{s,h,t-1} + \beta_5 \varphi_{s,h,t} * Post_t \\ & + \beta_6 \varphi_{s,h,t} * RAMR_{s,h,t-1} * Post] * \eta_i + \lambda_1 \tau_{i,s} \left(v_z PayorShare_{i,z,t-1} \right) + \varepsilon_{i.s.h} \end{aligned} \quad (6.10)$$

Individual choice is a function of observed (to the econometrician) patient and surgeon attributes as well as a set of unobserved factors: insurer network constraints, random tastes for and use of report cards, and the role of agency in this choice. I incorporate the unobserved terms as random coefficients (Berry et al. 1995; Nevo 2001; Train 2003).

Assume that the unobserved components of utility are distributed according to the distribution $\phi(\eta_i, \tau_i | b, W)$ that is known up to a mean and covariance, b and W, to be estimated. Thus the probability that a patient i chooses surgeon s is the closed form logit choice probability integrated over the distribution of the unobserved terms. The probability that patient i selects surgeon s given choice set j can be expressed as follows:

$$P_{i,s} = \int \left(\frac{e^{\beta X_{i,s}}}{\sum_j e^{\beta X_{i,j}}} \right) \phi(\eta_i, \tau_i | b, W) d\lambda \quad (6.11)$$

The integral over the unobserved components of utility does not have an analytical solution. However, I can estimate the model using simulated maximum likelihood (Train 2003). Estimates for the demand system are computed by solving analytically for the logit choice probabilities and integrating out the random taste distribution by taking draws from the joint distribution of unobserved terms. Using this numerical simulation, I compute the likelihood of observing the choices based on the observed and unobserved components of choice.

I draw n values of the unknown components η_i and τ_i from the normal distribution. For each draw I then compute the choice probability using Eq. (6.11). For n Halton draws, the simulated log-likelihood is

$$SLL = \sum_{i=1}^{I}\sum_{n=1}^{N} I[i = s]\ln \ddot{P}_{is} \tag{6.12}$$

where $I[i = s]$ is an indicator function taking the value 1 if individual i chose surgeon s and zero otherwise and $\hat{P}_{i,s} = \frac{1}{N}\sum_{n=1}^{N}\left(\frac{e^{\beta X_{i,s}}}{\sum_j e^{\beta X_{i,j}}}\right)$. The coefficient vector that maximizes (6.12) is the maximum simulated likelihood estimator (Train 2003).

6.5 Results and Discussion

6.5.1 Base Model

Table 6.2 presents results from estimating the model using a multinomial logit specification. I return to estimation that includes unobserved terms below. The first column presents estimates for the demand parameters when patients are assumed to respond only to the latent measure of surgeon quality, prior-quarter RAMR. Distance enters, as expected, negatively and significantly over most ranges. The significant coefficient on distance squared suggests a nonlinear cost of travel. Turning to consumer quality elasticity of demand, parameter estimate for β_1 suggests a significant negative response to RAMR in the absence of public quality reporting. Estimates of β_2, however, are small and insignificant, suggesting little response to the release of quality report cards.

I next incorporate controls for market-based information sources *(U.S. News and World Report)* as well as controls for insurance and physician agency. The estimates for this version of the model are in column 2 of Table 6.2. Incorporating these additional "omitted variables" to the version of the model in column 1 gives parameter estimates for β_1 and β_2 that are negative and significant. The coefficients on surgeon RAMR suggest a response to mortality in the prereporting period as well as after report cards were released. The introduction of report cards led to a significantly larger disutility from seeing a surgeon who has a higher RAMR. Comparing estimates of β_2 in columns 1 and 2 suggests that, after controlling for market-based learning about high-quality providers, patients do respond to surgeon quality more with publicly provided data.

Estimates for β_5 suggest that consumers respond more positively to *U.S. News and World Report*–ranked hospitals in the post-reporting period. This finding is consistent with complementarities between private information sources and public reporting. The same effect also holds for hospital teaching status. Consumers increasingly value

Table 6.2 Multinomial logit estimates of patient demand parameters

	Dependent variable: Log probability patient i select surgeon s					
	(1)		(2)		(3)	
Travel cost:						
Distance (Miles) i, s	–0.093	(0.002)***	–0.088	(0.002)***	–0.088	(0.002)***
Distance squared (Miles) i, s	0.001	(0.000)***	0.001	(0.000)***	0.001	(0.000)***
Distance (Miles) i, s*Post RC	–0.014	(0.004)***	–0.012	(0.004)***	–0.012	(0.004)***
Distance squared (Miles) i, s*Post RC	0.000	(0.000)	0.000	(0.000)**	0.000	(0.000)**
Mortality (Quality):						
RAMR s, t-1	–0.012	(0.001)***	–0.009	(0.002)***	–0.006	(0.004)*
RAMR s, t-1*Post RC	–0.001	(0.003)	–0.010	(0.004)***	–0.022	(0.009)**
Market based information:						
US News top hospital			–0.649	(0.023)***	–0.630	(0.027)***
US News top hospital * Post RC			0.219	(0.034)***	0.053	(0.040)
US News ranking h			0.020	(0.002)***	0.018	(0.002)***
Top ranking US New hospital m			0.565	(0.077)***	0.504	(0.077)***
US News top hospital *RAMR s, t-1					–0.005	(0.004)
US News top hospital *RAMR s, t-1* Post RC					0.083	(0.009)***
Teaching hospital h			–0.563	(0.020)***	–0.559	(0.023)***
Teaching hospital h*Post RC			0.161	(0.034)***	0.156	(0.039)***
Teaching hospital h* RAMR s, t-1					–0.001	(0.004)
Teaching hospital h *RAMR s, t-1* Post RC					–0.009	(0.008)
Agency:						
|Pr Mort i- Pr Mort s, t-1|			–2.370	(0.397)***	–2.349	(0.398)***
|Pr Mort i- Pr Mort s, t-1|* Post RC			–0.020	(0.330)	–0.012	(0.330)
Insurance network:						
Payer share s, t-1, i			–0.006	(0.061)	–0.019	(0.062)
Payer share s, t-1, i*HMO Contracts h			0.006	(0.003)**	0.087	(0.112)
Payer share s, t-1, i* Post RC			0.039	(0.107)	0.007	(0.003)**

Payer share s, t-1, i*HMO Contracts h*Post RC			0.000	(0.006)	0.001	(0.006)
Medicare FFS i* RAMR s, t-1					0.000	(0.006)
Medicare FFS i* RAMR s, t-1*Post RC					−0.009	(0.009)
Medicare HMO i* RAMR s, t-1					−0.022	(0.015)
Medicare HMO i* RAMR s, t-1*Post RC					0.014	(0.018)
Private HMO i* RAMR s, t-1					−0.003	(0.006)
Private HMO i* RAMR s, t-1*Post RC					0.000	(0.011)
Observations	1,048,706		821,793		821,793	
Log likelihood	−133652.05		−105839.30		−105787.50	

*, **, and *** denote statistical significance at the 10, 5, and 1% levels, respectively

teaching hospitals in the post-reporting period. These complementarities may pick up greater awareness of variation in quality due to the release of public report cards; the release not only increases individuals' use of the report cards themselves but leads patients to seek out other available private information sources.

Physician agency and insurance network effects also enter choice significantly on average. However, the demand shifters that capture physician agency do not differentially alter choice after the release of quality reporting. This is perhaps not surprising if physician agents have relatively good information even in the absence of quality reporting. Insurance network effects enter choice significantly only in the pre-reporting period and only on the interaction of the number of HMO contracts at hospital h with the share of payer type at the surgeon level. This provides weak evidence that surgeons at hospitals who are more willing to contract with HMOs care for patients with relatively more restrictive networks. This effect is invariant to the release of report cards and, given the lack of a significant estimate for the effect of a surgeon's share of a given insurance type alone, I do not emphasize this result as conclusive.

Column 3 contains estimates for the fully interacted model, which allows market-based information to interact with latent surgeon quality and report card–induced learning. In this specification the patient's response to surgeons' RAMR is significant both before and after the release of report cards.

The parameter estimate of β_2 in the fully interacted model is larger than the estimates in both columns 1 and 2. In fact, the differential response to surgeon quality after report card release is more than double the estimated response that controls for the average role of market-based learning and insurance in choice (column 2). These findings suggest that the interaction of market-based and public information alter consumer choice. As a result, models that do not control for prior consumer learning likely underestimate the effect of public reporting on choice.

The interaction of consumers' response to surgeons' RAMR with the type of insurance they have does not produce any significant effects. However, after incorporating these terms, there is a significant coefficient on the interaction of lagged payer share and the *Post* dummy variable, suggesting some increase in the constraints of networks after reporting. Because identification is coming from intertemporal changes, this finding may also be due to the rise of managed-care networks over the time period. This further underscores the need to account for the effect of insurance network constraints when estimating the effect of quality reporting on choice. Decomposing patients' response to RAMR across Medicare FFS, Medicare HMO, and private HMO patients suggests little differential effect of insurance on quality demand. Taken together, these results suggest that insurance network constraints play a relatively small role in a patient's choice of CABG surgeon, and to the extent that they do influence choice, this seems to be unrelated to surgeon quality.[6]

[6] In a set of unreported regressions I re-estimate the model allowing patient response to *U.S. News and World Report* to vary with the type of insurance a patient has. Consistent with the lack of response to RAMR, I find no differential response to private information provided by *U.S. News* between Medicare FFS patients and those in managed care (both Medicare and private).

The other relevant coefficient in the full model is β_6, the differential response to quality reporting given the market-based beliefs about a provider. The results in Table 6.2 suggest that *U.S. News and World Report* is a substitute for the information provided by PHC4. Teaching status and insurance do not differentially alter patients' response to quality information after the inception of reporting. The parameter estimates for the interaction of a *U.S. News and World Report* ranking and lagged surgeon RAMR is 0.083 and is significant at the 1% level. The response to new quality data within hospitals that have *U.S. News* rankings is substantially less after the release of report cards than among surgeons at hospitals that do not have such information available.

6.5.2 Incorporating Unobserved Effects

I next turn to estimating a version of the model that allows unobserved taste variation to enter as random coefficients in demand. This allows the effect of agency and the use of market-based information to vary in the population because of unobserved factors and the likelihood of a patient's being able to choose a given surgeon to vary because of unobserved insurance network constraints. The results of estimating Eq. (6.10) using maximum simulated likelihood are presented in Table 6.3. The first two columns contain mean and standard deviations for the parameters in the model, including market-based learning but with interacting private information with patient response to RAMR. Columns 3 and 4 present mean and standard deviation estimates for parameters in the fully interacted model.

The biggest change between the random coefficients and the base model is that estimates for the response to quality information (RAMR) are no longer significant in either specification. This is true both for the average effect and for the marginal increase due to quality reporting. However, in the fully interacted model (column 3) the estimate for the mean of β_2 is of a similar magnitude to the estimates in column 3 of Table 6.2, though the coefficient is not significant at conventional levels (p-value = .12). The estimates for the variance of the parameters on response to surgeons' RAMR are not significant in either version of the model.

I next turn to patients' response to a hospital's being ranked by *U.S. News and World Report*. These mean coefficient estimates are significant, both statistically and economically, in both specifications. The estimated standard deviation of patients' response to being ranked by *U.S. News and World Report* is large and significant in both versions. This lends further support to the idea that a subset of patients value *U.S. News* rankings highly while a larger group not only do not value (or access) this information but appear to avoid these hospitals, perhaps reflecting other rationing mechanisms or top hospitals' efforts to price-discriminate among patients.[7]

[7] I use the term price discrimination, but as discussed, this is more likely to be non-price-based efforts to ration care across patients with a different willingness to pay for *U.S. News* rankings.

Table 6.3 Random-coefficients estimates of patient demand parameters

	Dependent variable: Log probability patient i selects surgeon s							
	Mean		S.D.		Mean		S.D.	
Travel cost:								
Distance (Miles) i, s	−0.181	(0.004)***			−0.182	(0.004)***		
Distance squared (Miles) i, s	0.003	(0.000)***			0.003	(0.000)***		
Distance (Miles) i, s*Post RC	−0.022	(0.007)***			−0.019	(0.007)***		
Distance squared (Miles) i, s*Post RC	0.000	(0.000)***			0.000	(0.000)***		
Mortality (Quality):								
RAMR s, t-1	−0.004	(0.003)	0.000	(0.004)	−0.001	(0.004)	0.000	−0.007
RAMR s, t-1*Post RC	−0.003	(0.009)	0.001	(0.008)	−0.015	(0.010)	0.000	−0.017
Market based information:								
US News top hospital	−0.541	(0.033)***	0.543	(0.111)***	−0.535	(0.036)***	−0.606	−0.174***
US News top hospital * Post RC	0.396	(0.046)***	−0.005	(0.089)	0.273	(0.052)***	0.003	−0.013
US News ranking h	0.022	(0.002)***			0.021	(0.002)***		
Top ranking US New hospital m	0.657	(0.081)***			0.607	(0.081)***		
US News top hospital *RAMR s, t-1					−0.004	(0.005)	0.003	−0.177
US News top hospital *RAMR s, t-1* Post RC					0.059	(0.010)***	−0.002	−0.008
Teaching hospital h	−0.177	(0.026)***	−0.002	(0.052)	−0.159	(0.028)***	0.022	−0.028
Teaching hospital h*Post RC	0.034	(0.045)	−0.006	(0.085)	0.010	(0.049)	−0.016	−2.108
Teaching hospital h* RAMR s, t-1					−0.005	(0.004)	0.001	−3.134
Teaching hospital h *RAMR s, t-1* Post RC					−0.005	(0.009)	−0.003	−0.200
Agency:								
|Pr Mort i- Pr Mort s, t-1|	−4.980	(0.632)***	−0.093	(1.037)	−4.955	(0.633)***	−0.090	(1.030)
|Pr Mort i- Pr Mort s, t-1|* Post RC	2.266	(1.259)*	−0.064	(1.929)	2.314	(1.270)*	0.562	(1.886)
Insurance network:								
Payer share s, t-1, i	−0.045	(0.073)	0.018	(0.097)	−0.036	(0.073)	−0.011	(0.096)
Payer share s, t-1, i*HMO contracts h	−0.123	(0.150)	0.038	(0.271)	−0.107	(0.151)	0.034	(0.284)
Payer share s, t-1, i* Post RC	0.011	(0.033)***	0.002	(0.006)	0.011	(0.003)***	−0.001	(0.006)

Payer share s, t-1, i*HMO contracts h*Post RC	−0.003	(0.007)	0.001	(0.013)	−0.002	(0.007)	−0.002	(0.014)
Medicare FFS i* RAMR s, t-1					−0.001	(0.004)		
Medicare FFS i* RAMR s, t-1*Post RC					−0.007	(0.011)		
Medicare HMO i* RAMR s, t-1					−0.048	(0.018)***		
Medicare HMO i* RAMR s, t-1*Post RC					0.038	(0.022)*		
Private HMO i* RAMR s, t-1					−0.003	(0.007)		
Private HMO i* RAMR s, t-1*Post RC					0.000	(0.013)		
Observations	459,870				459,870			
Log likelihood	−65,398				−65,372			

*, **, and *** denote statistical significance at the 10, 5, and 1% levels, respectively

The estimated mean effect of patient-surgeon matching on choice—the role of physician agents—is significant not only before (as in the prior estimation) but also after the release of report cards. The variance in the population of agency is not significant for either the baseline effect or the marginal effect after reporting.

I find that allowing random coefficients to enter the model of insurance network constraints makes some difference in the estimated influence of network constraints. The estimated coefficients suggest a significantly larger mean effect after report cards, as before, though the variance of the estimate is not significant. In the fully interacted model the estimated response to quality by patients in Medicare HMOs is relatively larger (more negative) prior to the release of report cards. However, this effect is eliminated by the post-reporting period (I fail to reject the hypothesis that the sum of the coefficients on Medicare HMO* RAMR + Medicare HMO*RAMR*Post is equal to zero). Despite this finding, I do not interpret this as strong evidence for a role of Medicare HMO networks as agent in specialist choice based on quality. Given the volatility of payments and regulation as well as selection behavior in Medicare Part C markets and the fact that these factors were changing over time, I am concerned about omitted variables in my intertemporal identification.

Despite some changes to the findings in the random coefficients model, the basic results remain. I do note, however, that the effect of reporting on patients' response to quality is diminished. Given the assumptions on the form of the unobserved terms necessary to estimate Eq. (6.10), I focus on the base specification for the primary results and sensitivity analysis.

6.5.3 Sensitivity Analysis

The empirical strategy relies on two assumptions: that RAMR is a reasonable measure of surgeons' latent quality and that changes over time in response to RAMR are solely due to reductions in information asymmetries resulting from the release of quality report cards. If, however, other factors that influence patients' choice are correlated with RAMR or other changes between the pre- and post-report card period, the model is misspecified. To test for such a situation, I reestimate the model including only patients who received CABG after initially receiving angioplasty on the same day. This situation occurs when a patient has a complication during the angioplasty procedure and must be rushed to a CABG surgeon. Because these patients chose a surgeon in an emergency situation, I expect their choices not to respond to surgeon quality or to the release of information. If, instead, I find that surgeon quality affects the choices by these patients, particularly interacted with the post-reporting period, I will be concerned that measures of quality and reporting are correlated with the error term and that this may be driving the prior findings. Table 6.4 presents results for this specification.

Distance continues to enter significantly, reflecting the fact that angioplasty patients also prefer to receive care closer to their home. In all specifications,

Table 6.4 Multinomial logit estimates of patient demand parameters including only patients receiving both angioplasty and CABG on same day

	Dependent variable: Log probability patient i select surgeon s					
	(1)		(2)		(3)	
Travel cost:						
Distance (Miles) i, s	−0.101	(0.010)***	−0.082	(0.010)***	−0.083	(0.011)***
Distance squared (Miles) i, s	0.002	(0.000)***	0.001	(0.000)***	0.001	(0.000)***
Distance (Miles) i, s*Post RC	0.008	(0.038)	−0.014	(0.041)	−0.015	(0.041)
Distance squared (Miles) i, s*Post RC	0.000	(0.001)	0.000	(0.001)	0.001	(0.001)
Mortality (Quality):						
RAMR s, t-1	−0.007	(0.007)	0.000	(0.009)	0.003	(0.015)
RAMR s, t-1*Post RC	−0.006	(0.038)	−0.035	(0.043)	−0.082	(0.097)
Market based information:						
US News top hospital			−0.483	(0.118)***	−0.554	(0.135)***
US News top hospital * Post RC			0.000	(0.009)	−0.184	(0.394)
US News ranking h			0.031	(0.014)**	0.030	(0.014)**
Top ranking US New hospital m			1.001	(0.502)**	0.980	(0.502)*
US News top hospital *RAMR s, t-1					0.021	(0.020)
US News top hospital *RAMR s, t-1* Post RC					0.090	(0.096)
Teaching hospital h			0.402	(0.347)	−0.574	(0.113)***
Teaching hospital h*Post RC			−0.550	(2.058)	0.383	(0.395)
Teaching hospital h* RAMR s, t-1					−0.022	(0.019)
Teaching hospital h *RAMR s, t-1* Post RC					−0.022	(0.092)
Agency:						
ǀPr Mort i- Pr Mort s, t-1ǀ			−0.550	(2.058)	−0.588	(2.061)
ǀPr Mort i- Pr Mort s, t-1ǀ* Post RC			−0.150	(1.288)	−0.227	(1.294)
Insurance network:						
Payer share s, t-1, i			0.091	(0.321)	0.099	(0.322)
Payer share s, t-1, i*HMO contracts h			0.019	(0.014)	0.715	(1.270)
Payer share s, t-1, i* Post RC			0.772	(1.204)	0.019	(0.014)

(continued)

Table 6.4 (continued)

	Dependent variable: Log probability patient i select surgeon s					
Payer share s, t-1, i*HMO contracts h*Post RC			0.004	(0.067)	0.004	(0.069)
Medicare FFS i* RAMR s, t-1					−0.005	(0.019)
Medicare FFS i* RAMR s, t-1*Post RC					0.039	(0.114)
Medicare HMO i* RAMR s, t-1					−0.010	(0.092)
Medicare HMO i* RAMR s, t-1*Post RC					0.059	(0.142)
Private HMO i* RAMR s, t-1					0.022	(0.025)
Private HMO i* RAMR s, t-1*Post RC					0.036	(0.122)
Observations	28,017		28,017		28,017	
Log likelihood	−4391.29		−3433.69		−3431.27	

*, **, and *** denote statistical significance at the 10, 5, and 1% levels, respectively

surgeons' RAMR does not enter significantly either before or after the release of report cards. None of the variables for physician agency or the role of insurance networks enter the choice model significantly, either. Because a patient who requires CABG after receiving angioplasty is likely to be moved quickly to any available bypass surgeon, this also validates that these measure are not capturing unobserved variables that affect choice. The only market learning variables that enter significantly in Table 6.4 are the dummy for teaching hospital and *U.S. News and World Report* variables that are not interacted with surgeons' RAMR. This is not surprising, however, given that *U.S. News* rankings are for either cardiology or cardiac surgery overall at the hospital, not only for CABG. It thus appears that angioplasty patients also learn from market-based information and respond in a similar way to CABG patients (if anything, the premium on being the highest-ranked *U.S. News* hospital in a market is large even for angioplasty patients). The interactions between surgeons' RAMR and *U.S. News* information and teaching status are not significant in columns 2 or 3 of Table 6.4, suggesting that CABG-specific market-based learning is not picking up unobserved hospital-level observables that change over time.

6.6 Conclusion

This chapter considers the effect of privately and publicly provided information on patients' choice of a cardiac surgeon. *U.S. News and World Report* rankings significantly alter consumer choice, though this effect varies substantially in the population, and in some cases, patients prefer hospitals that are not ranked.

After the state of Pennsylvania introduced quality reporting for cardiac surgery, patients' response to quality increased. Patients' beliefs about quality due to *U.S. News and World Report* rankings significantly altered this response to the release of report cards. I find that the role of quality in patient choice was differentially smaller after the release of report cards among hospitals that were ranked by *U.S. News and World Report*. This provides evidence that private and public reporting substitute for each other.

The results also suggest that evaluations of reporting efforts should incorporate prior market-based learning into the model. Without this the estimated effect of reporting is likely to be biased down. Given that many studies of privately provided information find only a small or nonexistent effect of information release on consumer choice (see Kolstad and Chernew 2008 for a review of the evidence), these findings argue for continued investigation in this area.

Taken together, my results also underscore the importance of considering existing mechanisms for consumers to learn about quality when formulating information-based policy interventions. The distributional impact of such policies is also likely to vary depending on consumers' *ex ante* knowledge of providers' quality. If some markets have substantially more information available through market-based sources, the effect of reporting may differ from that in relatively less informed markets, where the effect is likely to be larger.

Without additional detail, it is difficult to make strong normative conclusions regarding the value of the introduction of quality report cards in Pennsylvania. This is true both in terms of a general evaluation and in trying to understand the normative effect of *U.S. News and World Report* rankings and physician and insurer agency. Future work that can evaluate the relative welfare gains from privately provided information and public reporting would be highly informative for policy.

In addition to studying information interventions, this chapter has estimated a demand model that separates the role of physician agency in demand for specialized care. Future work that considers the degree to which agents make optimal decisions for patients could inform appropriate information-based policy interventions as well as alternative incentive mechanism (e.g., payment policy) to improve choices in health care markets.

Finally, I have not considered the underlying normative value of the information contained in private and public report cards. This is important in interpreting these results for application, particularly in light of the substitution between *U.S. News* rankings and Pennsylvania's report cards. If *U.S. News* rankings are less correlated with socially desirable outcomes (e.g., increases in quality-adjusted life expectancy), then the results here suggest that market-based learning may undermine the value of public intervention. Of course, the converse may hold. Future studies that address this issue would be valuable in health care as well as in the many other markets in which information-based policy interventions have been applied and frequently overlap with private information provision.

Acknowledgments The data used in this analysis were obtained from the Pennsylvania Health Care Cost Containment Council (PHC4), which requests the following disclaimer: The Pennsylvania Health Care Cost Containment Council (PHC4) is an independent state agency responsible for addressing the problem of escalating health costs, ensuring the quality of health care, and increasing access to health care for all citizens regardless of ability to pay. PHC4 has provided data to this entity in an effort to further PHC4's mission of educating the public and containing health care costs in Pennsylvania. PHC4, its agents and staff, have made no representation, guarantee, or warranty, expressed or implied, that the data – financial, patient, payor, and physician specific information – provided to this entity, are error-free, or that the use of the data will avoid differences of opinion or interpretation. This analysis was not prepared by PHC4. This analysis was done by Jonathan T. Kolstad. PHC4, its agents and staff, bear no responsibility or liability for the results of the analysis, which are solely the opinion of the author.

6. Commentary: When Information Becomes Useful

Kenneth L. Leonard and Timothy Essam

"The Effect of Public and Private Quality Information on Consumer Choice in Health Care," by Jonathan Kolstad, tests whether patients are using information available from different sources to choose doctors of high quality. The central question, "Can patients adapt their behavior to information about quality and, in so doing, improve their own health?" is important in the health care literature, and it is one that we have also tackled, though with data from developing countries. In the case of this contribution, patients (together with their primary-care physicians) are choosing where to seek cardiac bypass surgery, using data collected by the state of Pennsylvania and published in *U.S. News and World Report*. In addition to tackling an important and difficult question, this chapter uses complex statistical methodologies to significantly improve the strength of its findings. It thus represents an important contribution to knowledge. Its primary contribution to this volume, however, is as an illustration of the complexities involved in measuring the value of information in general.

In this discussion of the chapter, we are not going to focus on the immediate question posed and answered by the author, since that is best left to him. Instead we will focus on how the chapter relates to the broader question of this volume. In the case of information about medical care, how should one present data on multiple stochastic outcomes to consumers? How much preliminary processing is required before the data can be shown to the public? Do data such as these lead to any change in behavior? Does this make the consumers of the information better off? More importantly, does the information make the average person better off? Some of these questions should seem surprising, given that people always have the option to ignore information; how can it be that accurately collected and reported information could ever make people worse off? As we shall see, this is all too easy, and proving that information has not caused harm is part of the reason that Dr. Kolstad's chapter is as complex as it is.

6.C.1. Information Asymmetry in Health Care

Health care has traditionally been seen as an interesting economic case study because of the imbalance of power, particularly with regard to information. Even in settings in which patients are empowered to choose treatments, hospitals, or

K.L. Leonard (✉) • T. Essam
Department of Agricultural and Resource Economics,
University of Maryland, College Park, MD, USA
e-mail: kleonard@arec.umd.edu; tessam@arec.umd.edu

physicians, they have little information or experience with which to make these decisions; the knowledge about medicine and the patient's condition lies with the physician. This means that patients cannot choose a physician and hospital and then demand high-quality services once they arrive, since they don't know what services are necessary and cannot evaluate whether high-quality services were ever delivered. For economists, this is a particular problem because there cannot be a "market for quality." Instead, patients attempt to ensure they get what they need by choosing the right physician and hospital.

However, in exactly the same setting where patients know so little, we can collect copious quantities of information, which should be useful to patients when they make decisions. Outcomes are an excellent example of how these data can be useful, and they illustrate some additional interesting features of health care. Whereas patients are unlikely to know much about what they need, or even what is happening to them, they know a lot about the outcome of treatment. So every patient gets one very high quality observation each time he visits a doctor. The chapter focuses on bypass surgery, which is not a particularly good example here, so we should imagine a mother taking her child in for an earache. The patient (and parent) can decide whether they liked the doctor's manner, how long they had to wait, the condition of the waiting room and how the child felt after they left the office, how he felt the next day, and even the day after that. However, this one piece of high-quality data is not particularly useful to the patient because doctor quality is related to health outcomes as only one piece of a complex set of variables. Outcomes are stochastic, meaning that a doctor can do everything right and the patient will not feel better, or she can do nothing right and the patient will feel better. It is possible to survive surgery from a terrible surgeon, and it is possible to die under the knife of the best surgeon. This means that although one piece of information is better than nothing, it is not an absolute guide for patients. In reality, patients can learn much more by looking at many points of low-quality data. It is more useful for the patient to know the outcomes for 100 patients operated on by a particular surgeon, or to hear the experiences of 100 children who went to a particular pediatrician for earaches.

This is the goal of these data collection efforts: to collect information from enough different cases to display a pattern. Thus, theory suggests that information from large data sets should be extremely useful to patients. However, the one thing that people often forget about health care is that what is hidden from patients is not hidden from other doctors. Patients cannot directly evaluate their doctors, and data collectors are in the same situation, but doctors can much more easily assess the skills of their peers without access to large data sets. This is a point that Kolstad recognizes and talks about explicitly, but we will come back to it in this review. In general, we need to remember that large data sets make local trends transparent to outside observers who otherwise would know very little, but that frequently there are people who have always known more about local conditions than the data can ever reveal. These well-informed actors may never be able to see the larger picture, but that does not stop them from acting in a way that could make data collection redundant.

6.C.2. Public Information Versus Expert Opinion

Kolstad is careful to recognize that patients have access to many forms of information, notably privately gathered private information, privately gathered public information, and publically gathered public information. Individuals, for example, can learn from their own experience, the experience of others in their social networks (parents talking with other parents about their experience with pediatricians), and even web searches (looking for information from other private sources). In general, individuals have much smaller networks but are willing to share whatever information they do gather. Thus, things like the Internet are valuable not just because we can find what others say, but because others are willing to post what they have learned. Other types of private organizations that may seek information include physicians and insurance companies. These private organizations may be less willing to share the information they have gathered, particularly if it was costly for them to gather. Interestingly, some private entities have incentives to invest significant resources in gathering information but then make them essentially public. *U.S. News and World Report* makes information available to its subscribers, but almost anyone could get access to this information at very low cost: it is effectively public. Then, of course, there are state and federal government entities, which can collect much more information (using legal mandates) and will deliberately make this information available. Kolstad's contribution shows that patients behave as if they had access to the publicly available public information and that they also value a private body's attempt to repackage this same information.

Kolstad finds that consumers are reacting to the information in *U.S. News and World Report* even before it is published; they have access to the information in some other form. However, they respond to this information even more when it is released in public form by a private entity. This suggests that there is some value to the manner in which the data are presented, either by the vehicle (a magazine) or by the format (discrete lists of recommended locations).

Does this mean that we can conclude the data are useful? Unfortunately not. Patients, individuals, and households react to all sorts of information, some of which is useful and some of which is not. Advertising, name brand recognition, and superstition are all forms of information that are publicly available but actually impede the flow of good information. This chapter stands on strong ground because it shows that patients react to information that is objectively useful, not just any form of information. What is interesting is that patients appear to partially incorporate such information even before it is easily available. We have shown similar results in work done in Tanzania (Leonard 2007; Leonard et al. 2009). There, households are learning by gathering information from the experiences of other households and making decisions as if they had access to high-quality information about physician quality. In these studies, the researcher has access to objective and correct information about quality that patients cannot access, but we can show that patients act as if they have access to this information. Thus, households are engaging in a private process that approximates the information that would be available in a public process. As in the U.S. study, the public information is more useful to households than the privately gathered information, but the private information does have some value.

When we know that the data accessible to patients are objectively correct, we have a stronger test of the value of information. In the data analyzed by Kolstad, the available information causes some people to choose better physicians. This must be better for the individuals making the better choices, but is it better for the collection of households? Does it improve overall welfare? This depends on the nature of the goods or services being provided. The way that Kolstad analyzes the data ensures that, for the sample he is studying, the average patient is seeking a better physician. Thus, in this setting, average quality should improve, but this finding is not automatically true in other settings.

If the good or service is inelastically provided (there isn't that much of it to go around), then one person's gain is likely to be another person's loss. Imagine there are only 20 operating theaters available and there are 20 patients waiting to undergo operations. Information about high-quality providers is likely to change who gets which operation, but it doesn't change the number of operations or the average quality of operations provided. If the people who switch to the better theaters are wealthy, then overall, nothing has been gained. If the people who switch to the better theaters are the people who need better surgery, then there may be overall gains from switching.

On the other hand, if the supply of the good or service is elastic (there is plenty to go around), then we might see larger improvements. If there are 25 available theaters for 20 patients, the worst 5 can be left empty and the average should improve. Thus, at the very least, we want to avoid a situation in which information simply leads to a shuffling of who gets what without any overall improvement in quality or productivity. Some people will be better off, but others will be worse off.

In health care the real hope for improvement comes when we think about the long-term supply of something like high-quality operating theaters. If information means that the best theaters are always full and the worst ones are empty, shouldn't we expect more good theaters (and surgeons) over time and fewer poor theaters and surgeons? Again, there is no reason to expect this automatic reaction to increases in demand. It must be the case that the additional revenue from attracting more patients is worth the cost of providing higher quality. This is not to say that hospitals and doctors do not want to have high ratings, but rather that it is not obvious that they would invest significant resources to improve their scores so as to attract more patients. This is even truer if they can use advertising to attract patients without having to increase quality: yes, some patients will seek quality as measured in things like *U.S. News and World Report*, but it is cheaper to simply advertise to attract more patients.

This leads us to what is probably the most important aspect of the information gathered by the state of Pennsylvania and published by *U.S. News and World Report:* doctors may care more about the esteem of their peers than they do about the opinion of prospective patients. Doctors are part of a profession, and the ideals of the profession of medicine are commonly held among doctors; they tend to care what other doctors think of them. So it is possible that, even though better patients are coming because of the available data, doctors are motivated by the opinion of other doctors. This is a good thing, but it illustrates that the data we collect might even help improve quality through a completely different mechanism than the one we are studying.

In fact, in many settings the expert opinion of peers may be much better than data collected by outside aggregators of information. For the data that Kolstad is analyzing, doctors have no reason to try to manipulate the results. When the data are collected, doctors and hospitals have little reason to believe the information could affect their bottom line. However, if patients continue to react to *U.S. News and World Report* and if doctors and hospitals care about how they react, there will come a time when manipulating the data is to the advantage of many hospitals and doctors. This has already begun to happen with other forms of rankings published by newspaper and magazines. In general, it is harder to manipulate the opinions of peers because opinions are not calculated by a formula (which allows people to see the flaws) and are therefore highly flexible and would respond to any long-term attempt at manipulation.

6.C.3. Implications

Jonathan Kolstad has examined an application in a setting where a large data set aggregates local events, and he has shown that presenting this information in a particular format is useful to some people. This is despite the fact that the information is gathered entirely from the experiences of other individuals and that patients by themselves, with their social networks and in collaboration with their doctors, appear to have access to some of this information already. In health care there are two sources of information available to patients: the information they can gather themselves (on outcomes) and the information available to their doctors (on outcomes and inputs). As we have seen, in this setting, doctors are useful as agents or aggregators of information because they know significantly more about the field of medicine than do patients.

This type of situation—aggregation of information that already exists, combined with the presence of a previously existing system for aggregating and processing information—is increasingly common, and therefore we should be able to draw some general lessons. Together with Molly Brown at NASA, we have been working on data using Normalized Difference Vegetation Index (NDVI) observations on fields in the Sahel region in Africa, trying to model the agricultural output from this area.[8] The satellite images are a poor representation of what each farmer knows

[8] NDVI data were obtained from the NOAA Advanced Very High Resolution Radiometer (AVHRR) archive and processed by the Global Inventory Monitoring and Mapping Systems (GIMMS) group at the NASA Goddard Space Flight Center (Tucker et al. 2005). Previous research has shown that NDVI can be used to detect deviations in production conditions and is correlated with net primary production and crop yields (Tucker et al. 1981; Prince 1991; Fuller 1998).

about his own crops, but we can quickly analyze data over a very large area and get a good idea of what is happening. In this case the aggregating agent is the market. Within about 6 weeks of the harvest, local and national markets in this area have absorbed all the relevant information about output, and prices reflect the balance between supply and demand and the costs of transporting food. The data are useful to us because it is difficult for researchers to collect the market data: we see local market prices only with a significant delay. However, since farmers can already see both their own high-quality data and the local information about prices, is there any possible gain from presenting this information to them? What would they learn from knowing how the average farmer is doing, given that they already know how they are doing and can observe market prices, which should already reflect what others are doing?

Two lessons from Kolstad's chapter are that the data need to be processed to be useful, and that information can be valuable if it allows people to learn things more quickly than they would be able to do with normal aggregation agents (their own doctors) or devices (markets).

If the data are to be useful, the information should be presented by someone who is planning on selling the analysis for profit. No farmer in Burkina Faso is going to pay for a glossy magazine, but something produced by a government ministry or international aid agency is likely to fail because workers in these offices aren't promoted on the basis of their ability to show sales of the information. *U.S. News and World Report* synthesizes a lot of information in a list of recommended doctors; what would we tell farmers? It would need to be something along the lines of picking from among four or five phrases: the total national harvest is well below average, below average, average, above average or well above average.

In addition, the marketed data can be better than the raw data or the market data if the information is timely. In the Sahel region of West Africa there is a short window (probably 8 weeks) between when the farmer knows he has a successful crop and when the markets have assimilated all the information from other farmer's crops. We are finding that a satellite image can potentially close this window, perhaps shortening it by 4 weeks. The farmer could use this information to decide how much of his crop to store, sell, or even leave in the field. In this case, although farmers are likely to benefit, traders and anyone with access to current information may suffer from the fact that their information is no longer private. Since traders use this information to move food around (potentially helping people) and they are more likely to move food if they can earn profits from their information, it is not clear that hurting traders is a useful strategy. It does seem likely that the farmers would benefit overall, and to the degree that the farmers are the poor people and that they have access to technologies to effectively store their crops (waiting for better prices they know will come), it is possible that, overall, such information would enhance development (poverty-fighting) objectives.

One interesting feature of satellite images is that it might be very hard to manipulate the data. A hospital has access to many tools for reclassifying patients and changing the definitions of services that could be used to make their outcomes seem better to agents like *U.S. News and World Report*. Farmers would have a much harder time trying to falsify the information that a satellite could observe from space. We cannot confidently say that it is not possible, but it seems to us to be very unlikely.

References

Arrow, K. (1963). Uncertainty and the welfare economics of medical care. *American Economic Review, 53*, 941–973.

Berry, S., Levinsohn, J., & Pakes, A. (1995). Automobile prices in market equilibrium. *Econometrica, 63*(4), 841–890.

Cutler, D. M., Huckman, R. S., & Landrum, M. B. (2004). The role of information in medical markets: An analysis of publicly reported outcomes in cardiac surgery. *American Economic Review, 94*(2), 342–346.

Dafny, L., & Dranove, D. (2008). Do report cards tell people anything they don't already know? The case of Medicare HMOs. *The RAND Journal of Economics, 39*(3), 790–821.

Dranove, D., & Sfekas, A. (2008). Start spreading the news: A structural estimate of the effects of New York hospital report cards. *Journal of Health Economics, 27*(5), 1201–1207.

Gaynor, M., Seider, H., & Vogt, W. B. (2005). The volume-outcome effect, scale economies, and learning-by-doing. *The American Economic Review, 95*(2), 243–247.

Gowrisankaran, G., Ho, V., & Town, R.J. (2006) *Causality and the volume-outcome relationship in surgery*. Mimeo: University of Pennsylvania.

Huckman, R., & Pisano, G. (2005). Turf battles in coronary revascularization. *The New England Journal of Medicine, 352*(9), 857–859.

Kolstad, J. T. (2010). *Information and quality when motivation is intrinsic: Evidence from surgeon report cards*. Mimeo: University of Pennsylvania.

Kolstad, J. T., & Chernew, M. (2008). Consumer decision making in the market for health insurance and health care services. *Medical Care Research and Review, 66*(1), 28S–52S.

McFadden, D. (1974). Conditional logit analaysis of qualitative choice behavior. In P. Zarembka (Ed.), *Frontiers in econometrics*. New York: Academic Press.

Mukamel, D. B., & Mushlin, A. I. (1998). An analysis of market share and price change after publication of the New York state cardiac surgery mortality reports. *Medical Care, 35*(1), 945–954.

Nevo, A. (2001). Measuring market power in the ready-to-eat cereal industry. *Econometrica, 69*(2).

Pennsylvania Health Care Cost Containment Council. (1998). *Coronary artery bypass graft surgery—1994–95 data: Research methods and results*. Harrisburg: Pennsylvania Health Care Cost Containment Council.

Pennsylvania Health Care Cost Containment Council. (2002). *Coronary artery bypass graft surgery—2000 data: Research methods and results*. Harrisburg: Pennsylvania Health Care Cost Containment Council.

Pennsylvania Health Care Cost Containment Council. (2004). *Coronary artery bypass graft surgery—2002 data: Research methods and results*. Harrisburg: Pennsylvania Health Care Cost Containment Council.

Pennsylvania Health Care Cost Containment Council. (2005). *Coronary artery bypass graft surgery—2003 data: Research methods and results*. Harrisburg: Pennsylvania Health Care Cost Containment Council.

Ramanarayanan, S. (2007). Does Practice Make Perfect: An Empirical Analysis of Learning-by-Doing in Cardiac Surgery. Mimeo: UCLA.

Schneider, E. C., & Epstein, A. M. (1996). Influence of cardiac-surgery performance reports on referral practices and access to care-A survey of cardiovascular specialists. *The New England Journal of Medicine, 335*(4), 251–256.

Steinbrook, R. (2006). Public report cards- beyond cardiac surgery. *The New England Journal of Medicine, 355*(18), 1847–1849.

Train, K. (2003). *Discrete choice methods with simulation.* New York: Cambridge University Press.

Wennberg, J. E., Cooper, M. M., & Dartmouth Atlas of Health Care Working Group. (1999). *The Dartmouth Atlas of Health Care 1999*. Chicago: American Hospital Publishing.

Fuller, D.O., (1998). Trends in NDVI time series and their relation to rangeland and crop production in Senegal, 1987–1993. *International Journal of Remote Sensing* 19, 2013–2018.

Leonard K.L., Adelman, S. and Essam T. (2009)"Idle Chatter or Learning? Evidence of social learning about clinicians and the health system from Rural Tanzania," *Social Science and Medicine*. 69: 183–190.

Leonard, K.L. Learning in Health Care: (2007)"Evidence of learning about clinician quality in Tanzania" *Economic Development and Cultural Change*, *55*(3), 531–555.

Pope, D. (2009) Reacting to Rankings: Evidence from "America's Best Hospitals" *Journal of Health Economics*, *28*(6), 1154–1165.

Prince, S.D., 1991. Satellite Remote Sensing of Primary Production: Comparison of Results for Sahelian Grasslands. *International Journal of Remote Sensing* 12, 1301–1311.

Tucker, C. J., Holben, B. N., Elgin, J. H.,&McMurtrey, (1981). Remote sensing of total dry matter accumulation in winter wheat. Remote Sensing of Environment 11, 171.

Tucker, C. J., Pinzon, J. E., Brown, M. E., Slayback, D., Pak, E. W., Mahoney, R., Vermote, E., El Saleous, N., (2005). An Extended AVHRR 8-km NDVI Data Set Compatible with MODIS and SPOT Vegetation NDVI Data. *International Journal of Remote Sensing* 26, 4485–4498.

Chapter 7
Adoption of Over-the-Counter Malaria Diagnostics in Africa: The Role of Subsidies, Beliefs, Externalities, and Competition

Jessica L. Cohen and William T. Dickens

Abstract Plans for the wide-scale distribution and subsidy of artemisinin combination therapies (ACTs), an antimalarial treatment, pose two problems for public health planning. First, many people seeking malaria treatment do not have the disease. If ACT subsidies could be targeted toward those with malaria, the cost of subsidies could fall. Second, the inappropriate use of antimalarial drugs may contribute to the emergence of drug-resistant parasites. Rapid diagnostic tests (RDTs) for malaria could help with both problems, but drug shop owners may have few financial incentives to sell them, given profits from overtreatment for malaria. A model of the provision of RDTs by profit-maximizing drug shops shows that if all parties know the probability of having malaria and if there are no subsidies for drugs and no external costs to inappropriate treatment, both monopolistic and competitive drug shop owners will provide RDTs under the same circumstances that a social welfare maximizing planner would. However, since drugs will be subsidized, customers overestimate their likelihood of having malaria, and since there are external costs to the misuse of antimalarials, profit-maximizing drug shops will likely underprovide RDTs. We show that a subsidy for RDTs can increase provision and, under adequate competition, induce everyone to use RDTs optimally. The results also highlight the importance of educating customers about the true prevalence of malaria and promoting competition among drug providers.

J.L. Cohen (✉)
Department of Global Health and Population, Harvard School of Public Health, Boston, MA, USA
e-mail: cohenj@hsph.harvard.edu

W.T. Dickens
Northeastern University, Boston, MA, USA

Brookings Institution, Washington, DC, USA
e-mail: wtdickens@gmail.com

R. Laxminarayan and M.K. Macauley (eds.), *The Value of Information: Methodological Frontiers and New Applications in Environment and Health*, DOI 10.1007/978-94-007-4839-2_7,

Keywords Malaria • Rapid diagnostic tests • Antimalarial subsidies • Markets for information • Retail sector • Drug resistance

7.1 Introduction

According to the 2009 World Malaria Report (2010), only 15% of children are treated for malaria with artemisinin combination therapies (ACTs)—currently, the only antimalarials that are effective against the disease.[1] The rest are treated with medicines to which the malaria parasite has acquired resistance. At \$6–\$7 for an adult dose, ACTs are considerably more expensive than the older, less effective antimalarials, the costs of which range from \$.20 to \$1 a dose (ACT Watch, PSI 2010).[2] The very low uptake of ACTs is attributed to the high price of these life-saving medicines in the private retail sector (where most Africans first seek treatment for malaria), combined with poorly functioning public sector facilities and supply chains. In an effort to increase access to ACTs, as well as crowd out artemisinin monotherapy and stem the development of resistance, efforts are under way to subsidize roughly 95% of their cost.[3]

Plans for wide-scale distribution and subsidies are likely to dramatically increase access to ACTs, but they are also likely to significantly increase the use of ACTs for nonmalarial illnesses. This is because most people seeking treatment for malaria either self-diagnose and purchase medicine at drug shops, or go to a public facility where they are diagnosed based on clinical symptoms but without a formal blood test.[4] Thus a large share of malaria treatment goes to people without malaria. In a recent randomized trial with rural Kenyan drug shops, Cohen et al. (2012) show that more than half of older children and adults purchasing subsidized ACTs do not have malaria. In another example, from Tanzania, only 46% of people receiving in-patient hospital care for "severe malaria" actually tested positive for malaria, the same rate as the general population (Reyburn et al. 2004). Because of acquired

[1] Artemisinin monotherapy is effective against malaria as well, but the World Health Organization and others in the global health community have pushed for artemisinin to be manufactured and sold in combination with other treatments with longer half-lives to preserve its efficacy (Arrow et al. 2004).

[2] ACT Watch Outlet Surveys, conducted by Population Services International, are available at http://www.actwatch.info/home/home.asp

[3] The Global Fund currently grants funds for ACTs in the public sector. The Affordable Medicines Facility–malaria (AMFm), funded by the Gates Foundation, the U.K. Department for International Development, and others (and hosted by the Global Fund), is being piloted in eight countries and will subsidize the cost of ACTs to first-line buyers (NGOs, wholesalers, governments, etc.) by roughly 95%. AMFm has negotiated the price of ACTs with manufacturers down to around \$1 a dose. Details about AMFm are at http://www.theglobalfund.org/en/amfm/

[4] According to the 2009 World Malaria Report, only 22% of suspected malaria cases that present at public health centers are confirmed with a test. In most African countries, more than 50% of people seek treatment for malaria outside the public sector (ACT Watch 2010).

immunity, the chances that a patient with a fever (the symptom most commonly associated with malaria) has parasites declines rapidly after age 5, and thus overtreatment is much more likely among older children and adults (Reyburn et al. 2004). In a Tanzania study with drug shop customers, only 18% of those five and over buying antimalarials were parasitemic (Kachur et al. 2006). Parasite prevalence in the area for this age group was 9%, suggesting that symptom-based self-diagnosis in this context was not much better than a random draw from the population.

Without improved targeting, such high rates of overtreatment mean that a large amount of ACT subsidy money will be spent on people without malaria. High rates of overtreatment have other downsides as well, including delaying proper treatment for the true cause of illness (a dangerous example is pneumonia in young children) and accelerating the development of drug resistance (Rafael et al. 2006; Perkins and Bell 2008). If people take ACT when they don't have malaria, it could also preclude learning about the effectiveness of ACTs over other antimalarials (Advaryu 2012).

A potential solution to those problems of ACT targeting would be to improve access to rapid diagnostic tests (RDTs) for malaria. Recent experimental results in Cohen et al. (2012) suggest that those seeking treatment for malaria are extremely interested in being tested for the disease and that drug shop customers may be willing to pay for an RDT. Combined with nonexperimental results on willingness to pay for RDTs (Uzochukwu et al. 2010), the results in Cohen et al. (2012) are an encouraging indication that, if RDTs are priced low enough and made available over the counter, consumer demand for malaria diagnostics may be substantial. We briefly describe the results of that experiment below.

The question then is whether drug shop owners would be willing to sell the tests. It is possible that, since sales of antimalarial drugs are a major source of revenue for drug shops, they would not want to offer the tests because they would not be able to sell antimalarials to customers who test negative. In this chapter we show that in the absence of subsidies or misperception of malaria frequency among drug shop customers, this is not the case. In fact, if there are no subsidies or externalities, if all those needing treatment are treated, and if both customers and drug shop owners correctly perceive the probability of malaria (conditional on symptoms), both monopolistic and competitive drug shops will provide tests in the same circumstances as would a social welfare–maximizing central planner.

We then explore how RDT provision is affected by ACT subsidies and by incorrect perceptions by consumers of the likelihood of malaria conditional on malaria-like symptoms. As noted above, ACTs are currently being subsidized in eight countries through the AMFm, and this could have major implications for the feasibility of RDT adoption in drug shops. We show that under ACT subsidies, there will be a tendency for underprovision of testing. Though not definitive, the fact that the majority of teenage and adult ACT buyers in Cohen et al. (2012) actually do not have malaria suggests that the probability of malaria infection conditional on symptoms is commonly overestimated. We show that if customers overestimate the likelihood that they have malaria, testing will not take place in circumstances where it should. Finally, we show that if there are externalities to mistreatment with antimalarials—for example, because it hastens the emergence of parasite resistance—provision of RDTs by drug shops will be suboptimal. All of these problems can be overcome to some degree by subsidizing RDTs.

7.2 Demand for RDTs: Consumers

The first experimental evidence on demand for RDTs among drug shop customers comes from a randomized controlled trial in western Kenya. Cohen et al. (2012) distributed vouchers to just under 3,000 households in the catchment area (4-km radius) of four rural drug shops. A sub-sample of households received vouchers for subsidized ACTs and for subsidized RDTs, and another sub-sample received vouchers for the subsidized ACTs only. ACT prices were randomly assigned and ranged from $.50 to $6, spanning the range of prices for alternative antimalarials available in drug shops. Households receiving RDT vouchers were randomly assigned to three treatment groups: free, $.20, or $.20 with the possibility of a refund. This last group had to pay $.20 for the RDT, but if they tested positive and went on to buy an ACT, they were refunded the cost of the test. The group receiving an offer for free RDTs or for $.20 RDTs with a refund had the strongest financial incentives to be tested for malaria prior to ACT purchase.

Among those with subsidized ACTs only (i.e., with no RDT voucher), a sub-sample of households were given "surprise" RDTs. That is, on purchase of ACTs, they were asked whether they would be willing to take a malaria test. Cohen et al. (2012) find that, although nearly all young children for whom ACTs were being purchased tested positive, less than 40% of older children and adults buying ACTs had malaria. Further, they find that the fraction of ACT buyers who have malaria diminishes as ACT prices go down (i.e., as ACT subsidies go up). This suggests that an ACT subsidy policy could exacerbate targeting problems. They then go on to show that subsidized RDTs, available over the counter alongside subsidized ACTs, can to some extent improve targeting.

Cohen et al. (2010, 2012) find some evidence that demand for RDTs is substantial. They find that, among those with RDT vouchers, more than 80% of people coming to buy ACTs took an RDT first. In other words, very few people choose to buy the medicine without first being tested. Further, they find that demand for RDTs was the same among those offered the test free and those who had to pay $.20. Although this RDT price is quite low, the study was conducted in an area where the daily wage is equivalent to $1.50, so the finding that demand for RDTs does not drop at all when the price increases from $0 to $.20 suggests that consumer valuation and willingness to pay for RDTs are notable.

Cohen et al. (2010, 2012) present encouraging evidence of significant demand for RDTs in drug shops. However, this study completely controlled the supply side, not allowing drug shop owners to choose whether RDTs were offered or at what price. Thus, the crucial next step in understanding whether RDTs can improve targeting of malaria medicine is exploring the conditions under which drug shops will find it profitable to make them available and affordable. We now turn to the supplier decision.

7.3 Supply of RDTs: Profit-Maximizing Drug Shops

Consider a simple framework where individuals periodically suffer from fevers, and some fraction m of fevers are caused by malaria. Drug shops have access to three products: an antimalarial drug, a rapid diagnostic test, and an alternative drug that is effective for nonmalaria-related fevers (e.g., an antipyretic or antibiotic).

7.3.1 *Monopolistic Drug Shop*

Define P^{NT} ("price no test") as the price a monopolist will charge for antimalarial treatment if RDTs aren't offered, and P^{WT} ("price with test") as the price the monopolist will charge if RDTs are offered. Define P^{T} ("price of test") as the price the monopolist will charge for the RDT if it is offered. We will assume that those who test negative for malaria will all purchase an alternative treatment at some price P^{A} (which we will treat as given).[5] The drug seller faces a constant unsubsidized cost for the antimalarial drug, the tests, and an alternative drug, which we denote C^{D}, C^{T}, and C^{A}, respectively.[6] Finally, the drug seller is assumed to expect that a fraction m^{D} of those seeking treatment for malaria will test positive.

We assume that the cost of antimalarials, whether or not tests are offered, is low enough relative to the expected value of treatment that all people who suspect they have malaria purchase an antimalarial. That is, we abstract (for now) from any potential effect of the tests on the decision to seek treatment at the drug shop, an assumption consistent with results in Cohen et al. (2010, 2012).[7]

Under these assumptions, when the test is not offered, we can write the expected profit per customer as

$$E\left(r^{NT}\right) = P^{NT} - C^{D}. \tag{7.1}$$

If RDTs are offered for sale in drug shops, the potential payoffs change. Individuals who test positive will be sold both the RDT and the antimalarial. Individuals who test negative will be sold both the RDT and the alternative treatment. Although the shop owner does not know the exact number of his customers who will test positive, the

[5] We treat the price of the alternative therapy as exogenously given because we assume that the market for it is much larger than those testing negative for malaria, so the cost of malaria medication and the availability of tests for malaria will have no effect on the price charged. We have in mind antipyretic drugs.

[6] If there is no alternative treatment, then $C^{A} = P^{A} = 0$.

[7] Cohen et al. (2010) find that people who are offered a subsidized RDT in addition to a subsidized ACT are no more likely to show up at the drug shop for treatment than those offered a subsidized ACT only.

expectation is that a fraction m^D will do so, and thus the expected payoff per customer if the test is offered is

$$\mathrm{E}\left(r^{WT}\right) = \left[P^T - C^T\right] + m^D\left[P^{WT} - C^D\right] + \left(1 - m^D\right)\left[P^A - C^A\right], \tag{7.2}$$

where the expressions in the squared brackets reflect the margins the shop makes on each of the three products sold.

Profit-maximizing drug shops will offer RDTs for sale if expected profits are higher with the sales of RDTs—that is, if $\mathrm{E}(r^{WT}) > \mathrm{E}(r^{NT})$. Combining (7.1) and (7.2) and rearranging terms, we can see that this is true as long as

$$m^D\left(P^{WT} - P^{NT}\right) + \left[P^T - C^T\right] + \left(1 - m^D\right)\left(\left[P^A - C^A\right] - \left[P^{NT} - C^D\right]\right) > 0. \tag{7.3}$$

From Eq. (7.3) we can see that three factors contribute to a monopolistic drug shop's willingness to offer the test. First, the shop could charge more for the drug when people are certain they have malaria. This is intuitive since the drug will be effective only when the person actually has malaria.[8] Even if this is not understood initially, over time, willingness to pay should increase as people discover that recovery is more likely when the drug is taken after a positive test. Second, the higher the markup on the test, the more likely the shop is to offer the test. Finally, if the margin on the alternative treatment is larger than the margin on the antimalarial if no test is offered, shops are more likely to offer the tests. This is unlikely to be the case, since the majority of alternative treatment purchases will be antipyretics, which are extremely inexpensive in Africa and are available widely in general stores, markets, and other outlets. To know when it will be in shops' interest to offer RDTs, we need to know what prices monopoly drug shops can charge. This requires an analysis of consumers' willingness to pay.

7.3.2 Consumers' Decision to Buy Test

The value to consumers of taking an antimalarial has two components. The first is the value of the improvement in health if they actually have malaria and receive the treatment for it. We designate that as W^M, where the W stands for willingness to pay for effective treatment. People know from experience that the treatment is not always effective, and they may understand that the reason is that other illnesses may appear symptomatically like malaria. Thus the second component of the value of treatment to a customer is the perceived probability that their symptoms are caused

[8] Some older antimalarials, such as chloroquine, have an antipyretic effect as well—so a person who had fever but not malaria and took an antimalarial might experience some benefit—but for the newer antimalarials, this is not the case.

by malaria, which we designate m^C. Thus their willingness to pay for the antimalarial drug in the absence of definitive test results is

$$\mathrm{E}\left(U^{NT}\right) = m^C W^M = P^{NT} \tag{7.4}$$

or the expected value to them of treatment when malarial infection is uncertain. Since drug shops want to maximize profits, a monopolist will charge the maximum price people are willing to pay for the drug if RDTs are not offered for sale ($m^C W^M$).

On the other hand, if a test is offered, consumers' expected value is the sum of the benefit if they test positive for malaria and if they test negative. Denoting W^A the willingness to pay for alternative treatment, the expected benefit if tested is

$$\mathrm{E}\left(U^{WT}\right) = m^C W^M + \left(1 - m^C\right) W^A \tag{7.5}$$

and customers will be willing to pay up to this amount in expected costs for treatment if tests are available. Their expected costs if tests are available and are purchased are

$$\mathrm{E}\left(C^{WT}\right) = P^T + m^C P^{WT} + \left(1 - m^C\right) P^A. \tag{7.6}$$

Even if tests are available, consumers may still choose to purchase the medicine without purchasing a test. Consumers will use the tests only if their expected welfare (benefits minus costs) is at least as great with the tests as without. That will be the case if

$$\begin{aligned} m^C W^M - P^{WT} &\leq m^C\left(W^M - P^{WT}\right) + \left(1 - m^C\right)\left(W^A - P^A\right) - P^T \\ &=> P^T \leq \left(1 - m^C\right)\left[P^{WT} + W^A - P^A\right]. \end{aligned} \tag{7.7}$$

If people choose not to be tested, they always pay for the drug but receive the benefit only a fraction m^C of the time. If they choose to be tested, they always pay for the test but pay for the antimalarial only if the test is positive. If the test is negative, they purchase the alternative treatment and receive consumer surplus $W^A - P^A$. Thus people are more likely to want to use the test (1) the lower the price of the test; (2) the less certain they are that they have malaria; (3) the more expensive the antimalarial drug is; and (4) the greater the consumer surplus from alternative treatment ($W^A - P^A$) if they do not have malaria.

Figure 7.1 portrays the actions consumers will take with different combinations of prices for the test and antimalarial drug. The consumer is choosing among being tested (and buying the appropriate drug conditional on test result), being presumptively treated (buying the antimalarial without the test), and doing nothing (buying no drug or test).

If the expected consumer surplus from buying the test and then the appropriate drug ($\mathrm{E}(U^{WT})$) is less than or equal to the expected cost ($\mathrm{E}(C^{WT})$), and the value of the test is above its price, then consumers will purchase the test and appropriate drug.

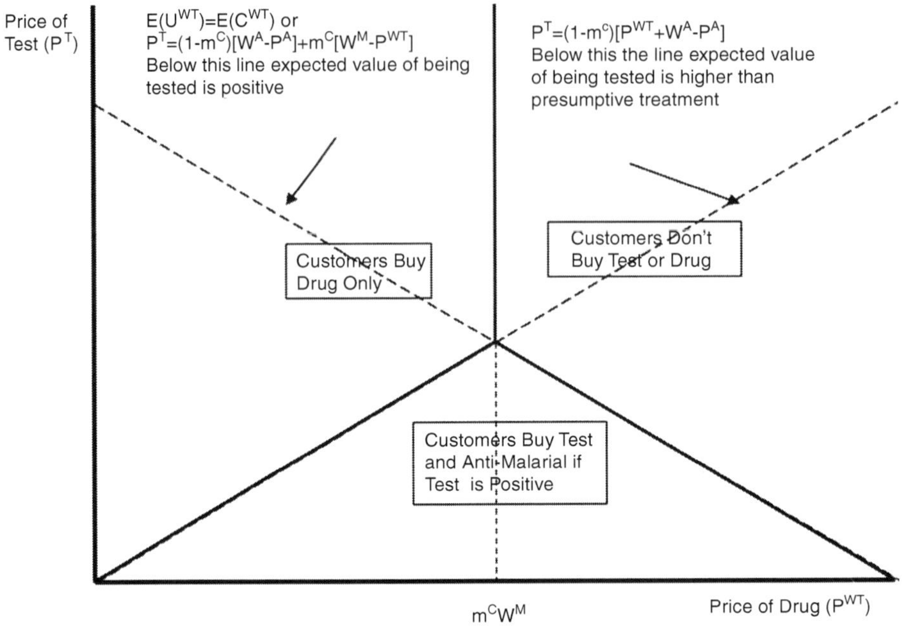

Fig. 7.1 Customer choices with different prices for tests and antimalarial drugs

This will be true anywhere in the lower triangle formed by the upward- and downward-sloping lines depicting the boundaries for the two conditions.

If the expected total cost with the test is above its value (the price of the test is above the upward-sloping line) but the price of the antimalarial is at or below the consumer's expected value of presumptive treatment with the antimalarial, then consumers will buy the drug but not the test. Finally, if the expected cost of the test and drugs is greater than the expected value of treatment with the test, and the cost of the antimalarial is above the expected value of taking it without knowing whether one has malaria ($m^C W^M$), then customers will buy neither the test nor the drug.

7.3.3 *Monopolist's Decision to Offer Test*

If the monopolist is going to offer the test, he will maximize profit by setting the prices for the antimalarial and the test such that customers' expected costs (from Eq. (7.6)) are just equal to the expected benefits (from Eq. (7.5)). Setting E(U^{WT}) = E(C^{WT}) and solving for P^T yields

$$P^T = \left(1 - m^C\right)\left[W^A - P^A\right] + m^C\left[W^M - P^{WT}\right]. \tag{7.8}$$

Substituting that into the equation for the firm's profit if the test is sold, we get

$$\begin{aligned}\mathrm{E}\left(r^{WT}\right) = \left(1-m^{C}\right)\left[W^{A}-P^{A}\right]+m^{C}\left[W^{M}-P^{WT}\right]-C^{T}+m^{D}\left[P^{WT}-C^{D}\right]\\ +\left(1-m^{D}\right)\left[P^{A}-C^{A}\right].\end{aligned} \tag{7.9}$$

That will be greater than profits without offering the test if

$$\begin{aligned}\mathrm{E}\left(r^{WT}\right)-\mathrm{E}\left(r^{NT}\right) = {} & \left(m^{D}-m^{C}\right)\left[P^{WT}-P^{A}\right]\\ & +\left[\left(1-m^{C}\right)W^{A}-\left(1-m^{D}\right)C^{A}\right]\\ & +\left[\left(1-m^{D}\right)C^{D}-C^{T}\right]>0.\end{aligned} \tag{7.10}$$

Note that if both the drug shop owners and the customers correctly perceive the probability that a sick person has malaria ($m^D = m^C = m$) and if there is no alternative treatment ($W^A = C^A = 0$), the monopolist offers the test so long as it saves on costs. That is, the monopolist will offer the test if the cost of the test (C^T) is less than or equal to the savings from the times the drug will not be purchased because the customer is not sick ($(1-m^D)C^D$). If an alternative treatment is available, then even if the cost of the test is greater than the savings from not having to buy the drug, the monopolist may still offer the test if the expected gain from being able to provide the alternative treatment when appropriate ($(1-m^D)[W^A-C^A]$) is sufficiently large. As we will see, this is the same condition under which the test will be provided by the competitive market and the same conditions under which a social welfare–maximizing planner would choose to make the tests available.

7.3.4 *Perfect Competition*

We've seen the conditions under which a monopolist will offer the test for sale, but under what conditions will profit-maximizing drug shops that face competition offer them? In a competitive market all prices are driven down to cost, and shops that don't offer the most attractive products to their customers will be driven out of business. Thus, RDTs will be offered so long as their costs are less than their benefits to consumers. With both drugs and the RDTs being offered at cost, customers of competitive drug shops will earn surplus

$$\mathrm{E}\left(U^{NT}\right) = m^{C}W^{M}-C^{D} \tag{7.11}$$

if they don't purchase an RDT. If they do purchase a test, their expected surplus will be

$$\mathrm{E}\left(U^{WT}\right) = m^{C}\left(W^{M}-C^{D}\right)+\left(1-m^{C}\right)\left(W^{A}-C^{A}\right)-C^{T}. \tag{7.12}$$

Thus in a perfectly competitive market, RDTs will be offered and purchased if and only if

$$\mathrm{E}\left(U^{WT}\right) - \mathrm{E}\left(U^{NT}\right) = \left(1 - m^{C}\right)C^{D} + \left(1 - m^{C}\right)\left(W^{A} - C^{A}\right) - C^{T} > 0. \tag{7.13}$$

RDTs will be offered and purchased so long as the cost is less than the expected savings from not buying the antimalarial when customers are not sick plus the extra benefit of getting a more appropriate therapy in that case. Note that if consumers' perceptions of the probability that they have malaria (m^C) are equal to the drug shop owners' perceptions (m^D), then the condition for the competitive market (Eq. (7.13)) is identical to that with a monopolist (Eq. (7.10)).

7.4 Optimal Provision of RDTs

We've seen that both monopolistic and perfectly competitive drug shops will sell RDTs under certain circumstances. How do those circumstances compare with what a social planner would deem optimal? The planner would want the tests to be sold and used if total social welfare was higher with use of the tests than without. We define social welfare or value when RDTs are not used as

$$V^{NT} = m\left(W^{M} + B^{M}\right) - C^{D} - (1 - m)C^{O} \tag{7.14}$$

where B^M is the external benefits of malaria treatment and C^O is the social cost of treatment of someone who is not sick with malaria with antimalarials, above and beyond the cost of the drugs. There are external benefits of malaria treatment to the extent that it reduces risks of infection to others. There are costs of treatment in excess of the cost of the drug if inappropriate treatment increases the rate at which malaria parasites become resistant to therapy. When tests are used, social welfare or value is

$$V^{WT} = m\left(W^{M} + B^{M} - C^{D}\right) + (1 - m)\left(W^{A} - C^{A}\right) - C^{T}. \tag{7.15}$$

A social planner would want the tests to be used when $V^{WT} > V^{NT}$ or when

$$(1 - m)\left(W^{A} - C^{A}\right) + (1 - m)\left(C^{D} + C^{O}\right) - C^{T} > 0. \tag{7.16}$$

Note that the existence of consumption externalities to taking the antimalarial if one is sick (B^M) has no effect on the optimal choice (since everyone who is sick is taking it under all conditions), but the social desirability of tests is higher if there are external costs to use of the drugs when they are not needed.

Comparing (7.16) with (7.13) and (7.10), we see that if there are no externalities to mistreatment ($C^O = 0$), and if there are no misperceptions of the likelihood of malaria ($m^D = m^C = m$), then both the competitive shop and the monopolist will supply the test in exactly the same conditions in which the social planner would provide them.

However, since neither the monopolist nor the consumer takes into account the costs of inappropriate treatment, the presence of such externalities can lead them to fail to provide tests in circumstances where the social planner would like to see them provided. Similarly, misperceptions of the true likelihood that a customer seeking treatment for malaria actually has malaria can lead to RDTs being provided in situations when they shouldn't be or not being sold in situations where they should be. We consider another possible source of this problem as well as a possible solution next.

7.5 Role of Subsidies, Beliefs, and Competition in Optimal Provision of RDTs

Consider now how the analysis changes if governments and NGOs want to make treatment for malaria and RDTs more affordable by subsidizing their prices. As noted in the introduction, there are many benefits to subsidizing ACTs (particularly in the context of credit constraints and disease externalities) that we don't consider here. Rather, our purpose is to ask whether subsidized RDTs, if made available alongside subsidized ACTs, would be sold by drug shops in a way that is welfare enhancing.

Define C'^T as the production cost of tests, which is equal to the subsidy plus the cost to drug shops, or $C'^T = C^T + S^T$, and define the production cost of the antimalarial drug analogously as $C'^D = C^D + S^D$. We can now rewrite the social planner's problem (Eqs. (7.14), (7.15) and (7.16)) as

$$\begin{aligned} V^{NT} &= m\left(W^M + B^M\right) - C'^D - (1-m)C^O \\ &= m\left(W^M + B^M\right) - C^D - S^D - (1-m)C^O, \end{aligned} \tag{7.14'}$$

$$\begin{aligned} V^{WT} &= m\left(W^M + B^M - C'^D\right) + (1-m)\left(W^A - C^A\right) - C'^T \\ &= m\left(W^M + B^M - C^D - S^D\right) + (1-m)\left(W^A - C^A\right) - C^T - S^T, \end{aligned} \tag{7.15'}$$

and the condition $V^{NT} < V^{WT}$

$$\begin{aligned} &(1-m)\left(W^A - C^A\right) + (1-m)\left(C'^M + C^O\right) - C'^T \\ &\quad = (1-m)\left(W^A - C^A\right) + (1-m)\left(C^D + S^D + C^O\right) - C^T - S^T > 0. \end{aligned} \tag{7.16'}$$

Note that even if there are no costs to inappropriate treatment ($C^O = 0$) and no misperceptions ($m^C = m^D = m$), subsidizing antimalarial treatment can create situations where both the competitive market and the monopolist will fail to provide the tests when it would be best to do so. This happens because the subsidy reduces the cost of antimalarials to the drug shops and thus lowers the cost-saving value of the test to them and to consumers, but it has no effect on the true social cost of the drug.

This problem, and the others previously described, could be overcome if it was possible to align the interests of private actors with the public purpose represented by the social planner's objective function. Is it possible to incentivize private drug shops to behave optimally? Yes, as long as private drug shops can be made to offer tests in the same circumstances as the social planner. To see whether this is possible, we look at the difference between the objective function of the social planner and that of the private drug shop.

A perfectly competitive drug shop will make the same choices as the social planner if the left-hand side of Eq. (7.13) is equal to the left-hand side of Eq. (7.16′), or if

$$\begin{aligned}&(1-m^C)(W^A-C^A)+(1-m^C)C^D-C^T\\&=(1-m)(W^A-C^A)+(1-m)(C^D+S^D+C^O)-C^T-S^{*T}\end{aligned}\tag{7.17}$$

where S^{*T} is the subsidy to the cost of the RDT that will cause the competitive drug shop to offer RDTs under the same conditions the social planner would. Rearranging terms, we see that this will be happen if

$$S^{*T}=(m^C-m)\left[W^A-C^A+C^D\right]+(1-m)\left[S^D+C^O\right].\tag{7.18}$$

If there are no errors in perception, then an optimal RDT subsidy will be equal to the proportion of customers without malaria times the drug subsidy plus the external cost of inappropriate treatment. If customers misperceive the probability that they have malaria, then there is an additional term.

Given how frequently people seeking treatment for malaria test negative for the parasite—studies noted in the introduction find this to be the case 35–80% of the time—if there are errors in perception, customers probably overestimate the probability they have malaria. If so, the RDT subsidy will have to compensate for this. To the extent the probability is overstated, the subsidy will have to be larger in proportion to the surplus from the alternative treatment plus the cost of the antimalarial drug.

The monopolistic drug shop will make the same choices as the social planner if the right-hand side of Eq. (7.10) is equal to the right-hand side of Eq. (7.16′), or if

$$\begin{aligned}&(m^D-m^C)\left[P^{WT}-P^A\right]+\left[(1-m^C)W^A-(1-m^D)C^A\right]+\left[(1-m^D)C^D-C^T\right]\\&=(1-m)(W^A-C^A)+(1-m)(C^D+S^D+C^O)-C^T-S^{*\prime T}.\end{aligned}\tag{7.19}$$

Rearranging terms, we see that this will be true if $S^{*\prime T}$ is set as

$$\begin{aligned}S^{*\prime T}=&(m^C-m^D)(P^{WT}-P^A)+\left[(m^C-m)W^A-(m^D-m)C^A\right]\\&+(m^D-m)C^D+(1-m)(S^D+C^O).\end{aligned}\tag{7.20}$$

Once again, in the absence of any misperceptions ($m^C = m^D = m$), a subsidy equal to the probability the customer does not have malaria times the cost of the antimalarial drug plus the external cost of inappropriate treatment will align the behavior of the drug shop with that of the social planner. If customers and drug shop owners misperceive the likelihood of malaria to the same extent ($m^C = m^D$), then (7.20) is identical to (7.18), and both the monopolist and the perfect competitor will behave like the social planner if the subsidy for the RDT is set optimally.

However, if drug shop owners and customers have different perceptions of the probability that customers are sick with malaria, monopolists will behave differently from competitive drug shops. Monopolistic drug shops will be less likely to want to offer tests to their customers if the owners think customers overestimate the probability that they are sick. Such customers will be willing to pay more for the drug than they would if they shared the drug shop owners' views, and the owners may not want to disabuse them of such views. Alternatively, if the customers view themselves as less likely to be sick than drug shop owners do, the drug shops will have an interest in promoting the test to increase the sale of the drug and the price they can charge for it (since people will consider the antimalarial more likely to be efficacious if they know they have the disease).

To see how much of a difference this will make for the optimal subsidy, we need to know what drug shops will charge for the antimalarial drug if they offer it with the test. Equation (7.9) shows what the monopolistic drug shop's profits will be as a function of the price of the antimalarial, assuming that the price of the test is set low enough that customers are just willing to seek treatment at the shop. Rearranging the terms in (7.9), we get that expected profits are

$$\begin{aligned} \mathrm{E}\left(r^{WT}\right) = {} & \left(1-m^C\right)\left[W^A-P^A\right]+m^C W^M-C^T-m^D C^D+\left(1-m^D\right) \\ & \times\left[P^A-C^A\right]+\left(m^D-m^C\right)P^{WT}. \end{aligned} \tag{7.9'}$$

We see from the last term that if the drug shop owner's perceived probability that the customer is infected is higher than that of the customer ($m^D > m^C$), then the drug shop will want to set the highest price for the antimalarial that it can (and thus the lowest price it can for the test).[9] On the other hand, if the shop owner sees the probability of a customer's being infected as lower than the customer does, they will want to set the price of the antimalarial as low as possible and the price of the test as high as possible if the shop is going to offer the test.

From Fig. 7.1 we can see what prices these will be. If the monopolist shop wishes to offer the test and to maximize profits, it will choose the price of the RDT and the antimalarial that is on the solid section of the downward-sloping line. If at the same

[9] If customers and drug shop owners have the same perceived probability of infection ($m^D = m^C$), then any choice of the price of the antimalarial and the RDT that satisfy the constraint that the customer expects that the test will save money (Eq. (7.7)) will maximize profit. In Fig. 7.1 this is any combination of the two prices on the solid section of the downward sloping line.

price for the RDT a lower price is charged for the antimalarial, profits will be lower. If the price of the RDT is increased beyond the maximum value on the solid part of that line, customers either won't purchase the RDT or won't seek treatment. From Fig. 7.1 we can see that the maximum price of the drug consistent with profit maximization corresponds to a zero price for the test. Thus if customers think it less likely that they have malaria than the drug shop owners, drug shops will give the tests away for free to identify those who have malaria and then charge as much as they can for the antimalarial—and still get customers to come to the shop.[10]

In the more likely case that customers perceive the probability that they have malaria to be higher than the drug shop owner does, the drug shop will want to charge as high a price as it can for the test. That price is given by the intersection of the upward- and downward-sloping lines in Fig. 7.1, and that can be found by solving Eq. (7.13) for P^T and setting it equal to the value for P^T given by Eq. (7.7). In this case $P^{WT} = m^C W^M$, and this is the value that should be used in computing the optimal subsidy in Eq. (7.20).

7.6 Discussion

We have shown that profit-maximizing drug shops have several incentives to offer their customers RDTs, and that in the absence of errors in perceptions, subsidies, or externalities, they will offer them in the same circumstances as would a planner who chooses whether to offer the test to maximize social welfare. However, there likely are externalities to inappropriate treatment, customers seem to perceive themselves as having malaria very frequently when they do not, and ACTs are being heavily subsidized in some countries. Thus in the absence of policy interventions, the private market will almost certainly under provide RDTs.

We have seen that a subsidy may be able to overcome the problem of under provision. How big would the subsidy have to be? Consider that the subsidy for the test that equates the interests of a drug shop owner and the social welfare–maximizing planner must be at least equal to the fraction of people seeking treatment who do not have malaria times the value of the subsidy to the antimalarial. Given that, on average, 65% of older children and 82% of adults seeking treatment for malaria test negative (Cohen et al. (2010, 2012); Kachur et al. 2006), and that an expected subsidy of 95% for ACTs with production costs of roughly \$1, just this one component of the optimal subsidy would be nearly the entire cost of the typical RDT (\$.60).

Thus a subsidy for RDTs may help but may not be a complete solution to the problem. From Eq. (7.13), we can see that if the price of the test is made low

[10] Drug shops would never pay people to take the test, since even those who did not think they might be ill would take the test just to get the payment.

enough with subsidies, consumers will always purchase them in a competitive marketplace.[11] The necessary subsidy may be quite large, and the best policy may be to give the tests to drug shops for free. However, in the absence of competition, monopolists may not offer the tests even if they are given to them at no cost. From Eq. (7.10), we can see that if consumers perceive the likelihood that they have malaria to be higher than the drug shop owners do, then profits from offering the test can be less than profits when the RDT is not sold, even if RDTs are given to shops for free. This observation suggests the importance of both educating customers about the prevalence of malaria and promoting competition among drug shops. The latter policy would have the additional benefit of reducing the cost of tests and drugs, thus making treatment accessible to more people (a factor we have not considered in our modeling). A full set of policies to maximize the benefits that RDTs might provide may require subsidies for the tests, education of consumers, and policies to promote competition among drug shops. These could be accomplished as part of a campaign to promote the use of ACTs.

[11] If Eq. (7.18) dictates a subsidy larger than the cost of the test to equate the behavior of the competitive drug shop and the social planner, the social planner would chose to offer the test at any production cost less than S^{*T}, and thus giving the tests away for free (in which case they will be used) is adequate.

7. Commentary: How to Solve One Problem Without Creating Another

Anup Malani

Two challenges motivate Cohen and Dickens in "Adoption of Over-the-Counter Malaria Diagnostics in Africa." First, individuals with malaria use the wrong malaria drug to treat their illness. They use monotherapies rather than combination therapies, specifically artemisinin combination therapies (ACTs), and the use of monotherapies is more likely to lead to drug resistance. Second, individuals use malaria treatments even when they do not have malaria. Specifically, individuals with fevers take malaria treatment even if they do not have malaria and either an antipyretic or antibiotic would be more effective. This too exacerbates drug resistance.

The favored policy response to the first problem, suboptimal malaria treatment, has been to subsidize the cost of ACTs. Unfortunately, this subsidy does not solve the second problem. Indeed, it may worsen it—a point to which I will return later. The proper policy response to the second problem, excessive malaria treatment, is to get individuals to take rapid diagnostic tests (RDTs) to verify that they have malaria before they take malaria treatments. Of course that is easier said than done. In a pair of papers, Cohen has taken up the question of how one can get individuals to take RDTs.

In a separate paper with Dupas and Schaner, Cohen reports on the results of an experiment in which individuals were randomized to subsidized ACTs and RDTs at different prices. The salient findings are two. First, subsidizing the price of ACTs appears to increase the degree of ACT use by individuals—especially older children and adults—who do not have malaria. Second, demand for RDTs is relatively inelastic. Specifically, demand is the same whether RDTs have zero price or a price equivalent to almost one-seventh of the subjects' daily wage. The results suggest that, if local pharmacies offer consumers RDTs for sale, those RDTs will be purchased, and the second problem—overuse—will be solved.[12]

[12] Although it is tangential to my main comments on the present paper, I am puzzled by this result in the predicate paper. For very few products in the world is demand truly inelastic. It is particularly surprising that demand for tests is inelastic given the high rate at which individuals take malaria medication even without verification they have malaria. Therefore, I suspect that some sort of crude Hawthorne effect may be responsible for the remarkable finding that price did not affect demand for RDTs. If I am correct, however, this means that the theory in the paper on which I am commenting is even more important. We must understand when drug sellers would also sell RDTs and when consumers would use them. The only change my suspicion would imply is that consumer demand for tests is more sensitive than the model in the theory paper assumes.

A. Malani (✉)
Lee and Brena Freeman Professor of Law, University of Chicago Law School, Chicago, IL, USA

Resources for the Future, Washington, DC, USA
e-mail: amalani@uchicago.edu

This volume's chapter by Cohen and Dickens takes up the natural question that follows: under what conditions will firms offer RDTs for sale, at least to the same extent that a social planner would want them to? The long answer is that it depends on several factors, including the beliefs of drug sellers and individuals about the prevalence of malaria and the externalities from excessive use of ACTs. But the useful normative policy proposal that emerges is that appropriate subsidies for RDTs may encourage RDT use and solve the problem that malaria drugs are overused.

In this comment I want to highlight two points that Cohen and Dickens make but do not stress and yet are very important for policymakers to understand. Moreover, I want to raise some more complications that they ought to consider in future research.

The first point I want to stress is that the policy designed to get people to use ACTs rather than monotherapies—ACT subsidies—exacerbates the second, overuse problem. By reducing the gap between the price of ACTs and the drug that individuals should take (antipyretics or antibiotics) if they know they do not have malaria, ACT subsidies also reduce the incentive of individuals to use RDTs and identify the proper drug to treat their illness. Indeed, to the extent that ACTs are more effective at treating malaria than monotherapies because they are less likely to be resistant, they will actually worsen the overuse problem after equating the price of ACTs and monotherapies. The implication is not that ACT subsidies are a bad idea. Rather, it is that the return to such subsidies is lower than expected.[13]

The second point is that a critical factor in evaluating the efficacy of any subsidy for RDTs is determining how they affect both sellers' and consumers' beliefs about malaria prevalence. As Cohen and Dickens acknowledge, if monopoly sellers think that malaria prevalence is lower than consumers think it is, then they would be reluctant to sell RDTs (or would require a higher subsidy to sell RDTs) because, through RDTs, consumers may learn that prevalence is lower and thus they may demand fewer ACTs. What I want to stress is that even if monopolist sellers were uncertain whether consumers thought prevalence was higher than it actually is, the risk that they might would actually encourage monopolists to at least delay selling RDTs. Once consumers learn that malaria risk is lower than they previously thought, that belief cannot be reversed. Thus the decision to sell RDTs has real option value.

The problem is even thornier if the monopolist seller starts wondering why an NGO or the government is subsidizing RDTs. If everyone who currently sought treatment actually had malaria, then there would be no need for RDTs. RDT subsidies are only required if individuals underuse ACTs or if they overuse it. If they underuse ACTs, an alternative solution is to further subsidize ACTs. If they overuse it, the RDT subsidies are required. Thus it is plausible that sellers will infer from RDT subsidies that malaria is lower than consumers suspect. But this very signal will

[13] To be even more clear, the blame ought to be placed not on ACT or ACT subsidies but on the low price of monotherapies. It is that low price that forces the use of subsidies for ACT to reduce the rate at which antimalarials generate resistance. However, if subsidies that equate the price of ACT and monotherapies increase use, then that too will generate resistance, a negative externality.

discourage monopolist sellers from offering RDTs in their stores. The one consolation, however, is that this should not affect the behavior of competitive sellers.

Beyond this point I want to recommend some topics for future research on RDT subsidies. The model that Cohen and Dickens present is purposely simplified to convey the basic intuition behind an RDT subsidy. All the comments that follow are meant to complicate that model to make it more realistic and help craft a more appropriate subsidy.

First, and most important, the present model assumes that individuals believe the RDT works. If they are uncertain of RDT accuracy, then they will have lower demand for RDTs. This has two consequences. One is that it is important to model how individuals update their beliefs about the accuracy of tests. From Gentzkow and Shapiro (2006), we know that individuals will judge tests partly by their priors and hence will be slow to learn about the accuracy of tests—at least without successful use of antimalarials to verify tests. Another consequence is that slow learning will require higher subsidies to encourage individuals to use RDTs.

A second topic for research is whether the subsidies for RDTs are so large that firms (or consumers) will face a negative price for RDTs. That raises the problem that governments and NGOs must monitor the use of RDTs; otherwise firms or consumers will simply order and dispose or take duplicative tests just to obtain income from the subsidy. That will increase subsidy costs without benefit.

Third, the present model assumes that individuals do not currently purchase diagnostic tests. But the fact is that they do. Buying an antimalarial is also the purchase of a diagnostic test. If the antimalarial does not work, people know either the antimalarial does not work or they do not have malaria.[14] As a result, the product choice they face is not an antimalarial or a test (the RDT). Rather, it is an antimalarial with a diagnostic test or a diagnostic test by itself (the RDT). This will change the equilibrium price for antimalarials, the demand for RDTs, and the magnitude of the subsidy required for the RDT.

Finally, the present model assumes that all individuals have identical beliefs about whether they have malaria and identical valuation for a cure conditional on having malaria. Of course both values will vary among the population. As a result, sellers face a downward-sloping demand for ACTs and RDTs even among people with fevers or with malaria. So a monopolist will sell fewer RDTs than the social planner desires and fewer than a competitive firm would sell, even if there were common knowledge about aggregate malaria prevalence and no externalities from mistreatment, contrary to the conclusion at the end of Sect. 7.4.

In summary, the chapter by Cohen and Dickens in this volume, combined with the companion piece by Cohen, Dupas, and Schaner, is an important step in addressing the problem of antimalarial overuse. The lesson—RDTs must be subsidized along with ACTs—is an important one for policymakers to learn. Further work is required to fine-tune the RDT subsidy amount, but that should not detract from the main lesson.

[14] If the individual does not have malaria but infers that the antimalarial does not work, one could say the antimalarial diagnostic suffered a false negative.

References

ACT Watch. (2010). *ACTwatch: Evidence for malaria medicines policy*. http://www.actwatch.info

Advaryu, A. (2012). *Learning, misallocation, and technology adoption: Evidence from new malaria therapy in Tanzania*, Mimeo. Yale School of Public Health. Available at http://www.yale.edu/adhvaryu/adhvaryu_learning.pdf.

Arrow, K., Panosian, C., & Gelband, H. (Eds.). (2004). *Saving lives, buying time: Economics of malaria drugs in an age of resistance*. Washington, DC: Institute of Medicine, National Academies Press.

Cohen, J., Dupas, P., & Schaner, S. (2010). *Prices, diagnostic tests, and the demand for malaria treatment*, Mimeo. Harvard School of Public Health.

Cohen, J., Dupas, P., & Schaner, S. (2012). *Price subsidies, diagnostic tests, and targeting of malaria treatment: Evidence from a randomized controlled trial*. National Bureau of Economic Research Working Paper #17943.

Gentzkow, M. and Shapiro, J.M. 2006. Media Bias and Reputation. *Journal of Political Economy*. 114(2).

Kachur, P., et al. (2006). Prevalence of malaria parasitemia among clients seeking treatment for fever or malaria at drug stores in rural Tanzania. *Tropical Medicine and International Health, 11*(4), 441–451.

Perkins, M., & Bell, D. (2008). Working without a blindfold: The critical role of diagnostics in malaria control. *Malaria Journal, 7*(Suppl I), S5.

Rafael, M., et al. (2006). Reducing the burden of childhood malaria in Africa: The role of improved diagnostics. *Nature, 444*(Suppl 1), 39–48. doi:10.1038/nature05445.

Reyburn, H., et al. (2004). Overdiagnosis of malaria in patients with severe febrile illness in Tanzania: A prospective study. *British Medical Journal*. doi:10.1136/bmj.38251.658229.55.

Uzochukwu, B., et al. (2010). Willingness to pay for rapid diagnostic tests for the diagnosis and treatment of malaria in southeast Nigeria: Ex post and ex ante. *International Journal for Equity in Health, 9*, 1.

World Health Organization. (2010). *2009 World malaria report*. Geneva: WHO Press.

Chapter 8
The Value of Determining Global Land Cover for Assessing Climate Change Mitigation Options

Steffen Fritz, Sabine Fuss, Petr Havlík, Jana Szolgayová, Ian McCallum, Michael Obersteiner, and Linda See

Abstract Land cover maps provide critical input data for global models of land use. Urgent questions exist, such as how much land is available for the expansion of agriculture to combat food insecurity, how much land is available for afforestation projects, and whether reducing emissions from deforestation and forest degradation (REDD) is more cost-effective than carbon capture and sequestration. Such questions can be answered only with reliable maps of land cover. However, global land cover datasets currently differ drastically in terms of the spatial extent of cropland distributions. One of the data layers that differ is cropland area. In this study, we evaluate how models designed to help in policy design can be used to quantify the differences in implementation costs. By examining these cost differences, we are able to quantify the benefits, which equal the loss from making a decision under imperfect information. Taking the specific example of choosing between REDD and carbon capture and storage under uncertainty about the available cropland area, we have developed a methodology on how the value derived

---

S. Fritz (✉) • S. Fuss • P. Havlík • I. McCallum • M. Obersteiner
International Institute for Applied Systems Analysis, Ecosystems Services and Management Program, Laxenburg, Austria
e-mail: fritz@iiasa.ac.at; fuss@iiasa.ac.at; havlikpt@iiasa.ac.at; mccallum@iiasa.ac.at; oberstei@iiasa.ac.at

J. Szolgayová
International Institute for Applied Systems Analysis, Ecosystems Services and Management Program, Laxenburg, Austria

Department of Applied Mathematics and Statistics, Faculty of Mathematics, Physics and Informatics, Comenius University, Bratislava, Slovakia
e-mail: szolgay@iiasa.ac.at

L. See
International Institute for Applied Systems Analysis, Ecosystems Services and Management Program, Laxenburg, Austria

School of Geography, University of Leeds, Leeds, UK
e-mail: see@iiasa.ac.at

R. Laxminarayan and M.K. Macauley (eds.), *The Value of Information: Methodological Frontiers and New Applications in Environment and Health*,
DOI 10.1007/978-94-007-4839-2_8,

from reducing uncertainty can be assessed. By implementing a portfolio optimization model to find the optimal mix of mitigation options under different sets of information, we are able to estimate the benefit of improved land cover data and thus determine the value of land cover validation efforts. We illustrate the methodology by comparing portfolio outputs of the different mitigation options modeled within the GLOBIOM economic land use model using cropland data from different databases.

Keywords Value of information • Land cover maps • Land use • Mitigation • GEOSS

8.1 Introduction

Activities within the land use, landuse change, and forestry (LULUCF) sector will play an increasingly important role in climate change mitigation in the future. Although LULUCF was a significant factor in the negotiations of the original Kyoto agreement, the protocol did not specify how emissions and reductions from this sector would be incorporated into the accounting system. Instead, this function was assigned to the Subsidiary Body for Scientific and Technological Advice (SBSTA) of the United Nations Framework Convention on Climate Change (UNFCCC) and the Intergovernmental Panel on Climate Change (IPCC), where a working group on LULUCF formulated a special report (Watson et al. 2000). The framework was then accepted at the seventh conference of the parties (COP-7) in Marrakech in 2001 (Schlamadinger et al. 2007).

Reductions in greenhouse gas (GHG) emissions can be achieved in the LULUCF sector in several ways, such as Reducing Emissions from Deforestation and Forest Degradation (REDD), increasing the area of land cultivated with biofuels, and improving agricultural practices. REDD is a multi-agency initiative that aims to establish a framework for the coordination of actions at the country level by creating a financial value for the carbon stored in forests, offering incentives for developing countries to reduce emissions from forested lands, and investing in low-carbon paths to sustainable development. However, within the Kyoto Protocol's first commitment period, 2008–2012, REDD in developing countries is not an allowable contribution, yet deforestation represents the main source of GHG emissions in, for example, Indonesia (Schlamadinger et al. 2007). The Bali Action Plan, an outcome of COP-13, held in Bali in December 2007, requires parties to include REDD in the post-2012 negotiations of the Kyoto agreement (FAO et al. 2008). At COP-15, in Copenhagen in December 2009, even though no overarching agreement was reached, leaders agreed to establish a "green climate fund," which is designed to mobilize $30 million on REDD + (which includes forest conservation and sustainable management) for mitigation, adaptation, technology, and capacity building, and further progress on this has been made at the previous COP-16 in Cancun in 2010.

Satellite remote sensing is an important potential source of data for determining initial conditions of land cover and forest cover for LULUCF and other land use models (Watson et al. 2000). Urgent questions exist about the land available for the expansion of agriculture to combat food insecurity, the extent of future competition for land between food and bio-energy, as well as how much land is available for afforestation projects. Moreover, questions arise about the cost-effectiveness of REDD policies versus bio-fuel targets. Such questions can be answered only with reliable maps of land cover. Recent data sets on global land cover are the MODIS land cover, based on the MODIS sensor and produced by Boston University (Friedl et al. 2002); the GLC-2000, based on the SPOT-Vegetation sensor and produced by the Joint Research Center of the European Commission (Fritz et al. 2003); and the GlobCover product, based on the MERIS sensor and produced by a consortium supported by the European Space Agency (Defourny et al. 2009). However, these data sets differ drastically in terms of cropland distributions and, especially, cropland area. Ramankutty et al. (2008) estimated that the cropland area is between 1.22 billion and 1.71 billion ha (at the 90 % confidence interval), which translates to a 40 % difference between land cover products. For example, using the maximum cropland area as the upper limit from the legend definition (e.g. for single classes 100 % cover and for Mosaic classes 50–70 % cover), we find that MODIS records 1,693 million ha, GLC-2000 records 2,201 million ha and GlobCover records 1,902 million ha (Fig. 8.1). At the same time there have been questions regarding the cropland extent reported by the UN's Food and Agriculture Organization (FAO), in particular for developing countries. For example, in the least developed countries, such as Malawi, the appropriate methods and tools to undertake reliable crop area estimates are simply not in place, and reported crop area contains a possible error of up to 30 % (World Bank, personal communication).

These large absolute and spatially distributed differences in cropland extent have implications for the GLOBIOM economic land use model used at the International Institute for Applied Systems Analysis because the data provide the initial conditions for the evaluation of mitigation options. To explore the value of this information, we construct a scenario with two mitigation options, REDD and the implementation of a new technology in the energy sector, carbon capture and storage (CCS). Each mitigation option has a different cost. However, the REDD mitigation option has increasing costs as less and less land is available. The uncertainty in these costs is also a function of which cropland extent layer is used as an input to the land use model. Uncertainty about whether the world is correctly represented by the figures reported by the International Food and Policy Research Institute (IFPRI) or the GLC-2000 land cover product or MODIS may carry substantial costs when choosing a mitigation policy portfolio. This is because the optimal mix of mitigation options under uncertainty might deviate substantially depending on whether IFPRI, GLC-2000, or MODIS reflects the true state of the world. This is also a function of the risk strategy of the decision maker. For example, a risk-averse strategy might typically be to accept higher portfolio costs to lower the overall risk. We acknowledge the potential importance of other sources of uncertainty, such as uncertainty in the economic land-use model and its underlying assumptions, as well as the exogenous drivers of the economic land-use model, such as the validity of population projections and assumptions about technological

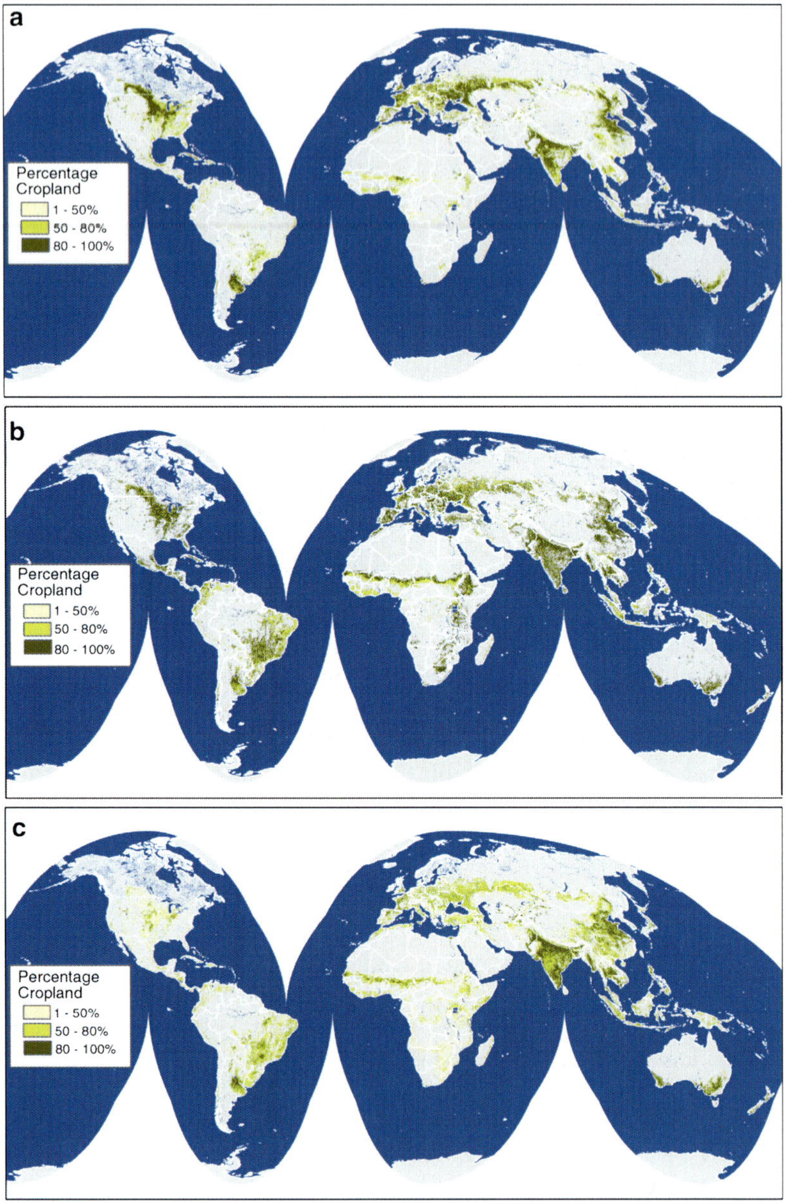

Fig. 8.1 Global distribution of cropland described by (**a**) MODIS, (**b**) GLC-2000, and (**c**) GlobCover

change. For example, higher than anticipated population would exert additional pressures on land, whereas unexpected technological breakthroughs that improve yields would reduce demand for agricultural land. Analyzing all sources of uncertainty is clearly beyond the scope of this paper. We therefore focus on land cover, since methods and tools are currently available to reduce this type of uncertainty.

Moreover, we acknowledge that the value of information derived in this paper has itself a particular uncertainty attached, since not all types of uncertainties can be considered simultaneously without obscuring the mechanisms in which we are primarily interested at this stage.

In this paper a methodology is presented that demonstrates how the value of reducing uncertainty can be assessed. A portfolio optimization model is implemented to find the optimal mix of mitigation options using different estimates of cropland from two land cover datasets as inputs to the model. We therefore created two land cover layers, one using the GLC-2000 cropland minimum (the cropland class is covered 50 % by cropland and 50 % by a noncropland class) and the other using the MODIS cropland maximum (where the cropland class is covered 100 % by cropland). This can still be considered a relatively conservative approach, since the maximum cropland extent reported by GLC-2000 would be even higher.

It can be shown that an increase in the probability that either GLC-2000 or MODIS is correct will lower the expected portfolio costs compared with the case where both have the same probability of being correct. This finding proves that there is added value in continuing to improve land cover information through better validation.

In the remainder of this paper we review the concept of the value of information along with applications in the existing literature. Subsequently, we present an analytical framework valuing the information from having better land cover data for two mitigation options under differing assumptions of the behavior of the cost function, which will be illustrated with an application. Finally, we consider the implications of this approach for the merit of global Earth observations (EO) and applications of this approach in future research.

8.2 Value of Information

The expected value of information (VOI) is a concept that has been used in stochastic programming for a long time (Birge and Louveaux 1997). Another term frequently used to describe this concept is the so-called willingness to pay (for information). The idea is that decisions taken on the basis of imperfect information can differ from those taken in a situation of complete or perfect information, and thus the decision-maker might be willing to pay the difference in costs or profits to be able to make a better-informed decision.

In the approach used in this paper, we compute the expected VOI for a portfolio model, where the optimal mitigation strategy depends crucially on the availability of information. The method that we use to optimize the decisions under perfect and imperfect information is standard portfolio theory (Markowitz 1952), using the variance of costs as a measure of risk. Even though both concepts are not original, the approach of using them to assess the VOI in the face of uncertainty about the availability of land, and thus the cost of one of the mitigation options, is worthy of demonstration, both in theory and with a practical example (using the case of avoided deforestation).

8.2.1 Review

Macauley (2006) provides a good overview of the literature from research using VOI and emphasizes that most models depend largely on the extent of the uncertainty of the decisionmakers, the cost of making a suboptimal decision in the light of better information, the cost of making use of the information and incorporating it into decisions, and the price of the next-best substitute for the information.

Macauley (2006) further demonstrates that the value of information is clearly 0 in the situation where a decision maker attaches a probability of 0 or 1 to a given event, which means that she considers the occurrence of the event no longer uncertain. The other case in which the information has no value is when no alternative actions are available, even if information could be obtained, or when a wrong decision will not result in any added costs. Similarly, information is most valuable when (a) the costs associated with a wrong action are high; (b) when many alternative actions are available; or (c) when the decisionmaker has no clear preference for one or more of the alternatives.

The expected VOI has been measured by two kinds of methods[1]:

- *Hedonic pricing*. These studies attempt to estimate the costs and benefits associated with environmental systems that have a direct effect on market values—for example, the use of wages or housing prices to infer the value of weather information or environmental quality as these affect wages or house prices.
- *Gains in output or productivity*. The VOI is generally found to be rather small in most of these studies. Macauley (2006) attributes this to the fact that people are willing to pay for information only beforehand. Often they are unaware of the severe consequences that imperfect information in the case of an uncertain event can inflict. In the same vein, people often attribute a very low probability to catastrophic events and then choose not to pay for information that may well be rather costly.

Finally, Macauley (2006) acknowledges that computation of the expected VOI is a suitable tool for the valuation of EO benefits. In this case, the availability of better information can save costs and lives and alleviate problems in the face of disasters.

In economics, the expected value of information has also been widely used. Looking at climate change policy analysis in particular, Peck and Teisberg (1993) and Nordhaus and Popp (1997) adopted a cost-benefit approach targeted at finding the optimal policy response to damages due to climate change. They then proceeded to estimate the extent to which the world would be better off economically if, for example, climate sensitivity and the level of economic damages were known.

[1] This review mainly applies to the Value of Information in the context of Earth Observations. The principles could equally have been applied within other scientific fields, but reviewing all of these is beyond the scope of this paper.

Most of these studies use multistage optimization, where all information about the correct level of the uncertain parameters arrives in one time instance. Others, like Fuss et al. (2008), have used stochastic dynamic programming allowing for a good description of the development of the uncertain parameters but with the disadvantage of having less scope in terms of controls and states. The VOI is derived by comparing profits and emissions when optimizing with stochastic prices to profits and costs when prices are deterministic (in which case the optimal decisions would be different from those in the stochastic setting).

8.2.2 Current Applications of Measuring Value of Earth Observations

The 10 Year Implementation plan of the Global Earth Observation System of Systems (GEOSS 2005) lists nine societal benefit areas (SBAs): weather, climate, ecosystems, biodiversity, health, energy, water, disasters, and agriculture. Despite the extensive literature on the costs and benefits of weather forecasts (Katz and Murphy 1997; Center for Science and Technology 2007), there have been few attempts to quantify the value and benefits of EO data in other SBAs. Studies that have addressed the benefit of EO for health and energy applications have been particularly scarce.

Moreover, to date, there have been few integrated assessments of the economic, social, and environmental benefits of EO and the GEOSS. A project funded by the European Union called GEOBENE (Global Earth Observation—Benefit Estimation: Now, Next and Emerging) developed tools and methodologies for studies of GEOSS benefit assessment. Some of those tools continue to be developed in the EUROGEOSS project, and two case studies are presented here. In the course of the GEOBENE project, a conceptual framework for assessing the benefits of GEOSS via a benefit-chain concept was developed. The basic notion is that an incremental improvement, and hence an incremental benefit in the observing system, must be judged against the incremental costs needed to build the observing system. Since it is not always easy to quantify the costs and in particular the benefits in monetary terms, an order of magnitude estimation is proposed. Moreover, it was shown that an understanding of the shape of the cost benefit curve can help guide rational investment in EO systems (Fritz et al. 2008).

An example of improved data for biodiversity conservation planning illustrates how the benefit chain concept can be applied. This case study, described in Fritz et al. (2008), demonstrates the benefits of replacing commonly available coarse-scale global data (the non-GEOSS scenario) with finer-scale data used in conservation decision making. The national land cover data set for South Africa was compared with the global GLC-2000 dataset, whose finer-scale data are like those expected from GEOSS and can thus be used to estimate the potential benefits of GEOSS data. When one compares the estimated cost of producing higher-resolution data for the South African case study (200 million €) with the cost of not having this information (1.2 billion €), it becomes clear that the improved data have real value.

A second example is demonstrated by Bouma et al. (2009), who used Bayesian decision theory to quantify the benefits if decisionmakers use EO data versus a scenario in which these data were not available. The authors examined the added value of EO for preventing potentially harmful algal blooms in the North Sea. Using expert elicitation to assess the perceptions of decisionmakers regarding the accuracy of the GEO-based algal bloom early warning system, the analysis indicated that the value (i.e., avoided damage) of an early warning system would be approximately 74,000 € per week. Since this is less than the costs of establishing and maintaining such an early warning system, investing in satellite observation for preventing potentially harmful algal blooms is an economically efficient investment. Increasing the accuracy of the information system substantially increases the value of information, where the value of perfect information was estimated at 370,000 € per week (Bouma et al. 2009).

A third example, in the field of the disaster SBA, is elaborated by Khabarov et al. (2008), who investigated, by means of simulation studies, how improvements in the spatial resolution of weather observation systems can help reduce the area burned by forest fires in Portugal and Spain. A fire danger index was computed on a daily basis, which was assumed to be used in decisionmaking. Official aircraft-based forest patrolling rules were applied. In the model, the total area burned and the total observed area were both considered, and the benefit of having fine- versus coarse-resolution data was assessed. By modeling the stochastic process of fire spread, the researchers estimated how much area burned could be saved if the fires were detected quickly through an improved patrolling pattern. This pattern could be designed using a finer weather grid. Simulations revealed that the use of finer-resolution data reduced the area burned by 21 % and the patrols could be reduced by 4 %. The cost-benefit ratio points towards a higher incremental benefit than the incremental cost of establishing a finer-grid patrol system.

An overall assessment of the GEOBENE project showed that in the majority of case studies, the societal benefits of improved and globally coordinated EO systems were orders of magnitude higher than the investment costs. A strong coordinating institution is required to ensure that an integrated architecture takes full advantage of the increased benefits and cost reductions achieved by international cooperation.

8.3 Analytical Framework: Portfolio Approach to Mitigation

8.3.1 Independent Constant-Cost Mitigation Options

In this paper we are interested in a situation where the decisionmaker can mitigate climate change either in the land use sector (e.g., through avoiding deforestation) or in industry (e.g., by introducing a new technology, such as carbon capture and storage).

The focus of our analysis is on the choice between these two options, where the land-use mitigation option exhibits increasing marginal costs, which differ between two scenarios depending on the land cover data set that is used. Scenario 1 is the GLC-2000 scenario, in which a substantial additional land resource is available for agriculture and cropland expansion; this is the "available land" scenario. Scenario 2 is the MODIS scenario, where most of the land is already in use and much less additional land is available for agriculture and cropland expansion; this is the "limited available land" scenario. The other mitigation option is assumed to be available at a constant cost at the beginning and is completely independent of the first option.[2]

We use standard portfolio theory (Markowitz 1952) to approach the problem of determining the optimal mitigation portfolio and derive the expected value of perfect information for the results obtained (Birge and Louveaux 1997).[3] In particular, the objective to be minimized is a weighted average of expected costs and variance:

$$\min_{x \in [0,1]} E[C(x)] + \omega Var[C(x)] \tag{8.1}$$

where the weight of the variance represents the level of risk aversion: the larger the weight of the variance in the objective, the more costs the decisionmaker will adopt to reduce this risk. E is the expected value operator; Var is the variance; ω is the measure of risk aversion and is larger than 0 for risk-averse decisionmakers and equal to 0 for risk-neutral decisionmakers; x is the share of emissions abated through avoided deforestation within the mitigation portfolio; and C is the mitigation costs.

Other studies analyzing mitigation strategies have also implicitly and explicitly incorporated risk-averse decisionmakers, but it is challenging to estimate the magnitudes of the risk aversion parameter for global decisionmakers, although much work has been conducted in eliciting farmers' degree of risk aversion using different types of utility functions (Lin et al. 1974; Binswanger 1980; Dillon 1971; Dillon and Scandizzo 1978).

At the global scale, integrated assessment models include damages from warming in their optimization of social welfare (see, e.g., Weyant et al. 1996 for an overview of the early literature). Anthoff et al. (2009) find high estimates for the social cost of carbon when explicitly including risk aversion, even with a model that incorporates relatively conservative damage estimates.

More closely related to our work, Springer (2003) suggests that diversification of mitigation activities allows for a reduction in risk exposure while taking advantage

[2] If two options in the land use sector were analyzed, these could be competing or complementary, so that costs would either decrease or increase as more of one option was chosen. This is not the topic of this particular study, but will be of interest in future research that will also consider biofuel policies.

[3] We acknowledge that this implies that we focus on the perfect information case, which will never materialize in reality. For this reason, the expected VOI derived should be interpreted as an upper bound of the VOI that can be attained by having increasingly accurate information.

of low-cost options. He uses the traditional Markowitz mean-variance approach to determine efficient international portfolios of carbon abatement options and derives information about expected returns from investing in emissions reduction from marginal abatement cost curves for CO_2.

Although the objective function used in this paper is different from the original Markowitz formulation of the expected value model, it is an alternative formulation proposed independently by Freund (1956), whereas both Markowitz's (1959) and Freund's (1956) formulations yield identical efficient frontiers (McCarl and Spreen 2007). Relating this back to the notation in Eq. (8.1), the expected value frontier for the optimal decisions across all $\omega > 0$ is identical to the one given by the Markowitz approach, with $\omega = 0$ giving the case where the decisionmaker is risk neutral. The limit case $\omega \rightarrow \infty$, on the other hand, represents the case where only variance is considered.

In Appendix 8.A we first derive analytically the optimal strategy and the expected VOI for a base case, where both mitigation options feature independently with constant costs. The expected value of perfect information (EVPI) is defined by the following equation:

$$EVPI(p, \omega) = E[C(\bar{x}(p, \omega))] + \omega Var[C(\bar{x}(p, \omega))] - pC_{12} - (1 - p)C_{21} \quad (8.2)$$

where p is the probability that the first land cover map is closer to reality, $\bar{x}$ is the optimal share of emissions abated through avoided deforestation within the mitigation portfolio, and C_{ij} is the mitigation cost either for a strategy or for option j in scenario i.

The derivations indicate that the optimal mitigation strategy is always a pure strategy in the case of perfect information, which implies that the decisionmaker never chooses a portfolio of the mitigation options. This result is independent of the level of risk aversion of the decisionmaker. Whether the first or second mitigation option is preferred thus depends on the scenario (i.e., which land cover map is a "truer" representation of reality).

In the case of imperfect information, and assuming that on average the cost of the first mitigation option is higher than the cost of the second option, there are some cases in which the decisionmaker (within a given interval of risk aversion) prefers a combination of the two mitigation options to a pure strategy. For this to be true, the probability of the scenario in which the first mitigation is cheaper must be sufficiently high. Otherwise, the decisionmaker will always prefer the option that is on average cheaper—independently of his risk aversion measure.

It can be shown that the case of perfect information is in fact a limit of the strategy in the case of imperfect information. Furthermore, the strategy is a decreasing function of the level of risk aversion (for a probability larger than the threshold). This means that the more expensive mitigation option enters the mitigation portfolio with a higher share if the decision-maker is more risk averse. In other words, the decisionmaker trades higher costs for a decrease in the variance—that is, risk.

8.3.2 Constant-Cost and Increasing-Cost Mitigation Options

The derivations for scenarios 1 and 2 and two mitigation options, where one has constant costs and the other one has increasing costs (Appendix 8.B), show that the introduction of imperfect information causes the decisionmaker to choose a mitigation strategy that is a compromise between the strategies optimal in the individual scenarios. This effect is independent of the risk attitude of the decisionmaker, so even without risk aversion, we get a mix of the two mitigation options as the optimal strategy.

Furthermore, for a given probability of the first scenario (i.e., a given land cover map is more correct than another), the optimal mitigation strategy of a risk-averse decisionmaker is the same as that of a risk-neutral investor, who perceives the probability attached to the first scenario differently. This probability is uniquely defined by the probability that the first scenario is correct and the level of risk aversion of the decisionmaker. It can be shown—independently of the level of risk aversion—that this probability is always closer to 0.5 than the probability that scenario 1 is true; that is, it is always closer to the solution where the decisionmaker is risk neutral, which is equivalent to the solution where she believes that the two land cover maps have equal probability of being correct.

Finally, the VOI is always found to be positive and it can be shown that there is a unique probability threshold below (above) which the VOI is increasing (decreasing) in the probability that the first land cover scenario is true. This implies that the decisionmaker's willingness to pay for having perfect information *ex ante* is highest at a given probability threshold, to the right of which the probability of scenario 1's being correct increases and to the left of which it decreases. That is, in both directions we move to a more informed situation, so that the marginal value of additional information decreases.

In the following section we present an empirical analysis, where we use the analytical model with two options, where one has increasing and the other one has constant costs (Appendix 8.B).

8.4 Mitigation Option Portfolio Example

Having defined the problem and the properties of the solution in a simple setting in Sect. 8.3, we now turn to an application of the second analytical model from Appendix 8.B using the GLOBIOM model to derive the function of the cost of the REDD mitigation option. The alternative mitigation option (a new technology, carbon capture and storage, in the industry and energy sector) is assumed to have constant costs. GLOBIOM is a global recursive dynamic partial equilibrium model that integrates the agricultural, bioenergy, and forestry sectors to give policy advice on global issues concerning land-use competition between the major land-based production sectors (Havlík et al. 2010). The global agricultural and forest market equilibrium is computed by choosing land-use and processing activities to maximize the sum of producer and consumer surplus subject to resource, technological, and political restrictions, as described by McCarl and Spreen (1980). Prices and international trade flows are endogenously computed for 28 world regions.

GLOBIOM enables one to estimate the marginal cost of REDD (C_{iREDD}) as the opportunity cost of activities that could take place on the deforested land, namely agriculture and biomass for bioenergy production. This cost is obtained from the dual value associated with a constraint that forces the model to respect a certain level of GHG emissions determined as a percentage of the business-as-usual emissions from deforestation. By varying the reduction level from 0 to 100 %, the entire marginal abatement cost curve can be uncovered.

The two alternative scenarios were differentiated by the underlying land cover maps; we used the GLC-2000 cropland minimum (the cropland class is covered 50 % by cropland and 50 % by a noncropland class) and the MODIS cropland maximum (the cropland class is covered 100 % by cropland). We calculated the ratio between the MODIS cropland maximum and the GLC-2000 cropland minimum area at the national level taking the GLC-2000 cropland minimum as the reference. To mimic the MODIS maximum cropland scenario, we multiplied the cropland reference area by this ratio and divided the crop yield level by the same ratio, assuming that total production of the reference year is known and valid for both scenarios. In those countries where the MODIS maximum cropland extent exceeded the GLC-2000 minimum cropland area, the additional cropland was assigned to the land category previously labeled "other natural land." This reduced the possibility of agricultural production expansion beyond forests. We consider the difference of cropland area chosen between the two land cover scenarios as relatively conservative, since we could also have modeled the difference between the MODIS cropland minimum and the GLC-2000 cropland maximum. Such scenarios would have increased the differences in cropland extent and consequently be more extreme.

We then test the sensitivity of the optimal mitigation strategy and the associated VOI to the cost of this "safe" alternative and the responsiveness to different levels of risk aversion, where we refer to a weight (ω) close to 0 as being risk neutral and then increase it to 0.002 in intervals of 0.0002.

For the latter, we fix the cost of the constant-cost mitigation option at $20 per tCO_2. Because the maximum potential from REDD between 2020 and 2030 is about 20 $GtCO_2$ with a price varying between $0 and $50 per tCO_2, the total amount to be mitigated by the combination of the constant-cost and REDD options is set equal to 20 $GtCO_2$.

In Fig. 8.2, the contribution of the REDD option to the overall mitigation contingent is shown for an increasing probability that the land cover map, for which REDD is relatively cheaper, is correct. In the risk-neutral case, where the decisionmaker minimizes expected costs irrespective of the variance (i.e., the risk aversion coefficient is equal to 0), we see that the red line rises from 13,000 million

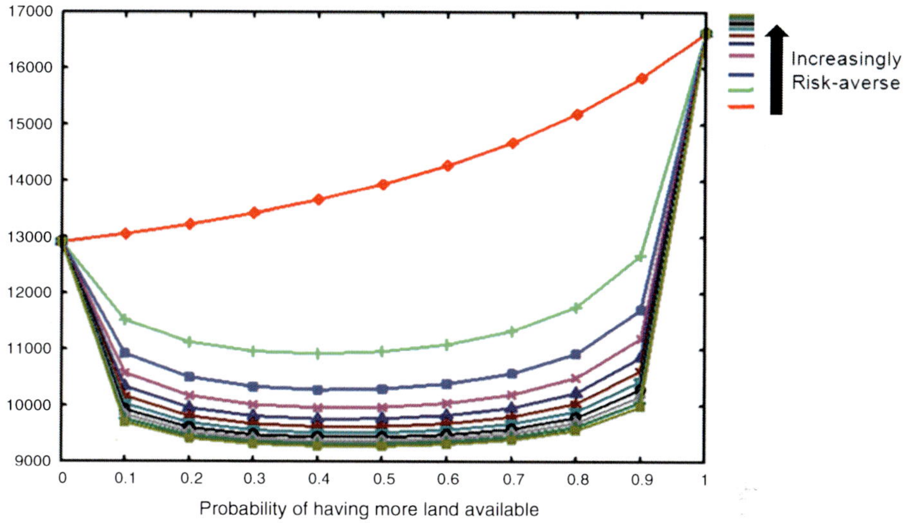

Fig. 8.2 CO_2 mitigated under REDD option for increasing probability that land cover data (scenario 1, available land) are true

tons (mt) of CO_2 to more than 16,000 mt as we become more confident that the land cover map with more additional land is correct. This implies that as the certainty that this map is correct increases, the REDD option becomes more and more attractive.[4] However, as the decisionmaker grows more risk averse (i.e., the lines in light green, blue, pink), we observe a different pattern: until a threshold of 40 % probability, the share of the REDD mitigation option actually decreases before it starts to increase. This can be interpreted in the following way: the points where the probability of having more land available is 0 and 100 % represent points with complete certainty. In the first case, the map with less available land is correct, whereas in the second case, the map for which REDD is cheaper is the true state of the world. These points thus also coincide with the risk-neutral results. At the probability threshold referred to above, the share of the REDD mitigation option is at a minimum, left of which the probability that the map with less available land is correct increases (and so the share of the constant-cost mitigation option increases at the expense of the current one). To the right of the minimum, the probability that the map with more available land is correct is higher, so the share of the REDD option is increased. Figure 8.3 shows the amount of mitigation using the second option (constant cost), which is clearly the mirror image of the first option's amount.

[4] Note that the optimal mitigation portfolio is never a pure strategy—not even in the case of risk-neutrality: it is always a mixture of both options, as explained in Sect. 8.3 and proven in the appendix.

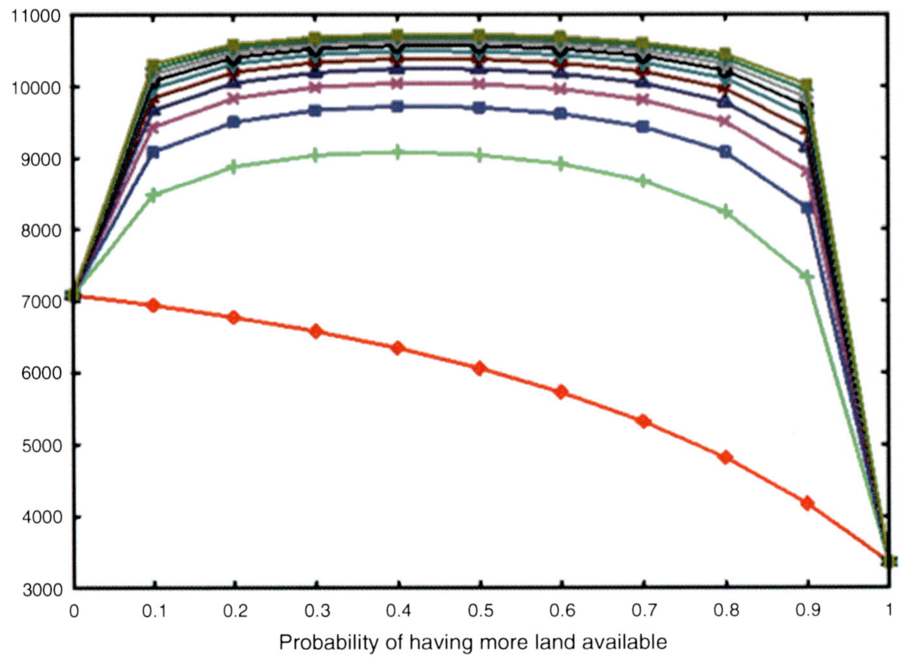

Fig. 8.3 CO_2 mitigated under constant-cost option for increasing probability that land cover data (scenario 1, available land) are true

As we move from the risk-neutral to the more and more risk-averse case, the curves change shape: they are sharply sloped at the ends, where certainty in the land cover data sets is highest, and flatter where uncertainty increases. This indicates that for increasing risk aversion, the patterns described above are reinforced, and more mitigation happens through the constant cost option. If we look at Figs. 8.4 and 8.5, which display the total expected cost of the mitigation portfolio and the variance, respectively, we can see that the decisionmaker will accept higher costs to reduce the overall risk (or the variance). These results are in line with the theoretical, general findings explained in the previous section and derived in the appendixes.

Finally, we compute the expected VOI according to the definitions presented in Sect. 8.3. In Fig. 8.6, the VOI is increasing to the left of 50 % and decreasing to the right of 50–60 % in the risk-neutral case. Only as risk aversion rises do the curves get skewed; that is, the maximum of the lines in Fig. 8.6 moves toward 70 % and then 80 %. This implies that the more risk averse you are, the more you value information, but after a certain probability threshold, you start to value additional information less because you are already relatively confident in the data. This probability threshold also increases for higher levels of risk aversion. In economic terms you see the marginal value of information switch signs at ever-higher probability levels, starting at 50 % for the risk-neutral case and ending around 80 % for higher levels of risk aversion. Once you are sure that the land cover map with more land available is correct, the VOI is 0 again.

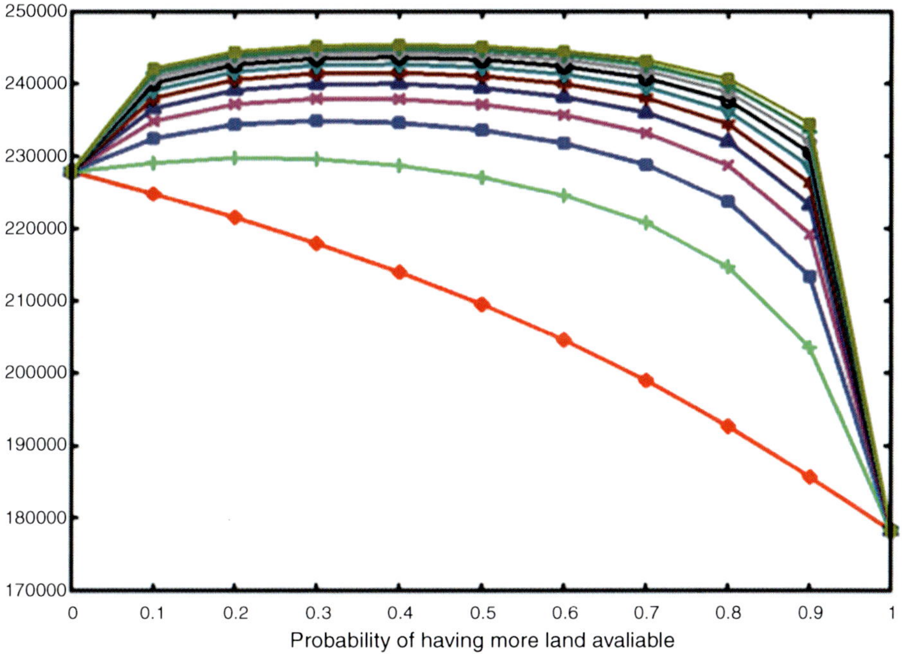

Fig. 8.4 Expected cost of mitigation portfolio (million $)

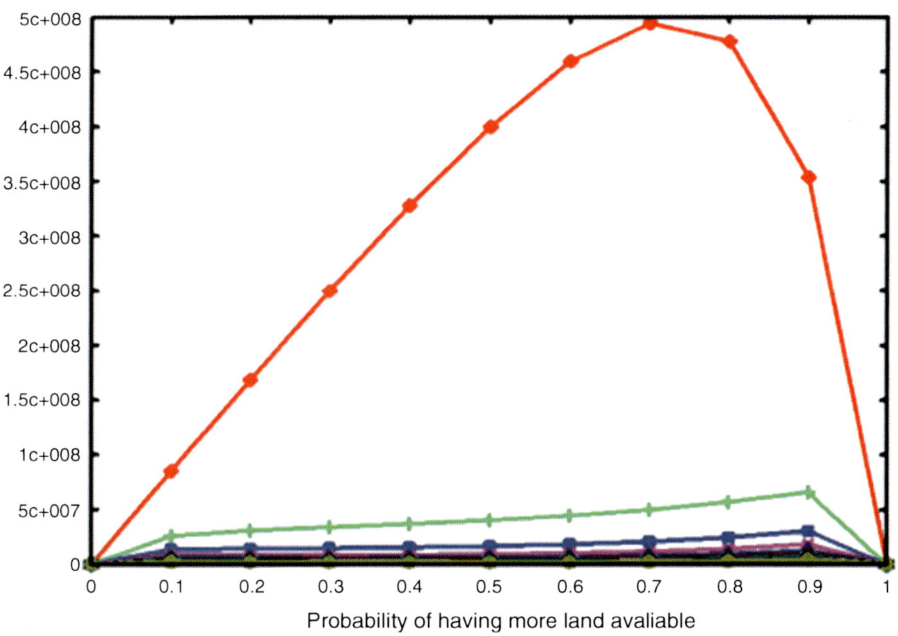

Fig. 8.5 Variance of mitigation portfolio costs

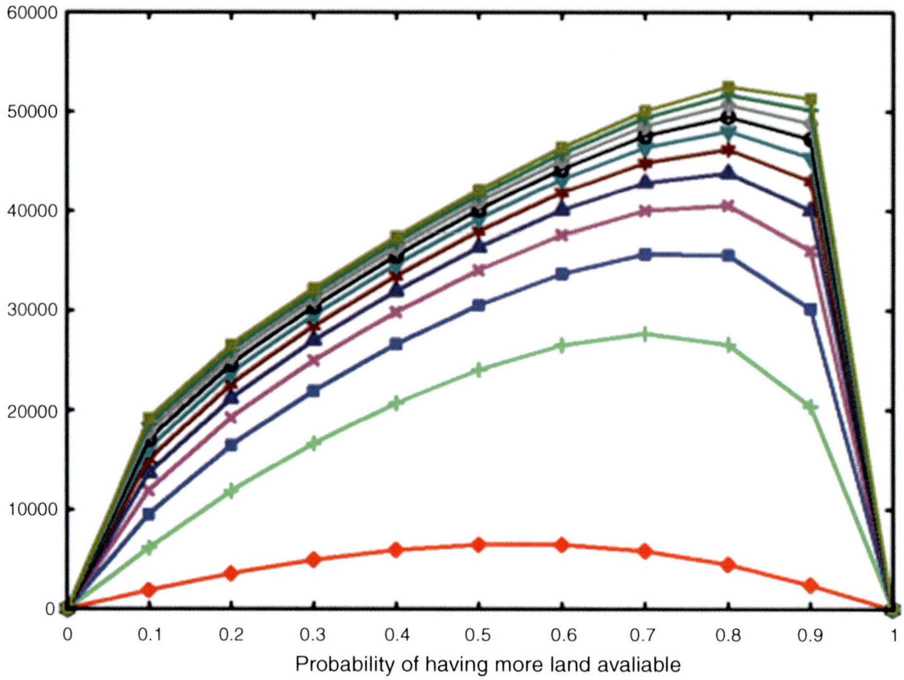

Fig. 8.6 Expected value of information (million $)

We then investigated the response of the expected VOI to changes in the assumption about the cost of the alternative mitigation option, carbon capture and storage—what we have called the constant-cost option above. In Fig. 8.7, the red line is derived for costs of this option of \$10/tCO_2, and the other lines (light green, blue, pink, etc.) represent progressively higher costs per ton of CO_2.

In this risk-neutral case, we see all the lines in Fig. 8.7 increasing, with more and more of the REDD option being adopted because the alternative (constant-cost) option is assumed to be comparatively more expensive. At \$60 per tCO_2, we observe a mitigation portfolio consisting of 100 % REDD, irrespective of the probability of having more land available being high. This is also why expected the VOI is equal to 0 in this case (see Fig. 8.8).

In Fig. 8.8, The VOI assuming the constant-cost option costs \$20 per tCO_2 is above the VOI assuming only a \$10 cost. This is because the decisionmaker regards the REDD option as less competitive if the alternative is so cheap. The VOI is then highest for the blue line, corresponding to \$30 per tCO_2, at around 50 % probability that the land cover data with more available land are correct. Beyond that, the alternative mitigation option gets less and less attractive than the REDD option, and the value of knowing with more certainty that this cover is correct decreases—that is, the pink line is underneath the blue one, followed by the dark blue and brown lines (at 0). Also, the maximum of expected VOI curves is to the left of the 50 %

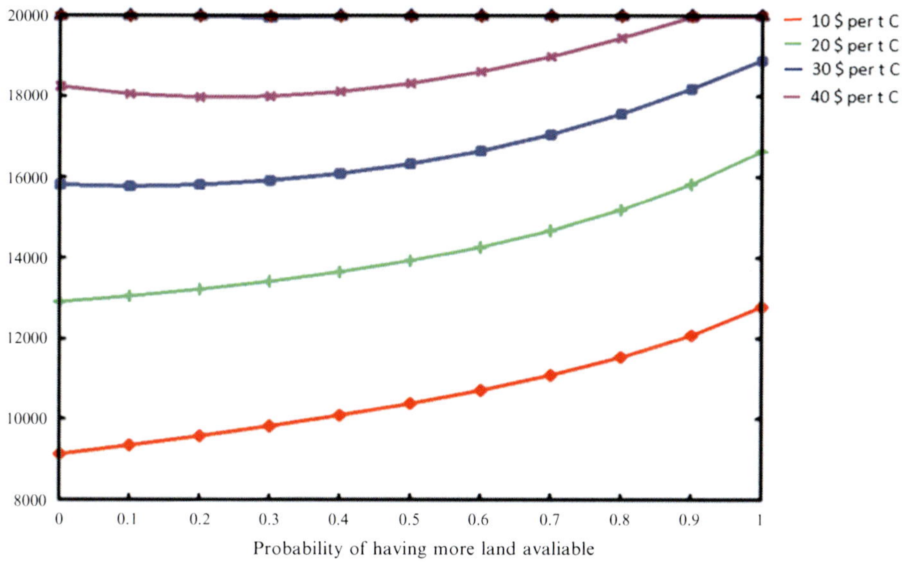

Fig. 8.7 CO_2 mitigated under REDD option for increasing probability that land cover data (scenario 1, available land) are true

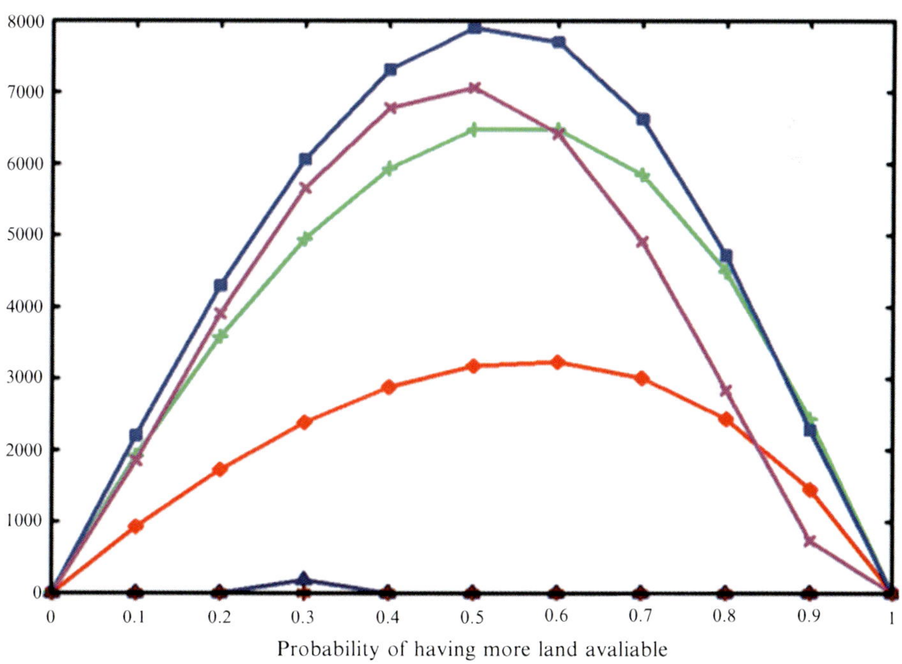

Fig. 8.8 Expected value of information (million $)

probability of having more land in the latter case (i.e. the constant-cost option is relatively more expensive) and to the right of 50 % in the case where the constant-cost option is cheap. This means that the marginal value of information decreases as the alternative option becomes more expensive, and vice versa, and the probability threshold beyond which additional information is valued at a decreasing rate gets lower and lower, too. In other words, the probability threshold required to commit more to the REDD mitigation option is not so high anymore.

8.5 Discussion and Outlook

This study has taken a very simple and straightforward approach to derive some powerful conclusions. We have used standard portfolio optimization to derive the optimal mitigation strategy, where one option to mitigate is avoided deforestation and the other option is outside the LULUCF sector, where we assume that a new technology, carbon capture and storage, becomes available at constant costs. The REDD option, however, displays increasing costs because other land uses compete with increasing land needs for REDD. In addition, the uncertainty surrounding the availability of land due to the inadequacy of existing land cover maps implies that the cost of REDD could be very different for different scenarios of land cover. In this study, we have used two land cover products, one of which shows more land available than the other, which means that the REDD option would be relatively cheaper if the first product is used. The purpose was to estimate the expected VOI and thus give us an idea about how much decisionmakers would be willing to pay to gain more certainty in the accuracy of land cover information.

An important conclusion (that can also be proven mathematically; see Appendix 8.B) is that even if the decisionmaker is risk-neutral, the existence of uncertainty leads to a portfolio of the two mitigation options and a positive expected VOI rather than a pure strategy using only the option that is on average cheaper. Only if the constant-cost option is so expensive that the REDD option is preferred under both land cover types do we find expected VOIs equal to 0 for any probability that more land is available. If we interpret the expected VOI as the willingness to pay for becoming more certain that a given land cover data set is true, then we can use Fig. 8.8 to provide the magnitudes of funds potentially involved: if the constant-cost option was \$30 per tCO_2 (the light blue line) and we were 30 % certain that the land cover map showing more land available was true, it would be worth more than \$1 billion to increase this probability to 40 %. For another 10 % improvement, we would still be willing to pay a little less than \$1 billion, and later this would decrease, since we are already relatively certain that this map is the right one. Since the "total" cost of the mitigation contingent is in the order of hundreds of billions of dollars, this represents only a small percentage. However, in absolute terms it implies huge funding potential for marginal improvements in the existing products. For example, the AFRICOVER project, undertaken for 12 countries in

Africa by FAO at a cost of several million dollars, provides a significant marginal improvement over coarser land cover maps such as GLC-2000 and improves our knowledge about land availability. The cost-benefit ratio points towards a positive return on investment.

Risk aversion increases the expected VOI even further. In addition, it also shifts the point at which the marginal change in VOI ceases to be positive. This implies that even for relatively high levels of certainty, a risk-averse decisionmaker would be willing to pay for further security, until the value of information falls to 0 in the case of complete certainty.

Future research should look more closely at the interaction between mitigation options arising within the LULUCF sector, since these options might have complementary features or compete with each other, thereby reinforcing the costs. Also, it will be of major interest to zoom into the properties of cost distributions when more options are considered and test other risk measures than just the variance, should potential losses not be normally distributed (i.e., if much could be lost in the tails of the distribution). Finally, uncertainties other than those arising from the costs surrounding the use of different land cover products should be analyzed. These include technological uncertainties, uncertainty about the correct stabilization target, and uncertainty about policy and regulation.

The example shown above has illustrated the tremendous value of information that reduces uncertainties in global land cover. It has shown that there is a high value in being able to map and quantify cropland extent accurately, in particular in Africa, where uncertainties are the highest. This demonstration involves one specific application. However, there are many other applications in which better land cover maps can help improve decisions, ranging from improved conservation planning for maximizing biodiversity on a local level to overall better land-use planning on a national level. This implies that there are co-benefits of having improved land cover information, and the VOI is probably much higher than what has been shown in this chapter.

Acknowledgments This research was conducted in the frame of the EU-funded EUROGEOSS (grant no. 226487) and CC-TAME (grant no. 212535) and GEOCARBON (grant no. 283080) projects.

8. Appendixes

8.A Independent Mitigation Options with Constant Costs

8.A.1 Problem Formulation and Assumptions

Let us denote C_{ij} as the mitigation costs for option j in scenario i $(i,j \in \{1,2\})$, representing the mitigation costs needed in the case where the observations from scenario j are correct and mitigation is carried out by option i. We will analyze the optimal mitigation strategy for both a case where the correct scenario is known beforehand and a case where this information is not available. The comparison between the two optimal mitigation strategies enables us not only to qualitatively assess the effect of uncertainty in the observations, but also to derive the value of information regarding which scenario is the correct one.

Let us assume that the mitigation cost is a linear function of the mitigation measures needed; that is, in the case where mitigation is carried out jointly by options 1 and 2 with options having shares of x and 1−x respectively ($x \in [0, 1]$), the cost in scenario i is given by

$$C_i(x) = xC_{i1} + (1 - x)C_{i2}. \tag{8.A.1}$$

Let us further assume that the observations represented by the scenarios are in principle diverse, such that neither of the mitigation options dominates the other; that is, without loss of generality, we can assume

$$C_{11} < C_{12}, \tag{8.A.2}$$

$$C_{21} > C_{22}. \tag{8.A.3}$$

Let us further assume that without loss of generality,

$$C_{11} + C_{21} - C_{12} - C_{22} > 0. \tag{8.A.4}$$

8.A.2 Model Formulation

Let us assume the optimal mitigation strategy is determined by the solution of the optimization problem

$$\min_{x \in [0,1]} E[C(x)] + \omega Var[C(x)], \tag{8.A.5}$$

where $E[.]$ and $Var[.]$ denote the expected value and variance, respectively. This formulation is a standard portfolio optimization approach, where the objective consists of the expected cost penalized by its variance. $C(x)$is the mitigation cost, which in our case is a random variable given by

$$C(x) = \begin{cases} xC_{11} + (1 - x)C_{12} & \text{with prob. } p \\ xC_{21} + (1 - x)C_{22} & \text{with prob. } 1 - p. \end{cases} \tag{8.A.6}$$

The parameter ω is the measure of risk aversion of the decisionmaker, where $\omega = 0$ models a risk-neutral and $\omega > 0$ a risk-averse behavior, with the level of risk aversion increasing with increasing ω. The probability p represents the information or belief of the decisionmaker about the reliability of individual scenarios. As already mentioned, we will investigate two cases:

- perfect information: the correct scenario is known, i.e., $p \in \{0, 1\}$
- imperfect information: there is no such information available at the decision moment, i.e., $p \in (0, 1)$

8.A.3 Solution

It is important to realize that the optimal mitigation strategy (i.e., the solution to the problem in Eq. (8.A.5)) is a function of the underlying parameters p, ω. Therefore, let us denote the optimal mitigation strategy by $\bar{x}(p, \omega)$.

Perfect Information

In the case where the correct scenario is known, there is no uncertainty concerning the mitigation costs, resulting in $Var(C(x)) = 0$.

If $p = 0$ (i.e., the second scenario is the correct one), then $E[C(x)] = xC_{21} + (1 - x)C_{22}$. Since (8.A.2) holds, the solution of (8.A.5) is attained for $x = 0$. Similarly, in the case where $p = 1$, the optimal strategy is $x = 1$.

This implies that the optimal mitigation strategy is in the case of perfect information always a pure strategy, never resulting in a portfolio of the mitigation options independently of the risk aversion of the decisionmaker. Whether the first or second mitigation option is preferred depends on the scenario: $\bar{x}(0, \omega) = 0, \bar{x}(1, \omega) = 1$ for any $\omega \geq 0$.

Imperfect Information

The solution of problem (8.A.5) is derived in Sect. 8.A.5. The most important result is summarized in the following Lemma:

Lemma 8.A.1. There exist $\hat{p} \in (0, 1)$, functions $\underline{\omega}(p), \bar{\omega}(p)$ and a function $\tilde{x}(p, \omega)$ such that

$$\underline{\omega}(p) < \bar{\omega}(p),$$

$\tilde{x}(p, \omega) \in (0, 1)$ for $p > \hat{p}$ and $\omega \in (\underline{\omega}(p), \bar{\omega}(p))$ and

$$\bar{x}(p, \omega) = \begin{cases} 0 & \text{if } p \leq \hat{p} \\ 0 & \text{if } p > \hat{p},\ \omega \geq \bar{\omega}(p) \\ \tilde{x}(p, \omega) & \text{if } p > \hat{p},\ \omega \in (\underline{\omega}(p), \bar{\omega}(p)) \\ 1 & \text{if } p > \hat{p},\ \omega \in [0, \underline{\omega}(p)] \end{cases}$$

for $p \in (0, 1)$ and any $\omega \geq 0$. The proof and analytical expressions for the probability threshold $\hat{p}$and risk aversion thresholds $\underline{\omega}(p), \bar{\omega}(p)$ together with the analytical expression for $\tilde{x}(p, \omega)$ are presented in Sect. 8.A.5.

Lemma A.1 discloses a quite natural but important conclusion. Assumption (8.A.4) states that, on average, the cost of the first mitigation option is higher than the cost of the second option. Lemma A.1 shows that, if p is high enough (i.e., the probability of the scenario where the first mitigation option is cheaper is high enough), then in some cases the decisionmaker (if his measure of risk aversion is within the given interval) prefers a combination of the two mitigation options to a pure strategy. On the other hand, if the probability threshold is not met, the investor prefers the option that is on average cheaper independently of his risk aversion.

As is proven in Sect. 8.A.5, the optimal mitigation strategy in the case of perfect information is a limit of the strategy in the case of imperfect information. In addition, the strategy is a decreasing function of ω for $p > \hat{p}$, i.e., the more risk averse the decisionmaker, the higher is the share of the second mitigation option in the optimal strategy, meaning that a risk-averse investor is willing to sacrifice some part of the expected costs for the benefit of a lower variance.

8.A.4 Value of Information

Using the notion of expected value of perfect information (EVPI) and the results derived in the previous section, we can quantify the value of the information on which scenario is the correct one.

EVPI (or often VOI) is a common term in decision theory used to quantify the maximum amount a decisionmaker would be ready to pay in return for complete (and accurate) information about the future (Birge and Louveaux 1997). The concept of EVPI was first developed in the context of decision analysis and can be found in classical references, such as Raiffa and Schlaifer (1961). The expected value of perfect information is, by definition, the difference between the value of the objective (i.e., costs) in the case where the information is unknown at the time of the decision and the expected value of the objective in the case where the information is known. In our case it can be expressed as

$$EVPI(p, \omega) = E[C(\bar{x}(p, \omega))] + \omega Var[C(\bar{x}(p, \omega))] - pC_{12} - (1 - p)C_{21} \quad (8.A.7)$$

with $E[\bar{x}(p, \omega)]$,$Var[\bar{x}(p, \omega)]$ given by Lemma 8.A.1. Since $\bar{x}(0, \omega) = 0$ and $\bar{x}(1, \omega)$ $= 1$, we have

$$C_{12} \leq \bar{x}(p, \omega)C_{11} + (1 - \bar{x}(p, \omega))C_{12} \quad (8.A.8)$$

and

$$C_{21} \leq \bar{x}(p, \omega)C_{21} + (1 - \bar{x}(p, \omega))C_{22}. \quad (8.A.9)$$

Since $Var[C(\bar{x}(p,\omega))] > 0$ for $p \in (0,1)$, the sum of p multiple of (8.A.8) and $(1-p)$ multiple of (8.A.9) yields

$$EVPI(p,\omega)>0 \tag{8.A.10}$$

8.A.5 Formal Proofs

To derive the results presented in Sect. 8.3, we first need to prove some preliminary lemmas:

Lemma 8.A.1.

$$E[C(x)] = p(xC_{11} + (1-x)C_{12}) + (1-p)(xC_{21} + (1-x)C_{22})$$

$$Var(C(x)) = p(1-p)[x(C_{11} + C_{21}) + (1-x)(C_{12} + C_{22})]^2$$

Proof: The first expression follows directly from the definition of the mean and (8.A). The second expression is obtained after some rearranging of terms by substituting (8.A) into the definition of variance, $Var(C(x)) = E[C(x)^2] - E[C(x)]^2$.

Lemma 8.A.2. The minimum of $Var(C(x))$over $x \in [0,1]$ is attained in $x = 0$ for any $p \in (0,1)$.

Proof: After rearranging the expression for $Var(C(x))$ derived in Lemma 8.A.1, we obtain the following:

$$Var(C(x)) = p(1-p)(x(C_{11} + C_{12} - C_{12} - C_{22}) + C_{12} + C_{22})^2,$$

$Var(C(x))$ is a quadratic function which is due to (8.A) increasing on $x \in [0,1]$; thus its minimum on the interval is attained in $x = 0$ independently of p.

Lemma 8.A.3. The minimum of $E[C(x)]$ over $x \in [0,1]$ is attained in

$$\hat{x}(p) = \begin{cases} 0 & \text{if } p \in [0,\hat{p}) \\ x,\ x \in [0,1] & \text{if } p = \hat{p} \\ 1 & \text{if } p \in (\hat{p},1]. \end{cases}$$

where

$$\hat{p} = \frac{C_{21} - C_{22}}{C_{21} - C_{22} + C_{12} - C_{11}}$$

Proof: Since $E[C(x)]$ is a linear function in x, its minimum on a compact interval is attained on its border, except for the case where $E[C(x)]$ is constant. Rearranging of terms in the expression for $E[C(x)]$ from Lemma 8.A.1 yields that this is the case if and only if $p = \hat{p}$. For $p < \hat{p}$, $E[C(x)]$ is decreasing in x by (A), (A), hence the minimum is attained in $x = 0$. $E[C(x)]$ is increasing for $p > \hat{p}$, thus its minimum over $x \in [0, 1]$ is attained in $x = 1$.

These results enable us to derive the solution to problem (A) for the case of imperfect information—that is, $p \in (0, 1)$.

Lemma 8.A.3 in combination with Lemma 8.A.2 yields that if $p < \hat{p}$, then the minimum of both $E[C(x)]$ and $Var(C(x))$ over $x \in [0, 1]$ is attained in $x = 0$. Since $E[C(x)]$ is independent of x if $p = \hat{p}$, the solution of (8.A) is attained in $x$minimizing $Var(C(x))$. Hence the solution of (8.A.5) is

$$\bar{x}(p, \omega) = 0$$

for $0 < p \leq \hat{p}$ for any $\omega \geq 0$.

In the following, we will assume $p > \hat{p}$, $p \in (0, 1)$. The objective of (8.A) is a quadratic function with a global minimum $\tilde{x}(p, \omega)$ which attained in x satisfying the first-order condition, which is a linear equation. After some rearranging, the first-order condition yields

$$wKx = L - wkK$$

with

$$K = 2p(1 - p)(C_{11} + C_{21} - C_{12} - C_{22})^2$$

$$k = \frac{C_{12} + C_{22}}{C_{11} + C_{21} - (C_{12} + C_{22})}$$

$$L = p(C_{21} - C_{22} + C_{12} - C_{11}) - (C_{21} - C_{22})$$

It should be noted that $K, k, L > 0$ by (8.A.2), (8.A.3), and (8.A.4) and $p > \hat{p}, p \in (0, 1)$. Thus the global minimum of the objective of (8.A.5) is attained in

$$\tilde{x}(p, \omega) = \frac{L}{Kw} - k.$$

It should be noted that $\tilde{x}(p, \omega) \in (0, 1)$ if and only if

$$\omega \in (\underline{\omega}(p), \bar{\omega}(p))$$

where

$$\underline{\omega}(p) = \frac{L}{(k + 1)K} < \frac{L}{kK} = \bar{\omega}(p).$$

This observation is crucial, since $\bar{x}(p,\omega)$ is equal to $\tilde{x}(p,\omega)$ if and only if $\tilde{x}(p,\omega) \in [0,1]$. If $\tilde{x}(p,\omega) \geq 1$, the objective of (3.5) is decreasing on $[0,1]$ and thus $\bar{x}(p,\omega) = 1$. On the other hand, the objective of (3.5) is increasing if $\tilde{x}(p,\omega) \leq 0$, hence $\bar{x}(p,\omega) = 0$. Therefore, we see that the set of (p,ω) on which $\tilde{x}(p,\omega) \in (0,1)$ is the same as the set of (p,ω) for which the optimal mitigation strategy is a portfolio of mitigation options. The derived results can be summarized as follows:

$$\bar{x}(p,\omega) = \begin{cases} 0 & \text{if } p \leq \hat{p} \\ 0 & \text{if } p > \hat{p},\ \omega \geq \bar{\omega}(p) \\ \tilde{x}(p,\omega) & \text{if } p > \hat{p},\ \omega \in (\underline{\omega}(p), \bar{\omega}(p)) \\ 1 & \text{if } p > \hat{p},\ \omega \in [0, \underline{\omega}(p)]. \end{cases}$$

The analytic expressions for both the probability and risk-aversion measure thresholds and the optimal mitigation strategy enable us to study their properties.

First of all, it is important to realize that for a given probability level p, the optimal mitigation strategy $\bar{x}(p,\omega)$ is a continuous function of ω. Second, comparing back to the results derived in Sect. 8.A.3 for the perfect information case, we see that

$$\bar{x}(0,\omega) = \lim_{p \to 0} \bar{x}(p,\omega)$$

and since $\lim_{p \to 1} \underline{w}(p) = +\infty$ also

$$\bar{x}(1,\omega) = \lim_{p \to 1} \bar{x}(p,\omega)$$

for any $\omega \geqslant 0$. In other words, the perfect information case is a limiting case of the case with imperfect information.

In addition, using the analytical expression for the global minimum $\tilde{x}(p,\omega)$, it can be easily shown that the optimal mitigation strategy $\bar{x}(p,\omega)$ is a decreasing function of ω.

8.B *Increasing-Cost LULUCF Mitigation Option*

8.B.1 Assumptions

Let us denote $C_i(x_1,x_2) : \mathrm{R}^2 \to \mathrm{R}_0^+$, $C_i \in C^2$ the mitigation cost function depending on the scenario i ($i \in \{1,2\}$) representing the mitigation cost depending on the extent of mitigation x_j performed by option j ($j \in \{1,2\}$). We assume that the mitigation cost function is scaled such that the total mitigation needed in both scenarios is equal to 1 and that no mitigation action results in 0 costs; that is, $C_i(0,0) = 0$.

We will analyze the optimal mitigation strategy for both a case where the correct scenario is known beforehand and a case where this information is not available.

The comparison between the two optimal mitigation strategies enables us not only to qualitatively assess the effect of uncertainty in the observations, but also to derive the value of information on which scenario is the correct one. First let us formulate the assumptions on the mitigation cost that are necessary for the analysis:

$$\frac{\partial C_i}{\partial x_1} > 0, \quad i \in \{1, 2\} \tag{8.B.1}$$

$$\frac{\partial C_i}{\partial x_2} = A > 0, \quad i \in \{1, 2\} \tag{8.B.2}$$

$$\frac{\partial^2 C_i}{\partial x_1^2} > 0, \quad i \in \{1, 2\} \tag{8.B.3}$$

$$\frac{\partial C_i}{\partial x_1}(0, 1) \leq A, \quad i \in \{1, 2\} \tag{8.B.4}$$

$$\frac{\partial C_i}{\partial x_1}(1, 0) \geq A, \quad i \in \{1, 2\} \tag{8.B.5}$$

$$\frac{\partial^2 C_1}{\partial x_1^2}(x) > \frac{\partial^2 C_2}{\partial x_1^2}(x) \text{ for } x \in [0, 1] \tag{8.B.6}$$

The first three assumptions are mathematical representations of the following situation: The marginal costs of the second mitigation option are constant and independent of the scenario (since the option is part of the analyzed industry) and the cost of the first function is assumed to be increasing and convex.

The fourth and fifth assumptions form necessary conditions, so there does not exist a dominant mitigation option; that is, the optimal choice of the investor is a combination of the two options. The last assumption states a relationship between the options, implying that the cost in the case of the second scenario is rising less steeply than in the first one.

Let us further introduce some simplifying notation and basic properties of the functions considered. Let $K_i(x) = C_i(x, 1 - x)$, i.e., $K_i(x)$ is a function of one variable only and denotes the mitigation cost as a function of the mitigation done in the first mitigation option, assuming the total mitigation is such that the mitigation target is met. $K_i(x)$ is an increasing (from (8.B.1) and (8.B.2)), convex (8.B.3) function of x. The first-order condition for minimization of $K_i(x)$ can be formulated in terms of function C_i as

$$\frac{\partial C_i}{\partial x_1}(x, 1 - x) = A \tag{8.B.7}$$

(8.B.1), (8.B.3), (8.B.4), and (8.B.5) imply that there exists a unique solution of (8.B.7), which we denote $\hat{x}_G^i$. Moreover, (8.B.3) ensures that the global minimum of $K_i(x)$ is attained in $\hat{x}_G^i$. (8.B.6) in turn implies that

$$\hat{x}_G^1 < \hat{x}_G^2 \tag{8.B.8}$$

8.B.2 Model Formulation

Similar to the baseline case, we are interested in finding the optimal mitigation strategy, which will be the solution of the same optimization problem as in the baseline case, which can be equivalently formulated in terms of functions $K_i(x)$ as

$$\min_{x \in [0,1]} E[K(x)] + \omega Var[K(x)], \tag{8.B.9}$$

where $E[.]$ and $Var[.]$ denote the expected value and variance, respectively. This formulation is a standard portfolio optimization approach, where the objective consists of the expected cost penalized by its variance. $K(x)$is the mitigation cost, which in our case is a random variable defined as

$$K(x) = \begin{cases} K_1(x) & \text{with prob. } p \\ K_2(x) & \text{with prob. } 1 - p. \end{cases} \tag{8.B.10}$$

The parameter ω is the measure of risk aversion of the decisionmaker, $\omega = 0$ modeling a risk-neutral and $\omega > 0$ a risk-averse behavior, with the level of risk aversion increasing with increasing ω. The probability p represents the information or belief of the decisionmaker about the reliability of individual scenarios. We analyze the following four cases:

I. *Perfect information.* The information on which scenario is correct is available prior to the decision point; that is, $p \in \{0, 1\}$.

II. *Imperfect information.* Such information is not available; that is, $p \in (0, 1)$. The three subcases represent different levels of risk aversion of the decisionmaker.

   (a) Risk neutrality. The decisionmaker does not care about the risk associated with the decision and is concerned only about the expected mitigation costs; that is, $\omega = 0$.
   (b) Absolute risk aversion. The decisionmaker cares only about the risk measured by the variance and neglects the expected cost.
   (c) Risk aversion. The decisionmaker prefers mitigation strategies leading to lower expected costs and a lower variance at the same time. The preference over them is measured by the risk aversion coefficient $\omega > 0$ present in

the target function. The risk-neutral and absolute risk aversion case are the limits of this case for $\omega \to 0$ and $\omega \to \infty$, respectively.

8.B.3 Solution

Perfect Information

Let us denote the optimal strategy for the cases analyzed here; that is, $p = 0, p = 1$ as $\hat{x}^2$, $\hat{x}^1$, respectively. In the case where the correct scenario is known, there is no uncertainty concerning the mitigation costs, resulting in $Var(K_p(x)) = 0$. If $p = 0$ (i.e., the second scenario is the correct one and $E[K(x)] = K_2(x)$), then if $p = 1$, $E[K(x)] = K_1(x)$. Thus the optimal mitigation strategies in the perfect information case are strategies in which the minimum of $K_i(x)$ is attained for$x \in [0, 1]$. We have already shown that $K_i(x)$ have global minima, which are attained in $[0, 1]$. Therefore

$$\hat{x}^1 = \hat{x}^1_G \tag{8.B.11}$$

$$\hat{x}^2 = \hat{x}^2_G \tag{8.B.12}$$

where $\hat{x}^i_G$ is the unique solution of (8.B.7), $i \in \{1, 2\}$ and from (8.B.8)

$$0 \le \hat{x}^1 < \hat{x}^2 \le 1 \tag{8.B.13}$$

Imperfect Information

Risk-Neutral Case

In this case we analyze a situation where $\omega = 0$. That is, the problem (8.B.9) is equivalent to

$$\min_{x \in [0,1]} pK_1(x) + (1 - p)K_2(x) \tag{8.B.14}$$

Let us denote the optimal strategy $\hat{x}^p$ as a function of p for which the minimum of (8.B.14) is attained. It can be proven that

$$\hat{x}^p \in (\hat{x}^1, \hat{x}^2) \tag{8.B.15}$$

and in addition

$$\frac{\partial \hat{x}^p}{\partial p} < 0 \tag{8.B.16}$$

where

$$\lim_{p \to 1} \hat{x}^p = \hat{x}^1 \tag{8.B.17}$$

$$\lim_{p \to 0} \hat{x}^p = \hat{x}^2 \tag{8.B.18}$$

Results (8.B.15) through (8.B.18) express that if the information about which scenario will turn out to be true in the future is not available at the decision moment, a risk-neutral decisionmaker will always prefer a strategy that lies in between the strategies that are optimal for each scenario if the information is available. They further show that the share of the first mitigation option in the mitigation strategy is decreasing with increasing probability, converging to the optimal strategy for the first scenario for $p \to 1$, and to the strategy optimal for the second scenario for $p \to 0$. The proof of results (8.B.15) through (8.B.18) is presented in Sect. 8.B.6

Absolute Risk Aversion

By definition of variance we have

$$Var(K(x)) = p(1-p)(K_1(x) + K_2(x))^2 \tag{8.B.19}$$

which means that for absolute risk aversion, the problem (8.B.9) can be formulated as

$$\min_{x \in [0,1]} p(1-p)(K_1(x) + K_2(x))^2 \tag{8.B.20}$$

In Sect. 8.B.6 we prove that the global minimum of $p(1-p)(K_1(x) + K_2(x))^2$ is attained on $[0, 1]$ in $\hat{x}^{0,5}$.

This discloses an interesting implication about the behavior of an absolutely risk-averse decisionmaker. We see that the optimal mitigation strategy is the same as in the case of a risk-neutral investor who believes that each scenario is equally probable.

Risk Aversion

As in the baseline case, let us denote the

$$\bar{x}(p, \omega) = \arg\min_{x \in [0,1]} E[K_p(x)] + \omega Var[K_p(x)], \tag{8.B.21}$$

$\omega > 0$, $p \in (0, 1)$. In Sect. 8.B.6 we prove that

$$\bar{x}(p, \omega) \in (\hat{x}^p, \hat{x}^{0.5}) \text{ if } p > 0.5 \tag{8.B.22}$$

and

$$\bar{x}(p,\omega) \in (\hat{x}^{0.5}, \hat{x}^{p}) \text{ if } p>0.5 \qquad (8.B.23)$$

This implies, in comparison with the risk-neutral case, that the decisionmaker prefers a more balanced mitigation strategy, choosing a strategy that is always closer to $p = 0.5$. More importantly, we saw in (8.B.15) through (8.B.18) that $\hat{x}^p$ is a continuous decreasing function mapping $[0, 1]$ on $[\hat{x}^1, \hat{x}^2]$. This means that for any $p \in (0, 1)$ there exists a unique $q \in (0, 1)$ such that

$$\bar{x}(p,\omega) = \hat{x}^{q} \qquad (8.B.24)$$

where $q \in (p, 0.5)$ for $p \in (0, 0.5)$ and opposite otherwise. In other words, the solution of the risk-averse case is equal to the solution of the risk-neutral case when the probability is equal to q. This shows that, in reality, the risk-averse decision-maker behaves in the same way as a risk-neutral investor, but in fact attaches a different probability to the scenarios, which is always closer to 0.5.

8.B.4 Value of Information

As in the baseline, we measure the value of information by EVPI. (8.B.24) implies that in a further analysis of the results it is sufficient to consider only a risk-neutral decisionmaker. Therefore, in this case EVPI is a function of probability only and can be expressed as

$$EVPI(p) = pK_1(\hat{x}^p) + (1-p)K_2(\hat{x}^p) - pK_1(\hat{x}^1) - (1-p)K_2(\hat{x}^2) \qquad (8.B.25)$$

As in the baseline case, we can show that

$$EVPI(p)>0 \qquad (8.B.26)$$

As we prove in Sect. 8.B.6.3, there exists $\hat{p} \in (0, 1)$ such that

$$EVPI(\hat{p})>EVPI(p) \text{ for any } p \in (0,1), p \neq \hat{p} \qquad (8.B.27)$$

Moreover, $EVPI(p)$ is an increasing function of p for $p \in (0, \hat{p})$ and decreasing for $p \in (\hat{p}, 1)$.

8.B.6 Formal Proofs

Risk-Neutral Case

First let us prove the following lemma:

Lemma 8.B.1. Let $q \in (0,1)$ and $F_i : R \to R, F_i \in C^2, i \in \{1,2\}$ satisfy

- $$\frac{\partial^2 F_1}{\partial x^2} > \frac{\partial^2 F_2}{\partial x^2} > 0$$
- The global minimum of $F_i(x)$ is attained in $x_i \in [0,1]$ with $x_1 < x_2$

  Then $F_q : R \to R$ defined by $F_q(x) = \alpha(qF_1(x) + (1-q)F_2(x))$ satisfies

$$\frac{\partial^2 F_1}{\partial x^2} > \frac{\partial^2 F}{\partial x^2} > \frac{\partial^2 F_2}{\partial x^2} > 0$$

and the global minimum of F_q is attained in $x \in (x_1, x_2)$. Moreover, for any $x \in (x_1, x_2)$ there exists $q \in (0,1)$ such that the global minimum of F_q is attained in x.

Proof: Since $F_i \in C^2$, also $F_q \in C^2$ and from the definition of F_q and $\frac{\partial^2 F_1}{\partial x^2} > \frac{\partial^2 F_2}{\partial x^2} > 0$ we have

$$\frac{\partial^2 F_1}{\partial x^2} > \frac{\partial^2 F}{\partial x^2} > \frac{\partial^2 F_2}{\partial x^2} > 0 \tag{8.B.28}$$

Hence if a global minimum of $F_q(x)$ exists, it is attained in x solving the first-order condition

$$0 = \frac{\partial F_q}{\partial x}(x) = q\frac{\partial F_1}{\partial x}(x) + (1-q)\frac{\partial F_2}{\partial x}(x) \tag{8.B.29}$$

x_1 and x_2 are the global minima of F_1 and F_2, respectively. Since $x_1 < x_2$ and $\frac{\partial^2 F_1}{\partial x^2} > \frac{\partial^2 F_2}{\partial x^2} > 0$, we have $\frac{\partial F_q}{\partial x}(x_1) < 0$ and $\frac{\partial F_q}{\partial x}(x_2) > 0$. Since $\frac{\partial F_q}{\partial x} \in C^1$, there exists a unique $x \in (x_1, x_2)$ such that $0 = \frac{\partial F_q}{\partial x}(x)$, which is thus the global minimum of $F_q(x)$. Furthermore, since $\frac{\partial^2 F_1}{\partial x^2} > \frac{\partial^2 F_2}{\partial x^2} > 0$, the implicit function theorem ensures, that (8.B.29) defines a unique smooth function $x(q)$ on $q \in [0,1]$ and thus $x_1 = \lim_{q\to 1} x(q)$ and $x_2 = \lim_{q\to 0} x(q)$, which in turn implies that for any $x \in (x_1, x_2)$ there exists $q \in (0,1)$ such that the global minimum of F_q is attained in x.

In the following let us denote $K_p(x) = E[K(x)] = pK_1(x) + (1-p)K_2(x)$. (8.B.15) is implied directly by Lemma B.1 for $q = p$ and $F_i(x) = K_i(x)$, (8.B.17) and (8.B.18) by the proof of Lemma B.1 (Note that the conditions of Lemma 8.B.1 are satisfied because of (8.B.6) and (8.B.8)). The implicit function theorem, as applied in Proof of Lemma 8.B.1, ensures that $\partial \hat{x}^p$ is a smooth function of p, which implies that $\frac{\partial \hat{x}^p}{\partial p}$ exists. For any $0 < p_2 < p_1 < 1$ we have, after some rearranging,

$$K_{p_1}(x) = qK_1(x) + (1-q)K_{p_2}(x) \tag{8.B.30}$$

for $q = \frac{p_1 - p_2}{1 - p_2}$, where indeed $q \in (0,1)$. Thus, by Lemma 8.B.1 for $q, F_1 = K_1$, $F_2 = K_{p_2}$ we have $\hat{x}^{p_1} \in (\hat{x}^1, \hat{x}^{p_2})$, which proves (8.B.16).

Risk-Averse Case

First let us realize that since $C_i(0,0) = 0$, (8.B.1) and (8.B.2) imply that $K_i(x) > 0$. If the global minimum of $L_p(x) = E[K(x)] + \omega Var[K(x)]$ exists and is attained in $\bar{x}(p,\omega)$, then $x = \bar{x}(p,\omega)$ must solve the first-order condition, which is equivalent to

$$0 = \frac{\partial L_p}{\partial x}(x) = \frac{\partial K_p}{\partial x}(x) + 4\omega p(1-p)(K_1(x) + K_2(x))\frac{\partial K_{0.5}}{\partial x}(x) \tag{8.B.31}$$

Note that $L \in C^2$ and $K_i(x) > 0$. For $p < 0.5$ we have $\hat{x}^p > \hat{x}^{0.5}$ and thus $0 < \frac{\partial L_p}{\partial x}(\hat{x}^p)$ and $0 > \frac{\partial L_p}{\partial x}(\hat{x}^{0.5})$. Hence there exists $\bar{x}(p,w)$ such that $0 = \frac{\partial L_p}{\partial x}(x)$ for $x = \bar{x}(p,\omega)$ where $\bar{x}(p,w) \in (\hat{x}^{0.5}, \hat{x}^p)$. Similarly for $p > 0.5$.

Value of Information

After rearranging terms and utilizing the first-order condition for $\hat{x}^p$, we obtain

$$\frac{\partial EVPI}{\partial p}(p) = K_1(\hat{x}^p) - K_1(\hat{x}^1) - K_2(\hat{x}^p) + K_2(\hat{x}^2) \tag{8.B.32}$$

which is a continuous function of p. From (8.B.15), (8.B.17), and (8.B.18) we obtain

$$\lim_{p\to 0}\frac{\partial EVPI}{\partial p}(p) < 0 \tag{8.B.33}$$

$$\lim_{p\to 0}\frac{\partial EVPI}{\partial p}(p) < 0 \tag{8.B.34}$$

Moreover, because of (8.B.6), after some rearranging, we obtain

$$\frac{\partial^2 EVPI}{\partial p^2} > 0 \tag{8.B.35}$$

Thus $\frac{\partial EVPI}{\partial p}(p)$ is an increasing continuous function of p, hence there exists $\hat{p} \in (0,1)$ such that $\frac{\partial EVPI}{\partial p}(\hat{p}) = 0$. Furthermore, (8.B.35) ensures that the global maximum of $EVPI(p)$ is attained in $\hat{p}$.

8. Commentary: The Uncertain Value of Reducing Uncertainty

Scott Farrow

Fritz et al. have an ambitious research agenda to determine the value of information about the extent of land cover in determining optimal mitigation choices for global climate change. The mitigation alternatives they investigate are an assumed industrial carbon capture technology and an alternative that reduces deforestation. They determine the optimal mix, including corner solutions, of the alternatives when there is uncertainty about the amount of land cover and the decisionmaker may exhibit varying degrees of risk aversion. Uncertainty about the amount of land cover determines the opportunity cost of the deforestation alternative and hence the cost of any mitigation policy.

The chapter is ambitious both because of its empirical modeling and in the development of a theoretical structure. The body of the text focuses on a general description of the models used to estimate the value of information and the quantitative results for various elements, such as mitigation costs and the value of information based on an economic land-use and impact model, GLOBIOM. The appendixes contain the mathematical development of two VOI models, one based on different but constant costs across the two mitigation options, and the other based on one increasing-cost option (reducing deforestation) while the carbon capture technology is assumed to be constant cost.

The careful development of the background mathematical model and its potential link to the empirical work is to be praised. However, improving the explanation of the linkages between the theory, specific equations, and the empirical model would be a substantial help to the reader. The authors state that the empirical results are based on the increasing-cost model, but some of the issues that could receive more attention are also apparent in the constant-cost model, with which I begin.

8.C.1 Core of Models Presented

The heart of the analysis is the expected value of perfect information. In the two constant-cost case, the authors in Eq. (8.A.7) define the value of information as

$$EVPI(p,\omega) = E[C(\bar{x}(p,\omega))] + \omega Var[C(\bar{x}(p,\omega))] - pC_{12} - (1-p)C_{21} \quad (8.C.1)$$

S. Farrow (✉)
Department of Economics, University of Maryland–Baltimore County, Baltimore, MD, USA

The Woods Hole Oceanographic Institution, Falmouth, MA, USA
e-mail: farrow@umbc.edu

Where (see their text for more detail)

- EVPI is the expected value of perfect information;
- p is belief, expressed as a probability, about the reliability of two measures of land cover;
- ω is a measure of risk aversion, ≥ 0;
- E and Var are expected value and variance, respectively;
- $\bar{x}$ is the optimal share of deforestation in the mitigation strategy; and
- C, C_{ij} are the mitigation costs either for a strategy or for option j with strategy i.

The first two terms on the right-hand side are the objective function when the decisionmaker chooses an optimal strategy under uncertainty; the last two terms constitute the expected value when the information is known at the time of decision.

Although the authors characterize the objective function as a portfolio optimization approach, in that the variance as well as the expected value linearly affect the objective, no empirical evidence or interpretation is provided to interpret ω other than 0 represents risk neutrality and there is an upper bound of infinity. Its units are the change in the objective function per unit change in the variance of costs in this problem. Is such a number very small? Or not? This is most obviously important in the two constant-cost models, where a mixed strategy requires various combinations of probability and risk aversion such that, as the authors present,

$$\bar{x}(p,\omega) = \begin{cases} 0 & \text{if } p \leqslant \hat{p} \\ 0 & \text{if } p > \hat{p},\ \omega \geqslant \bar{\omega}(p) \\ \tilde{x}(p,\omega) & \text{if } p > \hat{p},\ \omega \in (\underline{\omega}(p), \bar{\omega}(p)) \\ 1 & \text{if } p > \hat{p},\ \omega \in [0, \underline{\omega}(p)]. \end{cases} \tag{8.C.2}$$

Consequently, the scale of ω is important conceptually and empirically. Further, the authors derive the bounds in which ω leads to a mixed strategy as Eq. (8.C.3), where L, k, and K are functions of probability and cost:

$$\underline{\omega}(p) = \frac{L}{(k+1)K} < \frac{L}{kK} = \bar{\omega}(p) \tag{8.C.3}$$

Although generated by the later increasing-cost model, the authors report substantial increases in the value of information when there is risk aversion, as in Fig. 8.C.1 below; although the effect is presumably due to suddenly incorporating some weight on a large variance. However, we don't know the scale of ω and its plausibility in practice.

Moving on to the structure of the increasing-cost model based on avoided deforestation, the authors define an augmented expected value of information measure based on optimizing the linear mean and variance objective function. They find that the presence of risk aversion leads the decisionmaker to adjust risk-neutral probabilities toward a probability of one-half. However, the role of risk aversion is somewhat hidden by their focus on this probability adjustment.

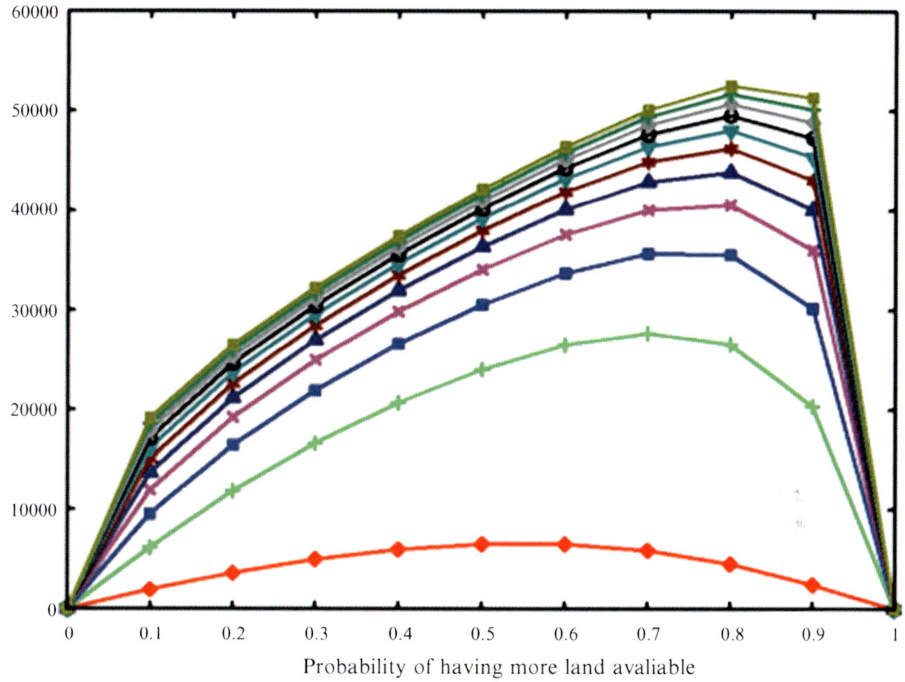

Fig. 8.C.1 Expected value of information (million $)

They find in Eq. (8.B.24) and related text that the optimal share of avoided deforestation is

$$\bar{x}(p, \omega) = \hat{x}^q \quad q \in (p, 0.5) \quad p \in (0, 0.5) \tag{8.C.4}$$

However, the degree of adjustment based on q and its link to the scale of ω are not made explicit. Consequently it becomes difficult to accept their empirical assertions about the value of information.

8.C.2 Alternative Framings and Extensions

The linear in mean and variance objective function plays an important role in the authors' results. Modelers of choice under uncertainty often specify a risk-averse utility function in income or wealth with various properties, often constant relative risk aversion. Such an approach seems an alternative here. Further, decisionmakers may be interested in some version of a marginal expected value of information or of partial information, since it is unlikely that space or any other technology will entirely resolve the uncertainty in land use as noted by the authors. It is more likely that there is a statistical distribution for land cover, and information may somewhat change that distribution. The "true" or "false" reductionist approach is useful here to get at ideas, but it is unlikely to represent the actual decisionmaking situation.

Finally, it may be useful to get some perspective by looking at the broader problem of uncertainty associated with making mitigation decisions for global climate change. There are clearly many sources of uncertainty in the model presented by the authors, such as the global economic model for the cost of the alternatives, including crop and other prices and behavior, the technology and costs of carbon removal, the statistical distribution of land cover, and the overall fit of the model. An approach that investigated uncertainty in the choice of mitigation options might involve determining the sensitivity of the model to a whole suite of uncertain variables and assessing the value of information of each from which to investigate a portfolio of research topics. The current approach, even if correct, does not give us any insight into whether the accuracy of the land-use cover maps is more valuable than partially resolving other aspects of uncertainty that are involved.

Ultimately, the authors are to be commended for their development of a conceptual model and for linking it to a large empirical model. Such linkages often require substantial simplification and explanation, which in the version they present makes their empirical results tantalizing but as yet highly uncertain to this reader.

References

Anthoff, D., Tol, R. S. J., & Yohe, G. W. (2009). Risk aversion, time preference, and the social cost of carbon. *Environmental Research Letters, 4*(2), 024002 (7 pp).

Binswanger, H. E. (1980). Attitudes toward risk: Experimental measurement in rural India. *American Journal of Agricultural Economics, 62*, 395407.

Birge, J. R., & Louveaux, F. (1997). *Introduction to stochastic programming*. New York: Springer.

Bouma, J. A., van der Woerd, H. J., & Kuik, O. J. (2009). Assessing the value of information for water quality management in the North Sea. *Journal of Environmental Management, 90*(2), 1280–1288.

Center for Science and Technology. (2007). *Weather and climate forecast use and value bibliography*. University of Colorado. http://sciencepolicy.colorado.edu/. Accessed 20 Aug 2007.

Defourny, P., Schouten, L., Bartalev, S., Cacetta, P., De Witt, A., Di Bella, C., Gerard, B., Heinimann, A., Herold, M., Jaffrain, G., Latifovic, R., Lin, H., Mayaux, P., Mucher, S., Nonguierma, A., Stibig, H-J., Bicheron, P., Brockmann, C., Bontemps, S., Van Bogaert, E., Vancutsem, C., Leroy, M., & Arino, O. (2009, May 4–8). Accuracy assessment of a 300 m global land cover map: the GlobCover experience. In: *Proceedings of 33rd international symposium on remote sensing of environment (ISRSE)*, Stresa, Italy.

Dillon, J. L. (1971). An expository review of Bernouillian decision theory in agriculture: Is utility futility? *Review of Marketing and Agricultural Economics, 39*, 3–80.

Dillon, J. L., & Scandizzo, E. L. (1978). Risk attitudes of subsistence farmers in Northeast Brazil: A sampling approach. *American Journal of Agricultural Economics, 60*, 425–435.

FAO, UNDP, & UNEP. (2008). *UN Collaborative Programme on reducing emissions from deforestation and forest degradation in developing countries (UN-REDD)*. URL: http://www.un-redd.org/

Freund, R. (1956). The introduction of risk into a programming model. *Econometrica, 21*, 253–263.

Friedl, M. A., McIver, D. K., Hodges, J. C. F., Zhang, X. Y., Muchoney, D., Strahler, A. H., Woodcock, C. E., Gopal, S., Schneider, A., Cooper, A., Baccini, A., Gao, F., & Schaaf, C. (2002). Global land cover mapping from MODIS: Algorithms and early results. *Remote Sensing of Environment, 83*, 287–302.

Fritz, S., Bartholomé, E., Belward, A., Hartley, A., Stibig H. J., Eva, H., Mayaux, P., Bartalev, S., Latifovic, R., Kolmert, S., Roy, P., Agrawal, S., Bingfang, W., Wenting, X., Ledwith, M., Pekel, F. J., Giri, C., Mücher, S., de Badts, E., Tateishi, R., Champeaux, J.-L., & Defourny, P. (2003). *Harmonisation, mosaicing and production of the Global Land Cover 2000 database (Beta Version)*. Luxembourg: Office for Official Publications of the European Communities, EUR 20849 EN, 41 pp. ISBN 92-894-6332-5*.

Fritz, S., Scholes, R. J., Obersteiner, M., Bouma, J., & Reyers, B. (2008). A conceptual framework for assessing the benefits of a Global Earth Observation System of Systems. *Systems Journal, IEEE, 2*, 338–348.

Fuss, S., Johansson, D., Szolgayova, J., & Obersteiner, M. (2008). Impact of climate policy uncertainty on the adoption of electricity generating technologies. *Energy Policy, 37*(2), 733–743. doi:10.1016/j.enpol.2008.10.022.

GEOSS. (2005). *The global earth observation system of systems (GEOSS) 10-year implementation plan*. See http://www.earthobservations.org/documents/10-Year%20Implementation%20Plan.pdf

Havlík, P. Schneider, U. A., Schmid, E., Böttcher, B., Fritz, S., Skalský, R., Aoki, K., De Cara, S., Kindermann, G., Kraxner, F., Leduc, S., McCallum, I., Mosnier, A., Sauer, T., & Obersteiner, M. (2010). Global land-use implications of first and second generation biofuel targets. *Energy Policy*, Corrected proof available online 7 April 2010.

Katz, R. W., & Murphy, A. H. (Eds.). (1997). *Economic value of weather and climate forecasts*. Cambridge: Cambridge University Press. 237 pp.

Khabarov, N., Moltchanova, E., & Obersteiner, M. (2008). Valuing weather observation systems for forest fire management. *IEEE, 2*, 349–357.

Lin, W., Dean, G., & Moore, C. (1974). An empirical test of utility vs. profit maximization in agricultural production. *American Journal of Agricultural Economics, 56*, 497–508.

Macauley, M. K. (2006). The value of information: Measuring the contribution of space-derived earth science data to resource management. *Space Policy, 22*(4), 274–282.

Markowitz, H. M. (1952). Portfolio selection. *Journal of Finance, 7*(1), 77–91.

Markowitz, H. M. (1959). *Portfolio selection: Efficient diversification of investments*. New York: Wiley.

McCarl, B. A., & Spreen, T. H. (1980). Price endogenous mathematical programming as a tool for sector analysis. *American Journal of Agricultural Economics, 62*, 87–102.

McCarl, B.A., & Spreen, T.H. (2007). *Applied mathematical programming using algebraic systems*. Available at http://agecon2.tamu.edu/people/faculty/mccarl-bruce/mccspr/thebook.pdf

Nordhaus, W. D., & Popp, D. (1997). What is the value of scientific knowledge? An application to global warming using the PRICE model. *The Energy Journal, 18*, 1–45.

Peck, S. C., & Teisberg, T. J. (1993). Global warming uncertainties and the value of information: An analysis using CETA. *Resource and Energy Economics, 15*, 71–97.

Raiffa, H., & Schlaifer, R. (1961). *Applied statistical decision theory*. Boston: Harvard University Graduate School of Business Administration.

Ramankutty, N., Evan, A. T., Monfreda, C., & Foley, J. A. (2008). Farming the planet: 1. Geographic distribution of global agricultural lands in the year 2000. *Global Biogeochemical Cycles, 22*(1). doi:10.1029/2007GB002952

Schlamadinger, B., Bird, N., Johns, T., Brown, B., Canadell, J., Ciccarese, L., Dutschke, M., Fiedler, J., Fischlin, A., Fearnside, P., Forner, C., Freibauer, A., Frumhoff, P., Hoehne, N., Kirschbaum, M. U. F., Labat, A., Marland, G., Michaelowa, A., Montanarella, L., Moutinho, P., Murdiyarso, D., Pena, N., Pingoud, K., Rakonczay, Z., Rametsteiner, E., Rock, J., Sanz, M. J., Schneider, U. A., Shvidenko, A., Skutsch, M., Smith, P., Somogyi, Z., Trines, E., Ward, M., & Yamagata, Y. (2007). A synopsis of land use, land-use change and forestry (LULUCF) under the Kyoto Protocol and Marrakech Accords. *Environmental Science & Policy, 10*, 271–282.

Springer, U. (2003). International diversification of investments in climate change mitigation. *Ecological Economics, 46*, 181–193.

Watson, R. T., Bolin, B., Ravindranath, N. H., Verardo, D. J., & Dokken, D. J. (2000). *IPCC special report on land use, Land-use change and forestry, UNEP*. URL: http://www.grida.no/publications/other/ipcc_sr/

Weyant, J., Davidson, O., Dowlatabadi, H., Edmonds, J., Grubb, M., Parson, E. A., Richels, R., Rotmans, J., Shukla, P. R., Tol, R. S. J., Cline, W. R., & Fankhauser, S. (1996). Integrated assessment of climate change: An overview and comparison of approaches and results. In J. P. Bruce, L. Lee, & E. F. Haites (Eds.), *Climate change 1995: Economic and social dimensions of climate change* (pp. 367–439). Cambridge: Cambridge University Press.

Chapter 9 Space Imaging and Prevention of Infectious Disease: Rift Valley Fever

David M. Hartley

Abstract Rift Valley fever (RVF) is a mosquito-borne viral disease causing febrile illness and death in domestic livestock (cattle, sheep, goats) and humans. In Africa, RVF erupts following abnormally high rainfall and flooding. Remote sensing surveillance of vegetative growth could provide early warning, weeks to months in advance of RVF emergence, and thus permit intervention strategies to ameliorate and prevent this infectious disease. To act on this advance notice, however, public health officials must quantify the economic cost associated with the disease (in terms of losses to agriculture and international trade as well as human morbidity and mortality) and weigh the averted losses against the diversion of financial and public health resources dedicated to other major ongoing health needs, such as malaria and HIV/AIDS. Other complications include the accuracy of the predictions, the shelf life of vaccines, and the effectiveness of vector control strategies.

Keywords Remote sensing • Infectious disease surveillance • Rift Valley fever • Satellite data • Mosquito-borne disease

9.1 Introduction

Infectious threats to human and agricultural populations are significant and diverse, ranging from emerging diseases to bioterrorism. Health care organizations and agencies routinely balance diverse considerations and competing priorities to best meet health needs. This process is often difficult; to maximize success, planning and decisionmaking must be guided by sound epidemiological data, methods, and practices.

D.M. Hartley (✉)
Department of Microbiology and Immunology, Georgetown University Medical Center, Washington, DC, USA

Fogarty International Center, National Institutes of Health, Bethesda, MD, USA
e-mail: Hartley@isis.georgetown.edu

R. Laxminarayan and M.K. Macauley (eds.), *The Value of Information: Methodological Frontiers and New Applications in Environment and Health*, DOI 10.1007/978-94-007-4839-2_9,

From a public health perspective, two maxims often guide approaches: (1) the importance of prevention; and (2) the concept of targeting scarce resources on those parts of a population most active in transmission of infectious disease. From an economic perspective, the loss in productivity due to morbidity or mortality plus the costs associated with treatment almost always outweigh the cost of preventing infection. Assuming that cost-effective interventions capable of reducing the burden of disease exist, resources are focused where they have the greatest impact. We can express these maximums more colloquially: (1) "an ounce of prevention is worth a pound of cure"; and (2) targeting interventions gives the "biggest bang for the buck."

In general, answering the question "To prevent infection, when and where do we intervene?" is difficult. However, for a large class of infectious diseases, those that are tied closely to climate and the environment, promising approaches have been described. These employ landscape epidemiology—an understanding of the relationships between ecology and infection—to predict the spatial and temporal distribution of pathogens, vectors, and/or hosts. Landscape epidemiology involves the integration of epidemiological data (e.g., collected from active or passive surveillance or field surveys) with climate, environmental, and ecological data (e.g., collected from space-based remote sensors or field measurements) within a geographic information system (GIS) (Clements and Pfeiffer 2009). The availability of massive data sets collected from specialized airborne and satellite sensors, coupled with the proliferation of computer and information technology, has placed the landscape epidemiology approach within the reach of researchers globally. The general idea is that when clear associations exist between variables such as temperature, precipitation, land cover, vector and host abundance, and vector competency, they can be exploited for public health purposes. If the occurrence of such determinants can be forecast prospectively,[1] then so much the better: it opens up the possibility of intervening to prevent disease transmission.

Because the landscape epidemiology approach gathers data about the spatio-temporal determinants of disease emergence and transmission, it provides the potential to cue public health agencies on both when and where to intervene. Studies have identified associations for many diseases, but it is critical to ask whether these studies have been put to use and whether the economic payoff was assessed. There are examples sufficient to answer the former question, but by and large, the literature is silent on the latter. In this chapter, we describe a body of research surrounding Rift Valley fever and see that much has been done. In this case, there is great promise for prospective forecasting of outbreaks, opening the door for public health intervention. However, much remains to be done to quantify the payoff.

[1] The word forecast is usually preferred over *predict*. As Neils Bohr is alleged to have once remarked, "Prediction is very difficult, especially about the future." Forecast is a slightly more forgiving concept than prediction because it implies a statistical skill or bounded uncertainty. If the meteorologist predicts rain for a given area on a particular day, she will be proved either right or wrong, but if she forecasts an 80% chance of rain, there's wiggle room. A forecast has meaning in terms of probability of occurrence, whereas a prediction is a categorical either-or proposition.

9.2 Rift Valley Fever and Remote Sensing Studies

9.2.1 Remote Sensing

Within the context of this chapter, remote sensing entails using sensors to gather information about Earth's surface or atmosphere from a distance. Data are typically collected from airborne sensors mounted on aircraft or from space-based sensors mounted on satellites. Herbreteau et al. (2005) describe the process of space-based remote sensing as the following four steps: (1) detection and measurement of electromagnetic radiation emitted, radiated, or reflected by objects on (or near) the surface; (2) recording of these data and transmission to ground stations; (3) reception of the data and processing into images; and (4) analysis and interpretation of the images through visual or computerized methods.

Sensors can be active or passive. Active sensors—for example, synthetic aperture radar (SAR)—send pulsed microwave signals to Earth and receive the returned signals. The resulting data are processed into images. SAR can "see" through cloud cover because clouds are transparent to microwave radiation. Since a transmitter on the sensor platform itself illuminates Earth's surface, SAR sensors can also produce images at nighttime. Passive sensors detect the reflected or emitted electromagnetic radiation from the surface. Reflected sunlight is the most common source of radiation measured by passive sensors. Sensors are often classified according to the region of the electromagnetic spectrum they measure (e.g., visible, infrared, and microwave). Data from remote sensors result from interactions between the primary sources of electromagnetic energy, atmospheric absorption and distortion of the radiation, and characteristics (e.g., sensitivity, spectral resolution) of the sensors. The details are complex and the reader is referred to the literature for details (e.g., Herbreteau et al. 2005).

An important term in remote sensing is resolution, of which there are four basic types. *Spatial resolution* refers to the area of Earth's surface corresponding to each pixel of an image. The pixels of high-resolution imagery represent small areas. For example, 10-m spatial resolution means each pixel image represents an area of 10 m^2. *Temporal resolution* refers to how often a remote sensor revisits a specified area at the same viewing angle. High-temporal-resolution satellite sensors have short revisit periods. *Spectral resolution* refers to the number and width of spectral bands (portions of the electromagnetic spectrum) measured by a sensor. High-spectral-resolution images contain many narrow bands.[2] Finally, *radiometric*

[2] Spectral sensors are said to be multispectral or hyperspectral. Multispectral sensors measure several wavelength bands, such as the visible green or portions of the near infrared region of the spectrum. Hyperspectral sensors measure energy in narrower and more numerous spectral bands.

resolution refers to the sensitivity of a remote sensor to variations in the surface reflectance. High-radiometric-resolution sensors are sensitive to small differences in reflectance values of Earth's surface.

Resolution and other characteristics of remote sensors combine to allow measurement of a wide range of observables of interest. In the case of infectious disease, there are meaningful meteorologic, climatic, and environmental observables that tell us important things about the likelihood of disease emergence and transmission. Beck et al. (2000) have elucidated physical factors that could be used for both infectious disease research and public health applications. Each factor is essentially an environmental variable thought to influence the survival of pathogens, vectors, reservoir species, and hosts. These factors are vegetation or crop type, vegetation green-up, ecotones,[3] deforestation, forest patches, flooded forests, general flooding, permanent water, wetlands, soil moisture, canals, human settlements, urban features, ocean color, sea surface temperature, and sea surface height. Precipitation, humidity, and surface temperature were not included in the list because they are difficult to derive from remotely sensed data.

Such factors have demonstrated relevance to a range of human and zoonotic diseases, including cholera (Lobitz et al. 2000), Lyme disease (Kitron et al. 1997; Dister et al. 1997), tickborne encephalitis (Daniel and Kolár 1990; Daniel et al. 1998), Q fever (Tran et al. 2002), dengue fever (Moloney et al. 1998), Sin Nombre virus (Boone et al. 2000), hantavirus pulmonary syndrome (Glass et al. 2000), Rift Valley fever (Linthicum et al. 1987, 1990, 1999; Pope et al. 1992; Anyamba et al. 2009), St. Louis encephalitis (Wagner et al. 1979), and others. Remote sensing has also proved useful in monitoring crop health and detecting nutrient deficiencies, disease, and weed and insect infestations (Hatfield and Pinter 1993). Vegetation indices derived from multispectral imagery are used to monitor the growth response of plants in relation to measured (or predicted) climate variables. Remote sensors can also quantify crop water stress.[4]

9.2.2 Natural History of Rift Valley Fever

Rift Valley fever (RVF) is a mosquito-borne viral disease causing febrile illness in domestic livestock (cattle, sheep, goats) and humans. Outbreaks of RVF are associated with widespread morbidity and mortality in livestock and morbidity in humans. Identified in Kenya in 1930, RVF is often considered a disease primarily

---

[3] An ecotone is a transitional zone between two ecological communities, such as between a forest and grassland.

[4] One application of this, for example, is the products provided by the Famine Early Warning System Network of the U.S. Agency for International Development, http://www.fews.net/ml/en/product/Pages/default.aspx

Table 9.1 Some outbreaks of Rift Valley fever in Africa, 1950–2010

Year	Region or country	Representative effects
1950–1951	South Africa	Estimated 100,000 sheep died and 500,000 aborted. Smaller losses in cattle. Approximately 20,000 human cases. Many survivors suffered blindness from retinal hemorrhage
1974–1976	South Africa	Outbreaks similar to 1950–1951 activity, producing 10,000–20,000 human cases
1977	Egypt	Affected estimated 25–50% of all sheep and cattle. Roughly 200,000 people fell ill, nearly 600 died
1987	Mauritania	Outbreak followed opening of Diama Dam, causing approximately 200 deaths
1997–1998	Kenya, Somalia	Large losses of domestic animals and human mortality
2000	Arabian Peninsula	Affected hundreds of people and thousands of livestock. First time RVF observed outside African continent
2006–2007	East Africa	Roughly 1,000 people diagnosed and 300 died
2010	South Africa	Estimated 50,000 farm animals infected, more than 1,500 died. More than 200 human cases reported, more than 20 deaths

of sub-Saharan Africa, though outbreaks have occurred in northern Africa (Egypt) and, recently, the Arabian Peninsula. Table 9.1 contains an illustrative list of outbreaks.

In Africa, RVF erupts aperiodically in 7- to 15-year cycles following times of abnormally high rainfall and flooding. RVF virus (RVFV) is spread by mosquitoes to livestock (and also wildlife hosts, though their importance in outbreaks and interepidemic maintenance is unclear). *Culex* mosquitoes are infected only directly, through feeding on infectious livestock, but floodwater *Aedes* mosquitoes also can be infected at birth by vertical transmission (i.e., mother-to-offspring passage of RVFV). RVF in livestock causes abortions in pregnant animals and mortality rates as high as 90% in neonates and 30% in adults. In humans, RVF is typically a self-limited febrile disease, though blindness and fatal hemorrhagic fever can result. Epizootics typically precede human disease in pastoral areas.

In times of heavy rainfall in East Africa, geological features knows as dambos flood. As they fill, desiccated floodwater *Aedes* mosquito eggs rewet and begin to develop. Infected and infectious adult mosquitoes then emerge, carrying viable RVFV, and bite nearby livestock. The livestock amplify the virus and develop high viremia (concentration of virus in their blood), sufficient to infect *Aedes*, *Culex*, and possibly other mosquitoes and biting insects. Humans can become infected through the bites of mosquitoes or by handling infected tissues (e.g., disposing of aborted tissues or slaughtering infected animals). The presence of heavy rains typically precedes large epizootics and epidemics, but a clear association between seasonal rainfall, vector abundance, and RVF serological prevalence has been demonstrated in western Africa (Bicout and Sabatier 2004).

9.2.3 Studies of RVF Using Remote Sensing

Roughly two decades of research on RVF provides the foundation for a public health early warning system.[5] Pope et al. (1992) studied central Kenyan RVF virus vector habitats with Landsat and evaluated their flooding status with airborne imaging radar. Landsat Thematic Mapper (TM; a multispectral imager) data were shown to be effective in identifying dambos in an area north of Nairobi. Positive results were obtained from a test of flood detection in dambos with high-resolution airborne SAR imagery. In the same year, Davies et al. (1992), studying patterns of RVF activity in Zambia, observed that animal serological conversion was associated with changes in vegetation. Both studies used a normalized difference vegetation index (NDVI), derived from data from multispectral remote sensors.[6] These studies established that NDVI tends to correlate with rainfall and RVF viral activity.

In 1999 Linthicum et al. published a study in which they found that Rift Valley fever outbreaks in East Africa between 1950 and 1998 tended to follow periods of abnormally high rainfall. Accounting for drivers of regional climate (abnormal rainfall in particular), they considered Pacific and Indian Ocean sea surface temperature anomalies in their analysis.[7] They concluded that a combination of such anomalies in both oceans and NDVI correlate well with RVF outbreaks, several months in advance of observed disease transmission. This study suggested that surveillance of Pacific and Indian Ocean sea surface temperatures and NDVI could support prospective forecasts of RVF activity.

A recent study documented the successful forecast of RVF activity in East Africa (southern Somalia, Kenya, northern Tanzania) based on these methods. Anyamba and coworkers (2009) derived a spatiotemporal RVF risk-mapping model based on climate-related data, forecasting areas of human and animal RVF in the Horn of Africa between December 2006 and May 2007. The forecasts compared favorably with subsequent entomological and epidemiological field observation. Disease was forecast 2–6 weeks in advance in the study region, a time potentially sufficient for prevention activities to be carried out—assuming that resources exist and that health authorities are primed to execute such activities. This is thought to be the first prediction of a RVF outbreak described in the literature.

[5] In what follows, we are not aiming to present a complete review of all the relevant studies.

[6] Based on plant reflectance, NDVI describes the relative amount of green biomass in the field of view of a multispectral sensor.

[7] Pacific Ocean sea surface temperature is related to the El Niño–Southern Oscillation. El Niño refers to the warming of the central and eastern Pacific Ocean, whereas the southern oscillation refers to changes in surface pressure in the tropical western Pacific.

9.3 Factors in Evaluating the Payoff of Remote Sensing

There is little in the research literature quantifying the economic dimensions of RVF or the cost-effectiveness of RVF interventions, making an investigation of the economic value of remote sensing–based surveillance problematic. In this section we describe some of the factors relevant for assessing the payoff.

9.3.1 Metrics

The studies above suggest that early warning, weeks to months in advance of RVF emergence, may be possible. Viewing such surveillance as a trigger for intervention strategies, payoff can be assessed in terms of losses averted. "Loss" can be measured in terms of reduction of human suffering or in terms of dollars corresponding to economic costs, as described below. Estimating costs associated with human RVF disease could be stated in terms of dollars or in terms of disability-adjusted life years (DALYs) averted. It is known, however, that DALYs capture only a fraction of the total costs associated with human disease. It is clearly important to consider veterinary and human public health costs when estimating the true consequences of RVF. Exclusion of the animal or the human dimension of RVF will lead to undervaluation of the damage. Recognizing this dual-burden nature of zoonoses is fundamental to any comprehensive assessment of RVF or similar disease (Perry and Grace 2009).

If it is clear that both the animal and the human dimensions of cost must be considered in any analysis of costs averted by RVF prevention or mitigation activities, it is also clear that analyses will be contextual. For example, some nations have large livestock industries that will be affected heavily by RVF. Some have the capacity to act on early warning of RVF activity weeks to months in advance by acquiring and/or distributing vaccine and instituting appropriate vaccination and vector control campaigns. Some have robust medical infrastructures that are able to deal with significant human morbidity. Other nations lack such capacities. The payoff of remote sensing in RVF prevention will vary between nations; inequalities can be substantial.

9.3.2 Costs of RVF

A variety of costs can be associated with RVF outbreaks. Recently, the National Agricultural Biosecurity Center (2010) has enumerated a list that includes costs seen by workers and consumers as well as those associated with control efforts, livestock morbidity and mortality, human morbidity and mortality, and international trade disruption.

9.3.2.1 Costs Associated with Control Efforts

There are three general types of intervention that are thought to be effective or partially effective for controlling RVF: vaccination of livestock, vector control, and livestock movement control (Davies and Martin 2003, 2006; Geering et al. 2002). Livestock vaccination is thought to be the most effective means to control RVF.[8] The most widely available vaccine is based on the modified live Smithburn strain of RVFV (WHO 1983). Although the vaccine is immunogenic, it can injure the fetus and cause abortion in up to 30% of pregnant sheep. A single dose results in long-lasting immunity. Inactivated vaccines often are poorly immunogenic, but they have the advantage of being suitable for use in pregnant animals and conferring maternal immunity (via colostrum) to offspring. An initial booster followed by annual injections is required. New vaccines are being developed but are not available currently (Lubroth et al. 2007).

Vector populations can be controlled via larvicidal treatment of mosquito breeding sites. Effective larvicide products are available commercially, though widespread flooding, typically seen leading up to and during RVF outbreaks, can complicate application. Ultra-low-volume adulticide sprays appear to have limited effect on RVF transmission (though they are used successfully to control transmission other arboviral infections) (Davies and Martin 2003, 2006; Newton and Reiter 1992). Estimated costs of purchasing and applying adult and larval control over large areas have been published recently; the overall costs can be significant (Anyamba et al. 2010).

Although strict control of livestock movement does not appear to affect transmission within outbreak areas, it is thought to be effective in preventing the long-distance translocation of RVF into nonenzootic and nonepizootic areas. The costs associated with sealing of transportation into and out of farms or agricultural areas include the lack of food delivery and milk collection.

There are other costs associated with epidemic and epizootic management. For example, deployment of public health workers to affected areas to conduct rapid assessment of the outbreak, procurement of personal protective equipment (e.g., gloves, masks, goggles, aprons) for farmworkers as well as veterinarians and field workers, and training of public health personnel on hygiene promotion and health education are all common activities in times of RVF transmission (International Federation of Red Cross and Red Crescent Societies 2010).

9.3.2.2 Costs Associated with Livestock Morbidity and Mortality

Livestock abortion and neonatal mortality can result in "lost" generations of animals following severe, widespread outbreaks of RVF. There are costs associated

[8] Interestingly, vaccination is not recommended once epizootic transmission is observed because campaigns can spread RVF virus by reuse of hypodermic needles.

with diagnosing animals as well as disposing of (e.g., via burial or incineration) and replacing dead animals. The effect of RVF infection on animal fertility and on milk production in aborting mothers is unknown.

9.3.2.3 Costs Associated with Human Morbidity and Mortality

RVF morbidity affects people in terms of medical treatment (hospitalizations, outpatient care, self-care), time lost from work, and reduced productivity related to long-term sequelae (e.g., blindness, neurological complications) (WHO 2010). Human populations at highest risk include farm residents and workers, animal health personnel, and abattoir workers.

9.3.2.4 Costs Associated with Regional or International Trade

RVF is an OIE (Office International des Epizooties) List A disease, meaning it has the potential for rapid spread, has potentially serious socioeconomic or public health consequences, and is of major importance in the international trade of animals and animal products (Bram et al. 2002). The presence of any OIE List A disease within a nation presents barriers to trade by providing trading partners a reason to impose embargoes, often compromising agricultural industries in the outbreak nation (Kitching 2000).

For example, in the 1997–1998 outbreak in East Africa, pastoral economies in Somalia suffered an export decline of more than 75%, following a Saudi Arabian embargo on animal products from the Horn of Africa (LeGall 2006). Trade bans in the 2006–2007 outbreak cost Kenya an estimated US$32 million in lost exports to the Gulf (Rich and Wanyioke 2010). Cessation of South Africa wool exports to China occurred in 2010. South Africa is the world's third-largest producer of wool used to make clothes, and China accounts for about 60% of the wool exports from South Africa (Lourens 2010). In the previous year, South Africa exported US$128 million worth of raw wool to China (Barrie 2010). In addition to embargoes, there are costs associated with establishing disease-free status and renewed trade (e.g., sustained, intensified surveillance; inspection measures at ports) following an outbreak of RVF. In the case of South African wool, China requires a 12-month period following cessation of disease transmission before wool from RVF areas can be allowed into the country.

9.3.2.5 Costs Seen by Consumers and Workers

In the Egyptian outbreak in 1977, losses of cattle and sheep led to shortages of red meat in the Cairo marketplace (Lederberg and Shope 1992, 71). Scarcity of meat and dairy products can lead to price changes as well as pressure to substitute foods for the products made scarce by RVF. In the case of trade embargoes, agricultural industry workers may be laid off or rendered unable to work. For example, the South Africa wool industry is thought to support approximately 18,000 farmers, and the national flock of 14 million sheep produces roughly 50 million kg of wool annually (Laurens 2010). Loss for a year of 60% of exports undoubtedly affects a substantial number of workers in this sizable industry.

9.4 Threshold for Public Health Response

As we have described above, costs associated with RVF can affect a society in diverse ways. There are preventive measures that are thought to be capable of ameliorating the damage of RVF if applied before virus circulation begins. Remote sensing–based early warning could, in theory, cue quick reaction activities to control the emergence and spread of RVF virus. It is pragmatic to ask: Beyond the simple questions of when and where to intervene, what do governments and public health authorities need to act on early warning forecasts? This is a complex question. Certainly, timely outbreak response requires effective early warning systems, but inertia must be overcome before response activities can be undertaken. Major RVF outbreaks are infrequent but when they occur can lead to substantial diversion of financial and public health resources normally dedicated to other major ongoing health needs, such as HIV/AIDS, TB, malaria, and diarrheal diseases (Breiman et al. 2010). The reticence to doing this is certainly understandable.

Lack of early response cued by remotely sensed RVF surveillance has been a cause of considerable outcry on the part of public and veterinary health authorities. In connection with the 2006–2007 outbreak, for example, an official of the UN Food and Agriculture Organization (FAO) noted,

> It is interesting, if rather disheartening, to watch another RVF epizootic emerge and evolve in eastern Africa and to note that it is such a close recapitulation of events that occurred in 1997/8 and decades before. It is a recapitulation not only with respect to disease evolution but also in terms of national and international preparedness—or lack of it. Those who followed ProMED in those days will be aware that the epizootic attracted intense international attention and was closely reported in postings, which contain much useful information. Despite seminal work on developing early warning systems based on remote sensing . . . it seems that the capacity to respond has not improved greatly in the high-risk countries in Africa. (Dr. Peter Roeder, Animal Production and Health Division, FAO Writing on ProMED, 12 January 2007, archive #20070112.0164)

At the same time, an African news source reported, under the headline "Kenya: NASA Gave Warning Over Deadly Fever,"

> The deaths from Rift Valley fever (RVF) could have been avoided if Kenya had heeded a warning by an American body that changing climatic conditions posed a risk. The UN Food and Agricultural Organization (FAO) says the US-based National Aeronautics and Space Administration (NASA) Goddard Space Flight center sounded the alarm way back in September [2006], 2 months before the 1st case was reported in Garissa. However, it is not clear whether the country received the warning or simply ignored it. . . . The center had warned that rising temperatures accompanied by heavy rains in the Central and Eastern

> Pacific Ocean and Western Indian Ocean could spark an outbreak of the disease. The warning was contained in FAO's September [2006] edition of the Emergency Prevention Systems Magazine, Empres Watch. The center had been monitoring climate in East Africa for several years. ... "The outbreak of Rift Valley Fever is another example that requires a quick and coordinated response," said FAO's New Crisis Management Centre manager, Karin Schwabenbauer ... (All Africa newswire, reported on ProMED, 12 January 2007, archive 20070111.0112) [9]

Despite such political pressure to act, it is important to realize that the situation is often complex. A recent, comprehensive set of case studies of the 2006–2007 outbreak in East Central Africa was published in the *American Journal of Tropical Medicine and Hygiene* (August 2010), and many of the nuances are described there. For example, current preparations of the Smithburn vaccine have a shelf life of approximately 4 years. Outbreaks in the Horn of Africa region occur aperiodically, with a mean of near 10 years between outbreaks. Veterinary health authorities cannot spend scarce resources on continually replenishing a stock of RVF vaccine when other needs are present continuously. Nor can manufactures maintain large stocks that are likely to expire before sale. Thus, vaccine may not be available at any given time. Nonetheless, waiting until there is a need to manufacture vaccine is problematic (Consultative Group for RVF Decision Support 2010).

Even if vaccine had been available in the Horn region in 2006, effective and safe administration triggered by the early warning described in Anyamba et al. (2009) would have been complicated; by the time the warning was issued, early outbreak areas had already been inundated by rains and were inaccessible. In fact, the Consultative Group for RVF Decision Support (2010) suggests that up to 141 days may be needed between a vaccine order and the successful acquisition of vaccine-associated herd immunity in a hypothetical target population of 100,000 animals. This is much greater than the 2–6 weeks' (14–42 days') advance notice permitted by existing early warning systems. Thus, to be actionable to decisionmakers, forecasts may have to provide longer lead times.

Forecasts also have to be accurate in terms of geographic specificity. One recent study demonstrated a range of accuracy in terms of observed human disease in at-risk areas (Anyamba et al. 2010). Comparing the locations of disease outbreaks among humans and the areas deemed at risk based on remote sensing–based forecasting between 2006 and 2008, the researchers found that in eastern Africa (2006–2007), 65% of human case locations were in at-risk areas; in Sudan (2007), 50% of human cases were in such areas; in Madagascar (2007–2008), 23%; and in southern Africa (2007–2008), 20%. Although the study does not estimate positive or negative predictive values for RVF disease, the observations suggest that the existing forecasting algorithm may be better suited for some areas than others. Whether public and veterinary health authorities in the Horn region will be compelled to act based on

[9] A search of the FAO EMPRES archive at the time of writing this chapter yielded only an EMPRESS Watch report entitled "Possible RVF activity in the Horn of Africa," dated November 2006, a few months before this article appeared.

such performance is unclear. However, as Breiman et al. (2010) observe, "Higher specificity of forecast models will be needed for them to be confidently used to activate action-steps, which require commitment of public health resources, especially when considering how those resources are often limited."

9.5 Discussion

Increased rainfall causes vegetation growth, which can be measured or inferred from orbiting remote sensors. In conjunction with epidemiological and other field observations, maps of vegetation indices have been used to estimate the occurrence of increasing vector populations and RVF viral activity in East Central Africa. Correlations between viral activity and satellite observations have been established. Correlations are significantly improved by the addition of Pacific and Indian Ocean sea surface temperature anomaly measurements. Such work has provided a strong foundation to forecast RVF emergence before an epidemic or epizootic activity is observed. Recent studies of the 2006–2007 RVF outbreak in East Africa suggest that about 2 months' advance notice is possible. With further experience and investigation, the accuracy of these forecasts can likely be improved and application can be generalized to additional regions.

How can we elucidate the value of such disease forecasts? Viewing an early warning system as a trigger for intervention, payoff can be assessed in terms of losses averted. Prevention and mitigation activities may be able to reduce the effects of RVF if implemented early. Preventing widespread disease, if possible, may avert a substantial fraction of losses that would occur in the absence of controls. However, it is unclear that 1–2 months is sufficient time to execute effective prevention measures. Quantifying the payoff of early warning is complex, and we are aware of no analyses in the peer-reviewed literature.

Additional studies developing a comprehensive understanding of the use of such forecasts should consider questions such as these:

For a given situation and region, how should decisionmakers balance competing needs of existing health concerns with the transient, multifaceted "one health" needs related to a potential RVF outbreak? What information is actionable? The Consultative Group for RVF Decision Support (2010) has recently described an analytic tool to guide decisionmakers in responding to future RVF emergencies in the greater Horn of Africa. The tool incorporates the concept that actions should be in proportion to an evolving risk profile and remote sensing–based forecasts. This tool will evolve as additional information becomes available; the degree to which it will be adopted and employed is unknown.

Given sufficient warning, to what degree is it possible to prevent RVF? Could preventive vaccination in combination with vector control substantially reduce human and animal RVF infections? The finite shelf life of existing vaccines, as described above, suggests that in many scenarios, vaccination cued by early warning may not be feasible (even if ground conditions are not already complicated by flooding and it is possible to distribute and administer vaccine). Yet in areas where RVF outbreaks are separated by decadal periods, vaccine-induced immunity may be the key to preventing virus emergence and circulation. Vector control can also be problematic, in terms of logistics and costs (Anyamba et al. 2010). For either vaccination or vector control to be an option, resources need to be ready if campaigns are to be administered preemptively. As Jost et al. (2010) observe, "Donors and international organizations must also reevaluate the policies that resulted in the bulk of financial aid being provided to affected countries only after human cases have been documented." Since livestock outbreaks typically precede human disease, holding back controls until transmission is well established is problematic.

Can appropriate data be collected to assess reduction in economic costs associated with reduced incidence of RVF due to prevention? Rich and Wanyoike (2010) analyze the economic effects of RVF in Kenya during the 2006–2007 outbreak and observe that "downstream impacts can often dwarf the impacts of the disease at the farm level, but public policy tends to concentrate primarily on losses accruing to producers." Combined epidemic-economic models capable of analyzing the economic benefits of RVF prevention and control activities may be helpful to elucidate the cost-effectiveness of early warning systems in relation to both types of losses.

Until such questions are answered, it is unclear how the payoff of remote sensing–based early warning systems can be assessed. Studies addressing these and related questions would not only suggest ways in which the use and application of RVF forecasting could be optimized, they may also illustrate approaches for other diseases. As alluded to above, modeling and simulation may be important tools for quantifying cost savings from instituting prevention measures early. Modeling of infectious disease outbreaks is recognized as an important tool for understanding the dynamics of the outbreak process, the effects of the disease, and the potential benefits of interventions (Hartley et al. 2011). Epidemic models could be used to drive economic analyses by providing estimates of disease incidence and mortality in both epidemic (short-term) and endemic (long-term) scenarios Gaff et al. (2007), Xue et al. (2012). They may also provide guidance on how quickly interventions are likely to have effect, and how early control measures need to be instituted to be effective (Gaff et al. 2011).

If the surveillance approaches described in this study are ultimately to have a demonstrated payoff, public and agricultural health decisionmakers must come to rely on and trust the products. Surveillance must be accurate year-in and year-out. The surveillance must also be capable of facilitating an effective response and triggering controls that can be executed during the window of opportunity afforded by the early warning. The research described in this chapter suggests that this is possible, though clearly much remains to be done to achieve the goal.

Acknowledgments The author is grateful to Resources for the Future (RFF) for supporting this work. Funding for research on Rift Valley fever came from the Center of Excellence for Foreign Animal and Zoonotic Disease Defense and the Research and Policy for Infectious Disease Dynamics (RAPIDD) program of the Science and Technology Directory, Department of Homeland Security, and Fogarty International Center, National Institutes of Health. The views expressed in this study are the author's own and not necessarily those of any funder.

9. Commentary: An Emerging Approach

Joshua Michaud

Public health has always faced the classic conundrum of valuing prevention. How do you assign worth when the desirable product of public health activity is the *absence* of disease in a population? Policymakers and the public in general are much more likely to notice and understand how important public health can be when it fails as opposed to when it works, since the results of failure—the societal and economic toll of disease and mortality—are highly visible and relatively easy to catalog. The value of cases prevented and costs avoided, on the other hand, is harder to appreciate. Often, public health practitioners lament that many proven and inexpensive health interventions are ignored or underutilized because their full benefits are not understood. Although this may be true to an extent, the problem of underutilization also derives from the continuing inability of public health practitioners themselves to fully understand, quantify, and communicate the value of their work. Public health methods for valuation are imperfect, and not used widely. Practitioners are understandably focused on the health consequences of their work and tend not to dwell on its economic value.

This valuation deficit is certainly present in the important area of disease outbreak prevention and response. It is into this "work-in-progress" valuation of disease outbreak information that David Hartley introduces his chapter, "Space Imaging and Prevention of Infectious Disease: Rift Valley Fever." Hartley's chapter provides an overview of a promising public health application of remote sensing, and it quickly becomes clear that it is an *initiation* of a discussion about assigning value to information useful for public health action rather than a proposed set of methodologies for doing so. The chapter proposes, introduces, outlines, and hints at the potential for using evaluation of costs and benefits of predicting and preventing infectious disease using remotely sensed data. Having read the chapter and having some sense of the other kinds of public health applications for which remotely sensed data could be used, I find it hard to argue with the chapter's primary conclusion—that the potential for linking satellite data with epidemiologic tools to design and implement predictive capabilities for disease outbreaks is excellent. Still, we are very much at the beginning of this conversation and the process of tool development in this area.

In this response to the chapter, my hope is to contribute to the dialog around remotely sensed data and its application to disease outbreak interventions by highlighting a few points from Hartley's chapter, building on and supplementing these points, and extending the points by outlining an even broader scope of application for

J. Michaud (✉)
International Development Program, Johns Hopkins University School of Advanced International Studies, Washington, DC, USA
e-mail: jmichau1@jhu.edu

this nascent interdisciplinary work. Further, I point out a few limitations of the approach and points for consideration to be kept in mind as we move forward.

9.C.1. Preliminary but Promising Field

The chapter indicates the existing set of methodologies and scientific literature on application of satellite derived data to disease outbreaks remains quite limited. Hartley's wording when reviewing the literature in the field is indicative: "no studies," "studies would provide," "new field," "literature is silent" and "great promise"; all of these phrases are present because, while there is reason to believe something of value could be extracted here, much more work is needed to understand and place a value on the information being generated.

As both a researcher in the field and a sometime consumer of the kind of predictive data on RVF that has been produced in recent years, through researchers like Hartley and through NASA and the U.S. Department of Defense's global emerging infection system program, I can personally attest that even the limited, preliminary kinds of predictions that have been made available in recent years have been valuable to U.S. policymakers, let alone those working in Africa and other locations to reduce the effects of RVF outbreaks in the field. The Department of Defense and others have worked with NASA to use remotely sensed data in East Central Africa, helping produce forecasts of likely RVF activity in the area that rely on monitoring the vegetation index to forecast areas where mosquito hatching, and possibly viral transmission, could occur over subsequent weeks and months. The Department of Defense analyses the remotely sensed data and makes intermittent forecasts of RVF transmission risks, making what might have been an academic exercise of providing proof of principle for the predictive capability of these models a true, operational program. I was a consumer of this information as part of my work with the National Center for Medical Intelligence, and I observed first-hand how outbreak risk forecasts for RVF had decisionmaking implications for Defense in regard to its deployments, field exercises, and force health protection efforts.

As Hartley has effectively outlined, RVF makes an excellent case study for this kind of predictive modeling because of the unique relationship between certain environmental variables. Transmission of disease is highly associated with microenvironments, locations where certain combinations of temperature, rainfall, and vegetation in areas with dambos and dambolike terrain together come together with the mosquito vectors and distributions of animal and human hosts to create ideal RVF transmission pockets. Fortuitously, many of the associated environmental variables can be characterized from space using remote sensing. So, as Hartley shows, there is little doubt that through existing technology, remote sensing can be useful in the prediction of RVF disease outbreaks in East Central Africa, and perhaps in other locations as well.

As interesting and compelling as it is to elucidate an epidemiological connection between satellite data and RVF outbreaks, some additional questions for the purposes of understanding the value of information are, "What are the economic implications of generating RVF (and other infectious disease) predictions?" and, "Are gathering and analyzing that information worth the expense?"

In theory, linking the release of a predictive assessment of a future RVF outbreak with the actions of policymakers and farmers in the region should be measurable, along with an assessment of the costs, but in reality, gathering the wide set of information needed to make accurate cost-benefit calculations, especially in an area such as East Central Africa, is tremendously problematic. The cost of the disease is borne most acutely, one could argue, in the economic sense through the loss of income when farmers' herds become infected. Certainly, the human toll can be significant, but much of the concern around RVF is centered squarely on its agricultural implications. For this reason, the United States is mainly concerned about RVF, I would say, as an agricultural importation threat. The costs of RVF to livestock farming and potential damage are very significant. For this reason, the Department of Agriculture considers RVF one of the most threatening "foreign animal diseases" out there, and that agency has performed some analyses of the potential economic repercussions should RVF be imported into the United States and cases of the disease be found in U.S. livestock. Local farmers in East Africa and other RVF-affected areas are no less cognizant of the potential losses associated with the disease. But, as Hartley touches on and I hope to indicate more fully in the next section, there is more than just the simple accounting of costs of illness and lost income to consider.

9.C.2. Economic Considerations for Prediction and Response

Rather than attempt to outline a full methodology for calculating the cost-benefit of disease prevention through RVF prediction, Hartley's chapter is only able to sketch how such a calculation might be done. This is mainly a reflection of the lack of prior academic work and methodology relevant to this particular area. Hartley ably reviews several categories of costs and benefits that would have to be included in valuation calculations, which I will not repeat here. Rather, in this section, I would like to highlight additional considerations that would have to be incorporated in a full valuation model for the kind of remotely sensed work that the chapter characterizes. As Hartley partially recognizes, attempting to determine whether RVF surveillance and prediction are worthwhile requires not only understanding the costs of the animal and human disease burden and the expense of the sensing platforms and public health interventions and the like, but also disentangling a larger set of questions about incentives and externalities that are inherent in disease outbreak prediction, detection, and response. The following discussion centers around three public health functions in this area: surveillance (initial detection—or accurate prediction—of an outbreak), reporting (communicating the presence of an outbreak once it is detected), and response (implementing public health actions to stop transmission and reduce cases of the disease). The discussion is not limited to RVF alone, since it draws lessons from other infectious diseases; the points are applicable for RVF, but also more broadly for many kinds of outbreaks.

Surveillance: Is More Information Always Better? It may come as a surprise to learn that in the context of disease outbreaks, more surveillance information is not always better. "Better" here refers to "economically rational" for the actors involved in conducting surveillance. The reason more surveillance might not be better involves the built-in economic disincentives to infectious diseases that potentially leave some people, industries, and countries worse off with more information. Take a Kenyan farmer with livestock at risk of being infected with RVF. Were he to discover that his animals had been infected, many (or all) might be put down in an attempt to control the disease, or access to markets where the farmer might sell his animals or derived products might be restricted. Such actions might lead to a significant loss of income or even destitution that he would wish to avoid. If some of the animals did become infected and the farmer was unaware (either by chance or by choice), then he might still be able to extract some gain from selling the animals and avoid the potential loss of his entire herd and income. Given the choice of knowing or not knowing, he might prefer not knowing—in other words, he is disincentivized to participate fully in surveillance. The same logic goes for a methodology of prediction using satellites or any other tools. An area's farmers may feel that by knowing about an impending epidemic in advance—one in which their livelihoods are guaranteed to suffer while the benefits of this knowledge are less certain—they could be worse off than not knowing.

Those kinds of disincentives for surveillance are not restricted to individual farmers in developing countries at risk for outbreaks. In the United States, for example, when birds illegally smuggled into the country, some of which might have come from geographic areas endemic for highly pathogenic H5N1 avian influenza (HPAI H5N1, another frightening "zoonotic" disease, or a disease animals that can affect humans), are intercepted, no laboratory testing of the birds is performed prior to culling them. The rationale for this is to avoid having to say that HPAI H5N1 has been found inside the borders of the United States. So far, the HPAI H5N1 virus has not been found in the United States, but detection of the virus would surely have major implications for the poultry and other industries because immediate trade restrictions and possibly panic might ensue. This is a missed surveillance opportunity put in place for economic concerns, and it indicates the power of an economic disincentive for more information about potentially deadly diseases.

In another example, there were similar difficulties in surveillance for bovine spongiform encephalopathy (BSE, or "mad cow disease") in the United Kingdom, since farmers had little incentive to report suspected cases in their herds. In fact, the United Kingdom's BSE inquiry report stated, "one reason why BSE was not picked up at a very early stage by the system was the lack of incentive for farmers to refer an isolated case of an unrecognized disease in their herd for laboratory investigation. Indeed, there was a positive disincentive, namely the cost of a post-mortem examination" (UK Government 2000). This obstacle often appears in the context of zoonotic infections that affect agricultural livestock because there are potentially large economic losses from the culling of sick and potentially sick animals. Such culling and destroying of livestock and the associated trade restrictions and lack of access to markets that usually coincide with an outbreak response can serve as a powerful disincentive for individuals (and sometimes whole towns or industries) from participating in disease surveillance. This disincentive for good surveillance information exists whether the surveillance is performed through diagnostic tests or through application of remote sensing data in a predictive climate-based model.

Reporting: Is Being Completely Transparent Always Rational? In a similar vein, there is commonly a disincentive to be fully open and honest about *reporting* detected outbreaks. Farmers who know their flocks are ill may avoid saying anything for fear of losing their income. Countries wishing to avoid economic damage sometimes downplay or fail to report disease outbreaks. In areas affected by HPAI H5N1, for example, poultry farmers are often reluctant to report cases of dead birds to health authorities for fear that officials will rob them of their livelihoods (and important sources of food) by culling their flocks. As one Nigerian poultry farmer stated, "If the government isn't able to compensate me [sufficiently], why should I bother to report if my birds become sick? Wouldn't I be better off just taking my chances?" (Bellagio Meeting 2006).

This reporting disincentive also plays out along international trade routes, motivating obfuscation by governments. A great hindrance to transparency and early disease detection internationally is the cost that an affected country faces when the rest of the world finds out about the outbreak. On learning about an outbreak, many times neighboring countries close borders, trading partners restrict or stop imports, and travel and tourism cease. These actions have real and sometimes very damaging effects on important industries or economic sectors within a country, and therefore there is a strong incentive for underreporting or not reporting at all (Cash and Narasimhan 2000). Economic costs can be significant, in particular if the infectious disease is linked to the agricultural export sector, as RVF often is.

There are many examples of this kind of negative trade consequence from reporting a disease outbreak. Some of the more commonly cited figures include the 1991 cholera epidemic in Peru, which is estimated to have cost the country more than $1.5 billion in lost exports and tourism (Knobler et al. 2006), and India's 1994 outbreak of suspected plague, which likely cost the country an estimated $1.7 billion (WHO 2005). Thailand initially denied it had H5N1 avian influenza in its chickens, and Indonesia delayed reporting its first bird outbreaks of H5N1 (CNN 2004). Burma failed to report its first H5N1 bird cases when they occurred in 2004 (Beyrer 2006). China has reportedly covered up H5N1 outbreaks in its flocks multiple times. In 2006, the World Health Organization (WHO) accused the Chinese Ministry of Agriculture of "selectively reporting" outbreaks of H5N1 in its chickens and refusing to send samples from infected birds out for testing. At that time, the chief WHO representative in China stated, "It's so sad that we haven't got that [outbreak] information or those [H5N1] viruses from the Ministry of Agriculture ... it's really beyond comprehension to us" (CBC 2006). Once reporting of these bird outbreaks does occur, the economic consequences can be very painful. When Thailand's troubles with bird flu became known, the resulting collapse in poultry exports cost it some $1 billion (*Economist* 2006). When Vietnam first reported the presence of H5N1 (the virus and the culling wiped out 17% of the country's chickens in 2004), the outbreak and subsequent trade bans resulted in a loss of more than $83 million for this developing nation (Vietnam News Brief Service 2004). In 2003, when another pathogenic avian influenza subtype (H7N7) was found in poultry in the Netherlands, Belgium, and Germany, 28 million birds were culled and restrictions on trade in both poultry and swine (Dutch pigs were found to harbor evidence of infection) were enforced (Kimball 2005).

Response: Can Anything Be Done About It? Even when a disease outbreak can be detected early and reporting does occur, there might exist a gap between what *should* be done to implement an ideal public health response, and what *can* be done given what a country, region, or local community can do or is willing to do. Information that is not "actionable" may not be valuable. It does no good for a country to know where and when an outbreak is occurring if it does not possess the ability, or the willingness, to respond. In such a case, the information would have been generated just for information's sake, not for policy action. Again, such a restriction on the value of outbreak information applies equally to confirmation in the form of a diagnostic test result, or a trusted prediction based on satellite data.

Therefore the links among surveillance, reporting, and response capacity are critical, and we should not emphasize more and better data when the relevant actors cannot implement or improve policies with that information. One of these activities without the others provides limited or no benefit; all must be provided. Clearly, disease outbreaks are prone to collective action problems, since "rational" action by individuals and governments protecting their own economic interests can lead to overall irrational outcomes, such as worse outbreaks, greater health consequences, and more interruptions of trade and economic activity. Valuing the information contained in the prediction of RVF outbreaks through remote-sensing data would have to take into consideration these characteristics. Could an accurate RVF outbreak prediction actually make farmers in the targeted area worse off economically because of preemptive trade bans or other damaging actions? Is this risk worth it if the outbreak likely cannot be contained, given weak public health capacity? What are the optimal outcomes for all parties involved, economically speaking? Hartley hints at these complications, but it is worthwhile to highlight them more clearly.

9.C.3. *Broader Potential for Environmental Observation and Disease Prediction*

Although Hartley's chapter focuses on RVF, a subtext here is that similar techniques and methodologies could perhaps be applied to other infectious disease threats. Certainly the literature is already relatively rich with studies examining the relationship between environmental variables and disease epidemics (Kelly-Hope and Thomson 2008; Harvell et al. 2002).WHO in 2005 identified 14 infectious diseases it classified as potential candidates for environmentally based "early warning systems," a list that includes RVF, malaria, dengue fever, cholera, meningococcal meningitis, and influenza (WHO 2005). All of the 14 diseases are affected to some extent by the environment, but each to a unique extent, such that variables strongly associated with one may not be associated with others. In addition, many factors besides the environment must be taken into consideration when judging the transmission of these pathogens—everything from geographic variations in endemicity to human and vector behavior, to varied and changing control measures, to dynamic immune states, and other measures that may be unknown or not measurable.

In the case of RVF, the disease's very direct link to the environmental conditions that favor mosquito breeding in dambos (precipitation, temperature, and other factors that can be measured through satellite monitoring) make it a good candidate for forecasting. For other diseases, environmental variables serve as drivers of disease transmission but are only relatively minor contributors to the overall set of factors that determine when and where disease outbreaks emerge and spread. Thus, among those infectious diseases linked to environmental factors, RVF in parts of East Africa may in fact be the lowest-hanging fruit of remote sensing–based disease prediction. Extending the prediction technique beyond RVF, while possible and worth pursuing, might be more involved, less accurate, and potentially more costly.

Dengue is sometimes referenced as a disease that might be predicted based on environmental factors. Just as in the case of RVF, breeding and activity of mosquito vectors are influenced by temperature and rainfall, but complications in the ecology of dengue transmission make it a bit more unpredictable. This is especially true in the case of the dreaded and explosive "urban" dengue outbreaks, because the drivers of these types of epidemics, which are becoming more common in many cities of tropical developing countries, are heavily based on human behavior rather than on strictly environmental factors. Urban dwellers who leave open containers of water or fail to clear stagnant puddles create accommodating habitats for mosquito breeding whether it has rained recently or not. The complexities of human immunity to dengue's multiple serotypes are not fully understood, also making clear prediction more difficult. These additional factors have made dengue a more difficult target for environmental modeling and linking to remotely sensed data. Other diseases bring their own complications: the link between the environment and plague, for example, is moderated through the activities not only of the vector that transmits the bacterium, but also the rodent hosts of that vector; this and plague's highly focal natures takes prediction from environmental observations several steps further away from a direct causality.

Perhaps the biggest prize (and the one with the largest potential benefit) in the outbreak prediction field is malaria. This mosquito-borne parasitic infection is highly endemic in many countries around the globe and remains one of the leading killers of children in low-income countries. Currently, the prediction of malaria outbreaks through use of environmental variables is fraught with complications and confounders. Nonclimatic factors such as population immunity levels, nutrition status of a population, the state of control measures at local levels, the use of antimalarial drugs, and the pattern of drug resistance in circulating malaria strains strongly influence the environment-malaria link. These difficulties have not prevented research and development of climate-based malaria predictions, however. In fact, multiple studies of the relationship between climate factors such as El

Niño–Southern Oscillation (ENSO) cycles or changes in temperature or rainfall and malaria transmission have shown there can be a relationship (Ebi 2009; Githeko and Ndegwa 2001), but to put it kindly, the evidence is mixed and highly contingent on the specific circumstances, location, and time.

In the background of those analyses of climatic variables and disease we have the looming shadow of global climate change and its possible effects on infectious disease. If prediction can be somewhat successful on the small scale of weeks and months, how successful can we be on a longer scale? Can we predict disease transmission patterns years in advance once we know what the climate will look like in the future? Current conventional wisdom in public health holds that many diseases that had previously been circumscribed to poor tropical areas of the world, driven by an ever-warming climate, will expand their reach into geographical areas and populations where they had previously not been found; malaria is typically held up as a prime example. But the link between climatic variables and disease transmission actually becomes more tenuous the larger the geographic and temporal scale over which one attempts to predict (Lafferty 2009). It is precisely because the nonclimate variables associated with transmission become so heterogeneous and difficult to model over large areas and longtime scales that the probability of accurate prediction becomes very small, and adherence to simple cause and effect becomes problematic.

Several examples indicate how far we have to go to make long-term predictions of disease transmission based on climate. A recent article by Gething et al. (2010) in *Nature* deftly points out the problems with blindly ascribing increases in malaria with climate change by comparing the best estimates for the effect size of climate change on malaria transmission compared with the effect sizes of different control and treatment measures. The authors' bottom line is that a warming planet over a long time frame has the potential to affect transmission, but the effects of available control measures and treatments dwarf the effects predicted from climate. In other words, climate effects are drowned out by control effects (not to mention other nonmeasurable effects, such as general development), and predictions are often based on the erroneous assumption that control and treatment measures and technologies won't change over the large time scales in these analyses. In another example, researchers in Australia concluded that as parts of that country become drier with climate change, the risk of dengue might actually increase as people hoard water in water tanks that would increase mosquito breeding (Kearney et al. 2009). This is a counterintuitive result, as one might assume that mosquito activity would most likely decrease—and disease transmission with it—as the environment becomes drier. Finally, it is worthwhile to note that two adjacent areas with equivalent climates can have dramatically different transmission patterns for some diseases that have been linked to climate. Researchers have examined the areas that straddle the U.S.-Mexico border and found that between 1980 and 1999 there were more than 62,000 reported cases of dengue on the Mexican side of the border (likely an underestimate), while on the U.S. side there were just 64 cases. The difference is mostly explained through differences in living standards between the United States and Mexico (Brunkard et al. 2007). So, making the link between observed information and disease is not linear and highly determined—many disease processes are complex and resist simple cause-and-effect explanations.

9.C.4. Concluding Remarks

Hartley's chapter is a valuable review of the possibilities and the obstacles of making predictions about infectious disease outbreaks using climate observations. Certainly, as the chapter indicates, there is reason to believe that real associations have been discovered and that true predictive associations can be made between earth observations and disease transmission. As this commentary has attempted to indicate, the strength of these associations is highly dependent on the disease in question, the geographic and temporal scales involved, and the available data and understanding of disease processes. What is lacking, as is made abundantly clear in the chapter, is proven methodologies or sets of tools that can be applied to valuing the predictive disease work and that take into consideration the complications and externalities associated with transmission of diseases like RVF. Collection of better data along with the development of more robust epidemiologic and economic models of disease prediction would go a long way to bringing us closer to understanding, and ultimately assigning the proper value to, these efforts.

References

Anyamba, A., Chretien, J. P., Small, J., Tucker, C. J., Formenty, P. B., Richardson, J. H., Britch, S. C., Schnabel, D. C., Erickson, R. L., & Linthicum, K. J. (2009). Prediction of a Rift Valley fever outbreak. *Proceedings of the National Academy of Sciences of the United States of America, 106*(3), 955–959.

Anyamba, A., Linthicum, K. J., Small, J., Britch, S. C., Pak, E., de La Rocque, S., Formenty, P., Hightower, A. W., Breiman, R. F., Chretien, J. P., Tucker, C. J., Schnabel, D., Sang, R., Haagsma, K., Latham, M., Lewandowski, H. B., Magdi, S. O., Mohamed, M. A., Nguku, P. M., Reynes, J. M., & Swanepoel, R. (2010). Prediction, assessment of the Rift Valley fever activity in East and Southern Africa 2006–2008 and possible vector control strategies. *The American Journal of Tropical Medicine and Hygiene, 83*(2 Suppl), 43–51.

Barrie, L. (2010, August 19). South Africa: Halts wool exports to China. Just-Style. Available online at URL http://www.just-style.com/news/halts-wool-exports-to-china_id108686.aspx. Accessed 7 Nov 2010.

Beck, L. R., Lobitz, B. M., & Wood, B. L. (2000). Remote sensing and human health: New sensors and new opportunities. *Emerging Infectious Diseases, 6*(3), 217–227.

Bellagio Meeting on Social Ethics and Influenza. "Ethics and Pandemics" [Blog post]. October 2006. <http://bellagio.typepad.com/main/2006/10/nigeria_update.html>.

Beyrer C (2006). Responding to AIDS, Tuberculosis, Malaria, and Emerging Infectious Diseases in Burma: Dilemmas of Policy and Practice. *PLoS Medicine* 3(10):e393.

Bicout, D. J., & Sabatier, P. (2004). Mapping Rift Valley fever vectors and prevalence using rainfall variations. *Vector Borne and Zoonotic Diseases, 4*(1), 33–42.

Boone, J. D., McGwire, K. C., Otteson, E. W., DeBaca, R. S., Kuhn, E. A., Villard, P., Brussard, P. F., & St Jeor, S. C. (2000). Remote sensing and geographic information systems: Charting Sin Nombre virus infections in deer mice. *Emerging Infectious Diseases, 6*(3), 248–258.

Bram, R. A., George, J. E., Reichar, R. E., & Tabaciinic, W. J. (2002). Threat of foreign arthropod-borne pathogens to livestock in the United States. *Journal of Medical Entomology, 39*(3), 405–416.

Breiman, R. F., Minjauw, B., Sharif, S. K., Ithondeka, P., & Njenga, M. K. (2010). Rift Valley fever: Scientific pathways toward public health prevention and response. *The American Journal of Tropical Medicine and Hygiene, 83*(2 Suppl), 1–4.

Brunkard JM et al (2007). Dengue Fever Seroprevalence and Risk Factors, Mexico Border, 2004. *Emerging Infectious Diseases* 13(10):1477-1483.

Canadian Broadcasting Corporation (CBC) (2006). "WHO slams Chinese Ministry of Health for not sharing bird flu info, viruses," 1 Nov. <http://www.cbc.ca/health/story/2006/11/01/flu-china.html>.

Cash R, Narasimhan V (2000). Impediments to global surveillance of infectious diseases: consequences of open reporting in a global economy. *Bulletin of the World Health Organization* 78(11):1358–1367.

Clements, A. C., & Pfeiffer, D. U. (2009). Emerging viral zoonoses: Frameworks for spatial and spatiotemporal risk assessment and resource planning. *Veterinary Journal, 182*(1), 21–30.

CNN (2004). "Bird flu: Global response sought." 27 Jan. <http://www.cnn.com/2004/WORLD/asiapcf/01/27/bird.flu/index.html>.

Consultative Group for RVF Decision Support. (2010). Decision-support tool for prevention and control of Rift Valley fever epizootics in the Greater Horn of Africa. *The American Journal of Tropical Medicine and Hygiene, 83*(Suppl 2), 75–85.

Daniel, M., & Kolár, J. (1990). Using satellite data to forecast the occurrence of the common tick Ixodes ricinus (L.). *Journal of Hygiene, Epidemiology, Microbiology, and Immunology, 34*, 243–252.

Daniel, M., Kolár, J., Zeman, P., Pavelka, K., & Sadlo, J. (1998). Predictive map of Ixodes ricinus high-incidence habitats and a tick-borne encephalitis risk assessment using satellite data. *Experimental and Applied Acarology, 22*, 417–433.

Davies, G., & Martin, V. (2003). *Recognizing Rift Valley fever* (FAO Animal Health Manual No. 17). Rome: Food and Agricultural Organization of the United Nations.

Davies, F. G., & Martin, V. (2006). Recognizing Rift Valley fever. *Veterinaria Italiana, 42*(1), 31–53.

Davies, F. G., Kilelu, E., Linthicum, K. J., & Pegram, R. G. (1992). Patterns of Rift Valley fever activity in Zambia. *Epidemiology and Infection, 108*(1), 185–191.

Dister, S. W., Fish, D., Bros, S. M., Frank, D. H., & Wood, B. L. (1997). Landscape characterization of peridomestic risk for Lyme disease using satellite imagery. *The American Journal of Tropical Medicine and Hygiene, 57*(6), 687–692.

Ebi KL (2009). Malaria Early Warning Systems. Chapter 4 in. Ebi KL et al. (eds.) *Biometeorology for Adaptation to Climate Variability and Change*. Springer, p. 49-74.

Economist. Editorial. "A shot of transparency." *Economist* 10 Aug 2006.

Gaff, H. D., Hartley, D. M., & Leahy, N. (2007). An epidemiological model for Rift Valley fever virus. *Electronic Journal of Differential Equations, 2007*(115), 1–12.

Gaff, H. D., Burgess, C., Jackson, J., Niu, T., Papelis, Y., & Hartley, D. M. (2011). A mathematical model to assess the relative effectiveness of Rift Valley fever countermeasures. *International Journal of Artificial Life Research, 2*(2), 1–18.

Geering, W. A., Davies, F. G., & Martin, V. (2002). *Preparation of Rift Valley fever contingency plans* (FAO Animal Health Manual No. 15). Rome: Food and Agricultural Organization of the United Nations.

Gething PW et al. (2010). Climate change and the global malaria recession. *Nature* 465, 342-345.

Githeko AK, Ndegwa W (2001). Predicting Malaria Epidemics in the Kenyan Highlands Using Climate Data: A Tool for Decision Makers. *Global Change and Human Health* 2(1):54-63.

Glass, G. E., Cheek, J. E., Patz, J. A., Shields, T. M., Doyle, T. J., Thoroughman, D. A., Hunt, D. K., Enscore, R. E., Gage, K. L., Irland, C., Peters, C. J., & Bryan, R. (2000). Using remotely sensed data to identify areas at risk for hantavirus pulmonary syndrome. *Emerging Infectious Diseases, 6*(3), 238–247.

Hartley, D.M., Rinderknecht, J.L., Nipp, T.L., Clarke, N.P., Snowder, G.D., and the National Center for Foreign Animal and Zoonotic Disease Defense Advisory Group on Rift Valley Fever. (2011). Potential effects of Rift Valley fever in the United States. Emerg Infect Dis. 17(8):e1. Available online at http://wwwnc.cdc.gov/eid/article/17/8/10-1088_article.htm. Accessed 10 June 2012.

Harvell CD et al. (2002). Climate Warming and Disease Risks for Terrestrial and Marine Biota. *Science* 296:2158-2163.

Hatfield, P. L., & Pinter, P. J. (1993). Remote sensing for crop protection. *Crop Protection, 12*(6), 403–413.

Herbreteau, V., Salem, G., Souris, M., Hugot, J. P., & Gonzalez, J. P. (2005). Sizing up human health through remote sensing: Uses and misuses. *Parassitologia, 47*(1), 63–79.

International Federation of Red Cross and Red Crescent Societies. (2010). *DREF Operation South Africa: Rift Valley fever* (DREF MDRZA003. EP-2010-000080-ZA). Available online at www.ifrc.org/docs/appeals/10/MDRZA003.pdf Apr 27, 2010. Accessed 29 May 2010.

Jost, C. C., Nzietchueng, S., Kihu, S., Bett, B., Njogu, G., Swai, E. S., & Mariner, J. C. (2010). Epidemiological assessment of the Rift Valley fever outbreak in Kenya and Tanzania in 2006 and 2007. *The American Journal of Tropical Medicine and Hygiene, 83*(2 Suppl), 65–72.

Kearney M, et al (2009). Integrating biophysical models and evolutionary theory to predict climatic impacts on species' ranges: the dengue mosquito Aedes aegypti in Australia. *Functional Ecology* 23(3):528-538.

Kelly-Hope L, Thomson MC (2008). Climate and Infectious Diseases. Chapter 3 in. M.C. Thomson et al (eds.), *Seasonal Forecasts, Climatic Change and Human Health*. Springer, 31-69.

Kimball AM (2005). *Risky Trade: Infectious Disease in the Era of Trade*. Ashgate, p. 54.

Kitching, R. P. (2000). OIE List A disease as a constraint to international trade. *Annals of the New York Academy of Sciences, 916*, 50–54.

Kitron, U., & Kazmierczak, J. J. (1997). Spatial analysis of the distribution of Lyme disease in Wisconsin. *American Journal of Epidemiology, 145*(6), 558–566.

Knobler S, Mahmoud A, Lemon S, Pray L, eds (2006). *The impact of globalization on infectious disease emergence and control: exploring the consequences and opportunities*. Workshop summary – Forum on Microbial Threats. Washington, DC: National Academies Press.

Lafferty K (2009). Calling for an ecological approach to studying climate change and infectious diseases. *Ecology*, 90(4):932–933

Lederberg, J., Shope, R. E., & Oaks, S. C. (Eds.). (1992). *Emerging infections: Microbial threats to health in the United States*. Washington, DC: National Academy Press.

LeGall, F. (2006, June 12). *Economic and social consequences of animal diseases* (World Bank Report). Available online http://go.worldbank.org/PMRI0CP0R0. Accessed 29 May 2010.

Linthicum, K. J., Bailey, C. L., Davies, F. G., & Tucker, C. J. (1987). Detection of Rift Valley fever viral activity in Kenya by satellite remote sensing imagery. *Science, 235*, 1656–1659.

Linthicum, K. J., Bailey, C. L., Tucker, C. J., Mitchell, K. D., Logan, T. M., Davies, F. G., Kamau, C. W., Thande, P. C., & Wagateh, J. N. (1990). Applications of polar-orbiting, meteorological satellite data to detect flooding in Rift Valley fever virus vector mosquito habitats in Kenya. *Medical and Veterinary Entomology, 4*, 433–438.

Linthicum, K. J., Anyamba, A., Tucker, C. J., Kelley, P. W., Myers, M. F., & Peters, C. J. (1999). Climate and satellite indicators to forecast Rift Valley fever epidemics in Kenya. *Science, 235*, 1656–1659.

Lobitz, B., Beck, L., Huq, A., Wood, B., Fuchs, G., Faruque, A. S., & Colwell, R. (2000). Climate and infectious disease: use of remote sensing for detection of Vibrio cholerae by indirect measurement. *Proceedings of the National Academy of Sciences of the United States of America, 97*(4), 1438–1443.

Lourens, C. (2010, March 24). *Fever outbreak may cost R35m in wool exports* (Business Report). Available online at http://www.busrep.co.za/index.php?fSectionId=561&fArticleId=5402283. Accessed 29 May 2010.

Lubroth, J., Rweyemamu, M. M., Viljoen, G., Diallo, A., Dungu, B., & Amanfu, W. (2007). Veterinary vaccines and their use in developing countries. *Revue Scientifique et Technique, 26*(1), 179–201.

Moloney, J. M., Skelly, C., Weinstein, P., Maguire, M., & Ritchie, S. (1998). Domestic Aedes aegypti breeding site surveillance: Limitations of remote sensing as a predictive surveillance tool. *The American Journal of Tropical Medicine and Hygiene, 59*(2), 261–264.

National Agricultural Biosecurity Center. (2010). *Rift Valley fever fact sheet.* Available online at http://nabc.ksu.edu/content/factsheets/category/Rift%20Valley%20Fever. Accessed 29 May 2010.

Newton, E. A., & Reiter, P. (1992). A model of the transmission of dengue fever with an evaluation of the impact of ultra-low volume (ULV) insecticide applications on dengue epidemics. *The American Journal of Tropical Medicine and Hygiene, 47*(6), 709–720.

Perry, B., & Grace, D. (2009). The impacts of livestock diseases and their control on growth and development processes that are pro-poor. *Philosophical Transactions of the Royal Society of London. Series B, Biological Sciences, 364*(1530), 2643–2655.

Pope, K. O., Sheffner, E. J., Linthicum, K. J., Bailey, C. L., Logan, T. M., Kasischke, E. S., Birney, K., Njogu, A. R., & Roberts, C. R. (1992). Identification of central Kenyan Rift Valley fever virus vector habitats with Landsat TM and evaluation of their flooding status with airborne imaging radar. *Remote Sensing of Environment, 40*(3), 185–196.

Rich, K. M., & Wanyoike, F. (2010). An assessment of the regional and national socio-economic impacts of the 2007 Rift Valley fever outbreak in Kenya. *The American Journal of Tropical Medicine and Hygiene, 83*(2 Suppl), 52–57.

Tran, A., Gardon, J., Weber, S., & Polidori, L. (2002). Mapping disease incidence in suburban areas using remotely sensed data. *American Journal of Epidemiology, 156*(7), 662–668.

UK Government (2000). *The BSE Inquiry Report.* Volume 2, Chapter 4. <http://www.bseinquiry.gov.uk/report/volume2/chaptee4.htm>.

Vietnam News Brief Service (2004). "Viet Nam: deaths of 5 children attributed to avian influenza H5 virus," 10 September.

Wagner, V. E., Hill-Rowley, R., Narlock, D. A., & Newson, H. (1979). Remote sensing: A rapid and accurate method of data acquisition for a newly formed mosquito control district. *Mosquito News, 39*, 282–287.

World Health Organization. (1983). The use of veterinary vaccines for prevention and control of Rift Valley fever: Memorandum from a WHO/FAO meeting. *Bulletin of the World Health Organization, 61*(2), 261–268.

World Health Organization (WHO) (2005). *Using climate to predict infectious disease epidemics.* Geneva.

World Health Organization (2007). *World Health Report 2007.* Geneva, p.3.

World Health Organization. (2010). *Rift Valley fever.* Available online at http://www.who.int/mediacentre/factsheets/fs207/en/ Revised May, 2010. Accessed 29 May, 2010.

Xue, L., Scott, H.M., Cohnstaedt, L.W., Scoglio, C. (2012). A network-based meta-population approach to model Rift Valley fever epidemics. J Theory Biol. May 4;306C:129–144

Chapter 10
Estimating the Benefits of Land Imagery in Environmental Applications: A Case Study in Nonpoint Source Pollution of Groundwater

Richard L. Bernknopf, William M. Forney, Ronald P. Raunikar, and Shruti K. Mishra

Abstract Moderate-resolution land imagery (MRLI) is crucial to a more complete assessment of the cumulative, landscape-level effect of agricultural land use and land cover on environmental quality. If this improved assessment yields a net social benefit, then that benefit reflects the value of information (VOI) from MRLI. Environmental quality and the capacity to provide ecosystem services evolve because of human actions, changing natural conditions, and their interaction with natural physical processes. The human actions, in turn, are constrained and redirected by many institutions and regulations such as agricultural, energy, and environmental policies. We present a general framework for bringing together sociologic, biologic, physical, hydrologic, and geologic processes at meaningful scales to interpret environmental implications of MRLI applications. We set out a specific application using MRLI observations to identify crop planting patterns and thus estimate surface management activities that influence groundwater resources over a regional landscape. We tailor the application to the characteristics of nonpoint source groundwater pollution hazards in Iowa to illustrate a general framework in a land use-hydrologic-economic system. In the example, MRLI VOI derives from reducing the risk of both losses to agricultural production and damage to human health and other consequences of contaminated groundwater.

R.L. Bernknopf (✉)
Department of Economics, University of New Mexico, Albuquerque, NM, USA

Western Geographic Science Center, United States Geologic Survey, Menlo Park, CA, USA
e-mail: rbern@unm.edu

W.M. Forney • R.P. Raunikar • S.K. Mishra
Western Geographic Science Center, United States Geologic Survey, Menlo Park, CA, USA
e-mail: wforney@usgs.gov; rraunikar@usgs.gov; saishruti@gmail.com

R. Laxminarayan and M.K. Macauley (eds.), *The Value of Information: Methodological Frontiers and New Applications in Environment and Health*,
DOI 10.1007/978-94-007-4839-2_10,

Keywords Integrated assessment • Landsat • Moderate-resolution land imagery • Remote sensing • Nonpoint source pollution • Value of information • Agricultural production • Land use and land cover • Joint production • Nitrate • Groundwater contamination • Hydrogeology • Ecosystem service • Environmental regulation • Agricultural policy • Renewable fuel standard • Ethanol • Economic loss • Risk

10.1 Introduction

Moderate-resolution land imagery[1] (MRLI) accrues benefits to society at large by providing a spatiotemporal land use and land cover (LULC) signal that can be linked to ground-based, land management activities and their effects on human and natural ecosystems. In this chapter, we develop a general framework for estimating the value of MRLI in an application to public policy issues related to agricultural production and its effects on ecosystem services.[2] An interdisciplinary analytical framework for making decisions is developed using the intrinsic heterogeneity of regional land characteristics (e.g. land cover transitions, soil characteristics, geomorphology, temperature, geology, transport and biochemical processes, and climatic regime), and external inputs (e.g., irrigation, nutrient application, crop management) to land parcels to produce corn and soy crops.

The societal benefits from MRLI information–that is, the value of information (VOI)–is the incremental value (i.e. cost savings) to public sector decision making (i.e. government regulation). Cost savings arise from avoiding costly errors in administering regulations for sustaining ecosystem services (e.g. groundwater contamination and compliance with the Clean Water Act). MRLI sensors and their data archives provide LULC information at a derived error rate in detecting farm land use that can be used as inputs to a probabilistic estimate of adverse change to an ecosystem service. For example, MRLI can provide a cumulative, multi-temporal accounting of crops that have differential effects on groundwater quality, which can be used to forecast critical levels of nitrate (NO_3^-) concentration in an aquifer. The data can inform decisions to regulate land use for mitigating a potentially irreversible loss of groundwater resources.

User surveys suggest that MRLI is used for decision making in land-use planning and management, water resources management, ecological forecasting, emergency and disaster management, national and homeland security, coastal zone

[1] Moderate resolution land imagery is defined in the spatial domain as having a pixel resolution between 30 and 250 meters.

[2] Ecosystem services are defined as the production of goods (such as timber, seafood, and industrial raw materials), life support processes (such as pollination, water purification, and climate regulations), life fulfilling conditions (such as beauty, cultural inspiration, and serenity), and preservation of future options of resource (such as biodiversity and genetic conservation for future use) (Daily 1997). In this particular case study, only the good of groundwater quality is considered.

management, and transportation management and infrastructure planning (Miller et al. 2011). Nelson et al. (2007) reviewed applications of Landsat—an MRLI sensor—in the agriculture sector, broadly defined to include production agriculture, water resources management, rangeland management, forestry, and environmental management. Early studies by Earth Satellite Corporation (1974) and ECON Inc. (1974) identified potential operational benefits of Landsat, in which both studies contained projections anticipating that agricultural applications would provide a significant share of the total benefit of the imagery. Results reported by Earth Satellite Corporation were $158 million[3] to $414 million for agricultural applications and those reported by ECON, Inc. (1974) would be in the range of $3.8 billion to $25.8 billion for agricultural applications. In both studies, the benefits from remotely sensed data were estimated by assuming that it resulted in improvements in production forecasts.

Other studies about the valuation of Global Earth Observation System (GEOS) information focused primarily on potential benefits of GEOS information (Macauley 2006, 2007; Williamson et al. 2002; Kalluri et al. 2003; Isik et al. 2005). Potential societal benefits from MRLI include cost savings in natural resource allocation, environmental regulation and reduced damage to public goods (Macauley 2007). Only a few studies have attempted to quantify the benefits of information from GEOS in monetary values (Macauley 2010; Bouma et al. 2009). Macauley (2010) developed an expenditure-based VOI estimation model to derive a value for Landsat data from the economic value of accurate estimates for forest carbon offsets. The study considers the scenario where a new hyperspectral sensor added to the platform can change the derived value of Landsat data for the period 2022–2026. This information is combined with climate policy scenarios and related to economic data about forest sequestration projected for the same time period. Bouma et al. (2009) used the stated-preference method to estimate the economic benefits of satellite-based information to be $2.68 million annually for managing water quality in the North Sea. Nelson et al. (2007) also reviewed the use of Landsat by the Risk Management Agency (RMA). This agency, as part of the United States Department of Agriculture (USDA), is responsible for monitoring compliance with the terms of the federal crop insurance program and assessing whether fraud has been committed. In testimony before Congress in 2006, USDA estimated the agency's return-on-investment in 2005 alone was 458 times the cost, based on $34.4 million in restitution and forfeiture and a $75,000 USDA image archive subscription fee. The Farm Security and Rural Investment Act of 2002 authorizes the secretary of Agriculture to issue rules the secretary considers necessary to ensure producers' compliance with programs that help farmers manage market risk and safeguard environmentally sensitive land. Some other, non-monetized uses of Landsat imagery are to approximate the extent and temporal dynamics of natural disasters such as flooding, hail storms and forest fires. Evaluation of cost

[3] All dollar values in this chapter are deflated to the 2009 price level.

savings attributed to each of the MRLI applications described above provides a limited indication of the partial VOI of MRLI.

This chapter is part of a larger study being conducted by the U.S. Geological Survey (USGS) to identify operational applications of Landsat data that have a quantifiable economic value. This chapter describes the development of a conceptual economic model that links remote sensing to physical processes. In the first section, the components of an integrated assessment approach (IAA) (Antle and Just 1991) that support the general framework are described. These components include profit maximization by producers, revealed preference for social risk, and cost effectiveness of regulation. The general framework is applied to a regional environmental externality problem. The next section contains an application of the economic model for agricultural production, nutrient loading, and the effects to an ecosystem service. This type of application expresses benefits as value-in-use, bequest, option, and existence values that result from the integration of remotely sensed observations and natural science process models and data. MRLI information is used to observe and help document the effect of agricultural production on ecosystem services to avoid costly errors in administering regulations to sustain those services. Satellite imagery is especially useful at the regional scale because it captures the temporal change of the population of land activities better than conventional methods of spatial and temporal sampling of representative locations (Wilkie and Finn 1996). Furthermore, MRLI observation and long-term datasets (archives) provide increased accuracy and precision for modeling biophysical processes such as LULC change related to groundwater dynamics, thereby improving the targeted response of decision makers and regulators as they seek to manage and maintain ecosystem services of the resources (groundwater quality). The example highlights the intersection—and potential conflict—of policies that encourage biofuels from corn production and incentivize reduced agricultural production via USDA conservation programs, and provisions of the Safe Drinking Water Act of 1974 (P.L. 107–377, 2002, amended). Depending on the quality of information available to decision makers, these policy and management tools may be employed more or less effectively. In other words, the greater sources of nonpoint source pollution can be more specifically addressed while locations that do not contribute to the nitrate loading and pollution of a given well could be less strictly regulated.

The model estimates the joint output of agricultural production and groundwater pollution as an example. On one hand, the model is used to estimate the economic value of agricultural production while minimizing the risk of a loss in groundwater quality. On the other hand, the model estimates the increased cost to producers from either increased input costs or reduced fertilizer application. The joint output model is followed by a loss estimation procedure that is used as the basis for comparing the value of sensor data sets. The example assumes the regulator needs to be cautious (reduce or eliminate fertilizer application for corn production) to avoid the loss of groundwater resources due to nonpoint source contamination (Lichtenberg 1991). The regulator's decision involves the implementation of regulation(s) in anticipation of a natural resource failure to avoid more costly alternatives later. The cost effectiveness of the MRLI is demonstrated when decisions are made less costly to society with the imagery rather than without it. Although the example presented here is preliminary, it is representative of the types of applications that are possible with MRLI. Finally, we summarize the model and express the need for empirical testing of the framework. At the time of the writing of this chapter, a pilot analysis is underway to evaluate the framework in Iowa.

10.2 Conceptual Framework

The conceptual framework for valuing MRLI is developed in an IAA within the context of agricultural, energy, and environmental policies. In the IAA, we combine two levels of decision making in an adaptation of the Antle and Just (1991) integrated framework (Fig. 10.1). The first level is the farmer and the second is the regional decision maker or regulator. The economic model is based on a farmer's choice among competing land uses in a geographic region (Antle and Valdivia 2006; Antle and McGuckin 1993; Antle and Just 1991). The upper part of

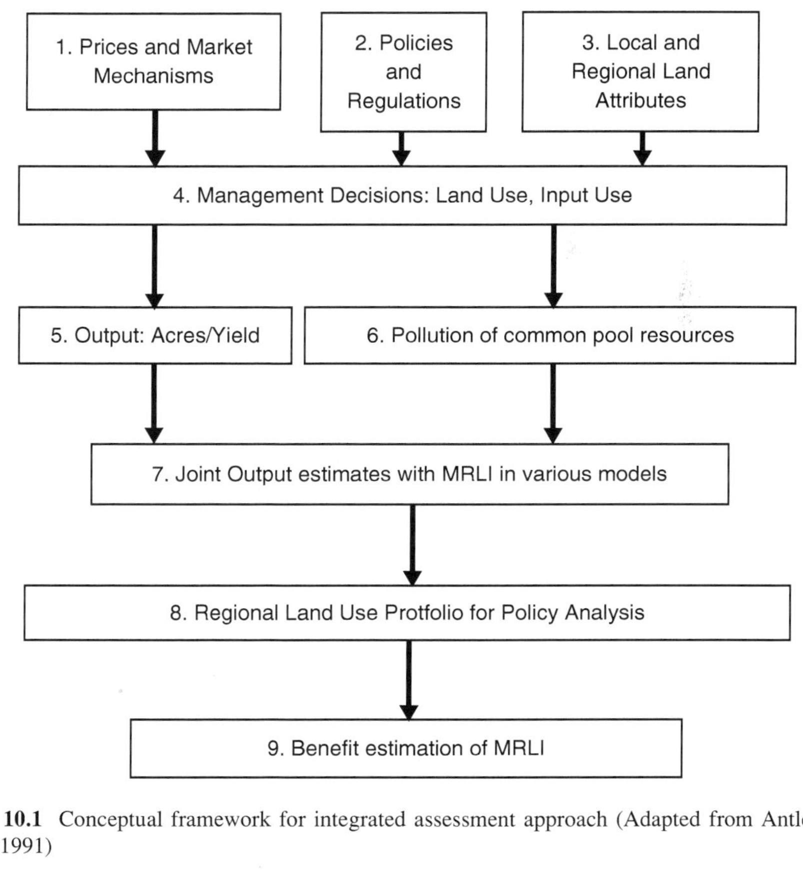

Fig. 10.1 Conceptual framework for integrated assessment approach (Adapted from Antle and Just 1991)

Fig. 10.1, boxes 1–6, relates to decisions by individual producers that result in a joint output of economic production and pollution. The lower part of figure, boxes 7 and 8, contains the observations of MRLI over multiple years and crop rotations to estimate temporal regional-scale production, the accumulation of agricultural inputs on the environment, and the risks to the decision maker. Box 9 of Fig. 10.1 integrates all previous boxes into a risk assessment associated with agricultural, energy, and environmental policies and the desirability of regulation.

The IAA incorporates two conceptual vectors, "Scale" and "Time," that are critical to consider for the analytical approach (Hong et al. 2007). Input variables are distributed spatially across the region and vary according to land characteristics and external factors that affect a given parcel. It is important to note that a parcel could include many fields, and farming operations could include multiple parcels.

Depending on the crop, its price elasticity, tendency for substitution, and the size and structure of the market, price data (Fig. 10.1, box 1) fluctuate from year to year and within a given year (*The Financials* 2010), whereas pollution (Fig. 10.1, box 6) potentially has a spatial and temporal cumulative effect on the resource over a number of years.

Environmental impacts of agricultural activities are controlled under several regulatory frameworks,[4] **R**, (Fig. 10.1, box 2). The degree to which these federal and state environmental and agricultural policies, regulations, and statutes apply depends on the particular farm and location of its fields (Fig. 10.1, box 3). Some locations, such as riparian corridors, will likely have more beneficial uses for a wider range of regulations (e.g. under the Clean Water Act, Endangered Species Act, and Safe Drinking Water Act). Others, such as highly permeable soils that allow surface water to infiltrate into the groundwater aquifer system, are likely to have fewer beneficial uses if the range of regulations is narrowed (e.g. under the Groundwater Protection Act, and total maximum daily loads, TMDLs). For those that do apply, the temporal interval that influences the decisions of an agricultural producer is assumed to be yearly, while the effect on natural resources is assumed to be monthly to seasonal. That said the creation, alteration, monitoring and enforcement of these regulations occur over longer time scales, which are assumed

[4] Regulatory frameworks include Farm Bill provisions, the USDA Conservation Programs, environmental policy, and energy policy. The Farm Bill legislation began in 1933. The Farm Bill is responsible for influencing many activities related to the decisions facing a farmer including crop insurance, credit programs and direct and counter-cyclical payment contracts. The Farm Bill also governs the USDA's Conservation Programs that provide voluntary, yet binding, cost-share programs. Depending on the contract and the program mechanism, the USDA's cost-sharing programs (Figure 10.1, box 2) have a duration of anywhere from three to thirty years. In addition to the agricultural programs, farmers in states such as Iowa face federal and state environmental policies and regulations such as the Federal Clean Water Act, Endangered Species Act, National Environmental Policy Act, Safe Drinking Water Act, 1990 Clean Air Act Amendments, Food Quality Protection Act of 1996; and the State Air Quality Code, Water Quality Codes, Groundwater Protection Act, Contaminated Sites (pesticides and fertilizers) Code, Pesticide Act of Iowa, Agrichemical Remediation Act, Agricultural Drainage Wells Code, Soil Conservation Districts Laws, Fishing and Game Hunting laws, Endangered Plants and Wildlife laws, Farmland Preservation Statutes, and Manure Management Plans and Tile Lines (Figure 10.1, box 2).

to be exogenous drivers of the producers behavioral changes and management decisions (Fig. 10.1, box 4).

As for energy policy, although the first energy policy that influenced ethanol production was the 1978 Federal Tax Credit, two more recent bills are of greater significance. The Energy Policy Act of 2005 set the renewable fuel standard (RFS) to increase ethanol levels from 4.0 billion to 7.5 billion gallons per year by 2012. In 2007, the Energy Independence and Security Act (P. L. 110–140) increased the RFS levels to require the use of at least 36 billion gallons of biofuel by 2022, 15 billion gallons being corn ethanol and the remainder being cellulosic ethanol (such as perennial grasses, biomass and municipal solid waste) and other advanced biofuels (Fig. 10.1, box 2). Spatially, their influence on the IAA is at the farm scale. As posited in this conceptual framework, the goals for ethanol production and the influence of the RFS on decision making of the producer will be across all their fields in response to expectations of higher corn prices.

Natural resource damage arises from agrichemical pollution and other effects of agricultural production, which range across many biogeochemical patterns and processes (Fig. 10.1, boxes 3 and 6). Some of the more salient ones are erosion and sedimentation, surface and groundwater nutrient loading and pesticide pollution, greenhouse gas fluxes, habitat and native vegetation loss, wetland dewatering and conversion, reductions in biodiversity, and take of at-risk or listed species (Fig. 10.1, box 6). For implementation of the IAA, essential considerations are: (1) the number of natural resource processes, pollutant loadings and their associated effects on ecosystem services (Fig. 10.1, box 6); (2) the degree of original modeling (or easily—adaptable existing models) to incorporate and adequately characterize the biophysical processes of the system related to the particular ecosystem services (Fig. 10.1, box 7); and (3) the temporal and spatial consistency of all the modeling components and their data availability. For our initial example in this study, we focus on groundwater pollution from agricultural nonpoint source pollution.

An economic model is developed for efficient allocation of resources from a regulator's perspective. Producers are expected to behave as profit maximizers under given regulatory constraints. A probabilistic estimate of pollution is based on a spatiotemporal agricultural production portfolio, which can be derived from a Cobb-Douglas production function. A forecast of the time to exceed a regulatory standard for resource consumption is made after the risk of contamination is determined. Whether and how much to regulate then depends on the regulator's risk preference.

10.2.1 Economic Model

Regulators seek to maximize the value of agricultural output while limiting the risk of resource damage. Given prevailing crop prices **P,** they choose regulations **R**:

$$\max_{\mathbf{R}} \mathbf{PQ}$$
$$s.t. \quad risks \leq \boldsymbol{\alpha}$$

where **P** represents prices of relevant crops, **Q** represents aggregate production of those crops, and **α** represents the probability of exceeding a regulatory standard that causes damage to a resource (here groundwater). Both the plot level and regional risks are related to the quality of information about crop production (**q**), variable inputs (**v**), farm management practices (**z**) and plot characteristics (**e**). The crop production **q** is the amount of each crop produced on a plot. It is a function of **v, z** and **e**.

In terms of **v,** by making decisions to maximize the production from their fields (Fig. 10.1, box 5), producers apply fixed and variable inputs such as irrigation, fertilizer, soil amendment, and pesticides (North Dakota State University 1997). These decisions and the rates and durations of application are made at certain times of the year, generally assumed to be within a given growing season. For example, during certain periods of the crop's growth cycle, fertilizer is applied to facilitate plant growth or insecticides are applied to limit the invasion of parasitic insects. The application of these two products, however, may not occur at the same time. This can hold true for other **v** used in production.

In terms of **z**, the methods for production, farm management practices are tempered by the biophysical characteristics of a given location. The method options available to farmers include weed and insect management techniques such as integrated pest management, seed selection (genetically engineered products, seed collected from previous years, and hybrid seed), crop rotation, tillage and biomass practices, mechanized- or hand- labor efforts, and others (Fig. 10.1, box 4) (North Dakota State University 1997).

In terms of **e**, properties of the plot that play crucial roles in the production of corn and potential damage to ecosystem goods are soil characteristics and variation in precipitation and temperature (Fig. 10.1, box 3). Physical and chemical properties of soils govern the quantity of agricultural inputs available to production as well as their fate and transport into groundwater resources. Additionally, rainfall intensity and duration and the use of irrigation are important factors in the fate and transport of inputs. These factors help explain crop yield in addition to resource pollution dynamics.

Next we describe more extensively and formulate models for each of the components that constitute the inputs needed for the regulator's maximization problem in Eq. (10.1).

10.2.2 Producer Behavior

The objective function and constrained risk in Eq. (10.1) depend on the behavior of the producer. That producer's behavior is one of profit maximization. **Q** is the aggregation of producer outputs **q** at the plot level and risk is a function of plot conditions $\mathbf{e}$ accumulating because of the actions of the producer. Given $\mathbf{R}$ the producers seek to maximize profit on each plot:

$$\begin{aligned}
&\max_{\{\mathbf{q}_t,\mathbf{v}_t,\mathbf{z}_t\}} \sum_{t=t_0}^{\infty} d^t \pi\left(\mathbf{q}_t, \mathbf{v}_t, \mathbf{z}_t, \mathbf{P}(\mathbf{R})_t, \mathbf{W}(\mathbf{R})_t, \mathbf{e}_t\right) \\
&s.t. \quad \mathbf{e}_{t+1} = \mathbf{e}_t + \Delta\mathbf{e}(\mathbf{q}_t, \mathbf{v}_t, \mathbf{z}_t) \\
&\qquad \mathbf{R}_t^{q\min} \leq \mathbf{q}_t \leq \mathbf{R}_t^{q\max} \\
&\qquad \mathbf{R}_t^{v\min} \leq \mathbf{v}_t \leq \mathbf{R}_t^{v\max} \\
&\qquad \mathbf{R}_t^{z\min} \leq \mathbf{z}_t \leq \mathbf{R}_t^{z\max}
\end{aligned} \tag{10.2}$$

Discounting is by factor d, π is the annual profit function, $\mathbf{W}$ is input costs vector and $\mathbf{R}$ is explicitly shown as minimum and maximum regulatory constraints on the decision variables and subsidies applied to the prices and costs the producer faces. Current choice variables affect the future by changing the properties of the plot by the function $\Delta e(\cdot)$. Time, t, is discrete corresponding to planting decisions made each annual growing season. In terms of aggregating the estimation of joint output and conducting statistical analyses for regional policy analysis, we assume an annual basis is appropriate (Fig. 10.1, box 8). Although many of the financial outlays for corn production practices, their pollution consequences, related biophysical phenomena, and environmental science measurements occur at finer time periods, the bottom line for a farmer is the amount of money she ends up with at the end of the year from the choices she has made during the prior year and in years past (Lence and Hayes 1995).

10.2.3 Revealed Preference for Social Risk

The regulator solves Eq. (10.1) acting as if $\boldsymbol{\alpha}$ is given, but this acceptable risk is also the result of an optimization process at the higher level of authority of the policy maker. A probability of exceeding regulatory standard, $\boldsymbol{\alpha}(\mathbf{R})$, is associated with policies $\mathbf{R}$. The expected present discounted value of the policies to society, $\pi^s(\mathbf{R}, \boldsymbol{\alpha}(\mathbf{R}))$, can be optimized by choosing $\mathbf{R}^*$ from possible policies $\boldsymbol{\Gamma}$:

$$\max_{\mathbf{R}\in\Gamma} \pi^s(\mathbf{R}, \alpha(\mathbf{R})) \tag{10.3}$$

The choices of policy makers reveal social preferences.[5] To the extent that π^s and $\boldsymbol{\Gamma}$ are stable, $\boldsymbol{\alpha}^* = \boldsymbol{\alpha}(\mathbf{R}^*)$ is also stable. We can infer $\boldsymbol{\alpha}^*$ from observed

[5] Others have used observed government actions to reveal social preferences. McFadden (1975) inferred the revealed value of indirect costs and benefits to highway route selectors and Ross (1984) shows how revealed preference can be applied to infer the implied social weights of regulators. We aren't using reveled preference to infer values, but rather to infer the optimal constraints implied by those values.

regulatory outcomes. This inferred $\boldsymbol{\alpha}^*$ is the risk constraint to which regulators have conformed, and we assume it is a reasonable risk constraint for the future to apply to marginal alterations of policy or to the case of improved information structure.

The risk constraint matrix, $\boldsymbol{\alpha}$, consists of the acceptable risk level for multiple resources (designated by the subscript) at multiple thresholds of failure (designated by the superscript):

$$\alpha = \begin{bmatrix} \alpha_1^{T1} & \alpha_1^{T2} & \cdots \\ \alpha_2^{T1} & \alpha_2^{T2} & \cdots \\ \vdots & \vdots & \ddots \end{bmatrix} \tag{10.4}$$

For example, the subscript 1 in Eq. (10.4) could designate nitrate pollution in unconstrained deep aquifers and the superscript T1 could designate threshold 10 mg/l, such that α_1^{T1} is the probability that nitrate pollution in an unconstrained deep aquifer will exceed 10 mg/l.

The conditional probability of exceeding the natural resource quality standard $\boldsymbol{\alpha}$ is derived from a cumulative probability distribution, which is a function of properties of a land plot that affect the regional groundwater resources and hydrogeologic processes of the aquifer.

10.2.4 Cost Effective Regulation with MRLI

The regulations described above with the additional information from MRLI ($\boldsymbol{\omega}(1)$) would be $\mathbf{R}^*(\boldsymbol{\omega}(1), \boldsymbol{\alpha})$ and without additional information ($\boldsymbol{\omega}(0)$) would be $\mathbf{R}^*(\boldsymbol{\omega}(0), \boldsymbol{\alpha})$ for the probability of exceeding the regulatory standard for resource damage $\boldsymbol{\alpha}$. The additional information may allow regulations to be better targeted so that the crop production will be different $\mathbf{Q}^*_{\mathbf{R}(\boldsymbol{\omega}(1),\,\boldsymbol{\alpha})}$ with the information than without $\mathbf{Q}^*_{\mathbf{R}(\boldsymbol{\omega}(0),\,\boldsymbol{\alpha})}$ at the same resource risk level. Therefore, the VOI to the regulator is stated explicitly as:

$$VOI_{\boldsymbol{\omega}(1)} = \mathbf{P}\left[\mathbf{Q}^*_{\mathbf{R}(\boldsymbol{\omega}(1),\boldsymbol{\alpha})} - \mathbf{Q}^*_{\mathbf{R}(\boldsymbol{\omega}(0),\boldsymbol{\alpha})}\right] \tag{10.5}$$

The application of the conceptual framework involves the use of the Landsat MRLI archive to estimate the joint outputs of agricultural production and pollution. Specific crop rotations are evaluated in a dynamic model that includes spatially explicit inputs in four dimensions: latitude, longitude, depth in the hydro-geologic system, and time. Our model characterizes relevant biogeochemical processes from the land surface through geologic strata to groundwater well extraction and quantifies the effect on farm income and potential damage to private and public well water supplies resulting from repeated applications of nitrogenous fertilizer (nitrogen) over the past several decades. As observed by MRLI, the pattern of the joint output of crops and pollution from the portfolio of land uses creates information to analyze locations by screening them on the basis of different criteria and to estimate an expected return on investment from farming practices in a socially-relevant context, namely their pollution and effects on common pool resources. Note that this approach is a close approximation of an alternative value of information based on the policy maker's optimization problem (Eq. (10.3)). Equation (10.5) will yield a lower VOI than this alternative, given diminishing returns of crop production, however this alternative approach would rely on a detailed understanding of the value of groundwater protection that is difficult to accurately observe.

10.3 Application

As a demonstration of the IAA, we forecast the risk of exceeding a regulatory standard for groundwater pollution in Iowa. As mentioned before, earlier studies on the application of MRLI show significant use in the agricultural sector. MRLI provides a time series of LULC signals that are linked to levels of inputs in crop production and their potential consequences to natural resources. In predominantly agricultural states such as Iowa, where approximately 80 % of the population depends on groundwater for drinking water, regulation of nitrate contamination associated with agricultural production is an issue of serious concern. Therefore, application of the conceptual framework to agricultural production and its effects on the ecosystem service of groundwater quality is selected.

10.3.1 Agricultural Production and Its Ecosystem Service Effects

Farmers may overuse or underuse fertilizers, especially nitrogenous fertilizers (Mortensen and Beattie 2005). The improper choice of fertilizer application rate can be costly to the farmer in terms of lower yields as a result of excessive vegetative growth, susceptibility to storm and insect damage, and poorer crop quality (Brady and Weil 2002) as well as higher fertilizer cost. We model producers' fertilizer application decision (1) as part of the dynamic optimization of profit. This decision hinges on the uncertainty faced by producers about the growing season's weather conditions. Given the policies and market conditions they face, producers might address uncertainty by choosing to err on the side of under- or over-application (Sheriff 2005). Fertilizer runoff and leaching-often a result of over fertilization and lack of uptake by vegetation-into streams or groundwater is an unintended consequence that can cause ecological and environmental damage (Martinez and Albiac 2006). Groundwater pollution occurs as a result of the interaction of several factors at the land surface including fertilizer application and its interaction with the soils and hydrogeology below the surface. Furthermore, under-fertilization limits yields. Both overuse and underuse are a misallocation of resources. Although it is assumed that a farmer must meet all the requirements of the laws and that the ecological structure and function of some locations provide multiple benefits and ecosystem services, a catastrophic social loss of clean groundwater is risked when over-application of nitrogenous fertilizer is widespread and persistent (Heal 1991). A regulator's informed intervention could help internalize externalities and reduce the losses associated with a misallocation of resources.

We apply a simple production function to model crop yield with a response plateau beyond which the marginal product is zero (Hall 1998). Further, we assume that corn producers operate at constant returns to scale and display diminishing marginal productivity (Livanis et al. 2009). We apply the production function (Eq. (10.6)) to many parcels in a regional-scale analysis, so the response function must be representative of the physical processes while being tractable region-wide. A Cobb-Douglas production model with a plateau meets both criteria (Mortensen and Beattie 2005):

$$\begin{aligned} q &= \zeta_0 \prod_{i=2}^{n_e} e_i^{\zeta_i^e} \prod_{i=1}^{n_z} z_i^{\zeta_i^z} (\zeta_1 + v_1 + e_1)^{\zeta_i^e} \text{ for } (v_1 + e_1) < NIT^{\max} \\ &= \zeta_0 \prod_{i=2}^{n_e} e_i^{\zeta_i^e} \prod_{i=1}^{n_z} z_i^{\zeta_i^z} (\zeta_1 + NIT^{\max})^{\zeta_i^e} \; \textit{otherwise} \end{aligned} \tag{10.6}$$

where q is one element of $\boldsymbol{q}$—production in tons per hectare of a crop, v_1 is the nitrogenous fertilizer application rate, z_i is one of the n_z relevant elements of **z** (e.g. no till, irrigation), e_i is one of the n_e relevant elements of **e** (e.g. e_1 is residual nitrogen, and other e_i include moisture content, slope, soil type, and depth to water), and $NIT^{\max}$ is the amount of nitrogen beyond which marginal production is 0. The profit-maximizing behavior of individual farmers will neglect the overall welfare of the region. Here we hypothesize that a mismatch of parcel characteristics and crop production can lead to an increased likelihood of exceeding the regional regulatory standard for nitrate concentration in groundwater.

In this problem, given the land characteristics and the particular crop that the farmer is trying to grow, we assume there is a continuum of land parcel types, and an associated optimal application of land uses, fertilizer and irrigation and other inputs as costs of production. The combination of individual production decisions and observation of the producers at regional scale provides a mechanism to evaluate pollution issues at both intensive and extensive agricultural margins over time. Decisions at the intensive margin include management decisions such as chemical application rates for a given unit of land (Antle and McGuckin 1993; Antle and Just 1991). These types of decisions are usually short run input decisions including how much nitrogenous fertilizer should be applied during a specific growing season. Decisions at the extensive margin determine what land is used for production versus other purposes (e. g., USDA conservation programs), and thus determine the environmental characteristics of land in and out of production. These decisions entail long-run concerns such as crop rotations over different growing seasons. Thus, some parcels of land ought to be used in a different way (e.g. marginally productive lands close to streams ought to be in a USDA conservation program) and the parcels of land in a given LULC assignment can be mismanaged (i.e. fertilizers can be over applied). Both of these problems increase the magnitude of the pollution problem. The magnitude of the problem is a function of the history of use *j*. Depending on such characteristics as application rates, watershed size and connectivity, slope position, nitrate mobility, dispersion and travel times, aquifer and well depth, and groundwater age, the history of use can influence the groundwater pollution levels on the order of years, decades or centuries (Tomer and Burkart 2003; Meals et al. 2010).

Evaluation of the accumulation of regional environmental effects begins with application of a multi-period, spatial model of agricultural production for corn-soybean crop rotations (Lambert et al. 2006), where the land use and crop rotations are observed using MRLI. The first part of the model is the agricultural production and response function in Eq. (10.6). The planting pattern dictated by Eq. (10.6) will—in steady state—reduce either to some rotation pattern between corn and soybean or to fallow (i.e., in forage, conservation program, or non-agricultural use) We should also note that, to a relatively minor extent in our study area, the choice is sometimes to grow crops other that corn and soybeans. The choice to plant a crop in a given season depends on relative prices of the crops and costs of inputs. The decision to produce corn during a particular rotation means that a specific quantity of nitrogen is applied and if the decision is to produce soybeans, a different quantity of nitrogen (if any at all) will be applied (Lambert et al. 2006). Fallow land requires no nitrogen, but later planting of the site will affect retained nitrogen.

The longest ongoing studies of corn-soybean sequencing in the northern Corn Belt (Lauer et al. 1997) indicate that in a typical rotation, the production of corn is 13 % higher than with continuous corn and soybean production is 10 % higher than with continuous soybean. Transitioning from fallow, first year corn is 15 % more productive, but no improvement is apparent for second-, third-, or later-year corn crops, compared to simply growing corn continuously. Soybean production is 18, 8, and 3 % more productive for first-, second-, and third-year crops, respectively, compared to continuous soybean, and no improvement is found after the third consecutive year of soybean planting. These data average results from specific experimental plots and indicate the crops' typical production trends over time; the exact effect of a rotation cycle for a given crop's production depend on the particular site's characteristics, rotation history, and other management practices.

The producer re-calculates the dynamic optimization in Eq. (10.2) with all available information at time $t = t_0$ when preparations for the next planting are made. The producer knows site characteristics, past planting on the site, cost of inputs, and expectations for the price of the alternative crops this season and future costs and prices. At that time, the producer chooses the next crop to plant that maximizes the profit for the current season plus the discounted expected profit from all future seasons (that also depend on the current choice). This choice is a transition from the past planting to the next. Because of uncertainty about each producer's knowledge and expectations, this transition is a probabilistic transition from the crop of one season to the crop for the next. The state of the crop planting at time t is represented by $\boldsymbol{\delta}_t$:

$$\boldsymbol{\delta}_t = \begin{bmatrix} \delta_{1,1,t} & \delta_{1,2,t} & \delta_{1,3,t} & \delta_{1,4,t} \\ \delta_{2,1,t} & \delta_{2,2,t} & \delta_{2,3,t} & \delta_{2,4,t} \\ \delta_{3,1,t} & \delta_{3,2,t} & \delta_{3,3,t} & \delta_{3,4,t} \end{bmatrix} \tag{10.7}$$

where $\delta_{j,\,l,\,t} = 1$ for land use j at time t for the lth consecutive year and 0 otherwise,$\delta_{j,\,4\,,t} = 1$ if land use j is planted at time t for the fourth or higher consecutive year and 0 otherwise.

The crop choice is mutually exclusive; therefore:

$$\sum_j \sum_l \delta_{j,l,t} = 1 \tag{10.8}$$

The state of the crop planting can transition to either the first year of another crop or the next year of the same crop. The probability $\boldsymbol{\gamma}$ describes the probability that $\boldsymbol{\delta}_t$ will transition to some $\boldsymbol{\delta}_{t+1}$. The transition probability from the lth year of crop j to the $l1$th year of crop $j1$ is $\gamma_{j,l}^{j1,l1}(\mathbf{z}_t, \mathbf{P}_t, \mathbf{W}_t, \mathbf{e}_t)$. Thus, given each possible crop state, we have a discrete choice of three possible outcomes that depend on prices and characteristics known at time t. For first-year corn in the prior year, the next crop state could be second-year corn with probability $\gamma_{1,1}^{1,2}$ or first-year soybean with probability $\gamma_{1,1}^{2,1}$, or first-year other with probability $\gamma_{1,1}^{3,1}$. But, for first-year corn all other transitions are not possible, e.g. the crop state could not be third-year soybean next season or fourth-year other, so those transition probabilities are by definition zero. We relate the transition probabilities to known factors using a multinomial logit:

$$\mathrm{In}\left(\frac{\gamma_{1,1}^{1,2}}{1-\gamma_{1,1}^{1,2}-\gamma_{1,1}^{2,1}}\right) = {}_{1,1}^{1,2}\beta_0 + {}_{1,1}^{1,2}\beta_1 p_1 + {}_{1,1}^{1,2}\beta_2 p_2 + {}_{1,1}^{1,2}\beta_3 p_3 + {}_{1,1}^{1,2}\beta_4 w_1 \\ + {}_{1,1}^{1,2}\beta_5 w_2 + {}_{1,1}^{1,2}\beta_6 w_2 z_1 + {}_{1,1}^{1,2}\boldsymbol{\beta}^e \mathbf{e} + \varepsilon_{1,1}^{2,1} \tag{10.9}$$

$$\mathrm{In}\left(\frac{\gamma_{1,1}^{2,1}}{1-\gamma_{1,1}^{1,2}-\gamma_{1,1}^{2,1}}\right) = {}_{1,1}^{2,1}\beta_0 + {}_{1,1}^{2,1}\beta_1 p_1 + {}_{1,1}^{2,1}\beta_2 p_2 + {}_{1,1}^{2,1}\beta_3 p_3 + {}_{1,1}^{2,1}\beta_4 w_1 \\ + {}_{1,1}^{2,1}\beta_5 w_2 + {}_{1,1}^{2,1}\beta_6 w_2 z_1 + {}_{1,1}^{2,1}\boldsymbol{\beta}^e \mathbf{e} + \varepsilon_{1,1}^{2,1} \tag{10.10}$$

$$\gamma_{1,1}^{3,1} = 1 - \gamma_{1,1}^{1,2} - \gamma_{1,1}^{2,1} \tag{10.11}$$

where p_1 is the real price of corn, p_2 is the real price of soybean, p_3 is the real subsidy per hectare for conservation program lands, w_1 is the real cost of nitrogenous fertilizer, and w_2 is the real cost of diesel fuel for operations including irrigation. Similar equations apply to the transition from every other crop state.

10.3.2 Loss Estimation

A societal choice involves decisions at regional scale and has an overarching effect on a variety of individual land-owners and firms. Over long periods of time, production of corn, its associated pollution, and its cumulative environmental effects at regional scale can cause substantial changes to groundwater. The policy issue is whether this ecosystem service becomes stressed and could be compromised, or even lost, because of changes in the land use pattern. The social risk is the under- or over-regulation pertaining to groundwater contamination. However, the true state of contamination in space and time at the regional scale is unknown and can be only estimated with uncertainty. Thus, there is a need for a probabilistic model of groundwater vulnerability.

A probabilistic estimate of where and when to regulate the crop producers in a region can be based on the relevant circumstances above the groundwater over t years or by way of a health standard, both of which can be assembled into the vector α in Eq. (10.4). Thus, for our analysis we assume there exists some conditional probability distribution of an adverse environmental effect in a region from land use j that is known to the regulator (Nelson and Winter 1964) and consists of three elements (Philips 1988): (1) the set of possible states of the environment $\{D_k\}$ that a nonpoint source groundwater contamination incident of resource damage type for all $j, j \in J$ land uses involved in the transition to state k, $k \in K$, possible states of concentration; (2) an observation about land use j with information ω from a specific type of MRLI; and (3) the conditional probability that a time series of land uses was observed, suggesting that a particular state of the environment will prevail in the future.

Like the production function in Eq. (10.6), the probability of a particular level of groundwater resource damage attributed to a land use at any one point in time in Eq. (10.12) is related to the intensity of crop production $\mathbf{q}$, the variable inputs $\mathbf{v}$, the methods $\mathbf{z}$, and the properties of the plot $\mathbf{e}$:

$$p_k = f(\mathbf{q}, \mathbf{v}, \mathbf{z}, \mathbf{e}) \tag{10.12}$$

p_k depends on nitrogenous fertilizer related activities on the soil surface and other nitrogen sources (e. g., atmospheric deposition, land use and cropland type, population density, and manure management), management practices including irrigation and tillage, properties of the soil in the plot, temperature and precipitation, movement of nitrate (converted from nitrogenous fertilizer) to aquifers (depending on soil biological, chemical, and physical properties and the presence of clay), movement and conductivity of the lithology (depending on characteristics of sub-surface geology such as hydraulic conductivity and bedrock such as limestone), denitrification of accumulated nitrate in aquifers (presence or absence of denitrification facilitating layers in the soil and at depth), texture of aquifers (especially unconsolidated sand and gravel), and drainage and recharge rate of and well extraction from the groundwater system (Nolan and Hitt 2006; Nolan et al. 1997, 2002; Canter 1997).

The contamination problem is one of a rate of accumulation of long-term nitrogen application in crop production. Near-surface, soil nitrogen-cycling processes such as fixation, mineralization, immobilization, nitrification, denitrification and plant uptake are often costly and difficult to measure accurately and precisely (Nolan et al. 2010). That said, a model focusing on the near-surface soil processes is:

$$e_{1,t+1} = e_{1t} + v_{1t} - le_t - \eta_t - u_t \tag{10.13}$$

where v_{1t} is the quantity of nitrogen applied as fertilizer e_{1t} is the nitrogen stock in the soil, and le_t is the quantity of nitrogen that has leached from the soil. Of the remaining variables u_t is the amount of nitrogen uptake by land use j, and η_t is the nitrogen volatilization. Nitrogen uptake and volatilization are a proportion of yield $(u_t = \upsilon q_t)$ and fertilization $(u_t = \tau v_{1t})$, respectively, where τ is the volatilization rate into the air, υ is the nitrogen uptake rate by plants, and t is the number of time periods of observation by MRLI.

The nitrate leached le_t, is the principal component of nitrogen applied that accumulates in groundwater over the period of time. Nitrogen fertilizer leached in the form of nitrate le_t is explained by:

$$le_t = f\{Y, NIT, PPT, TEM, Si, LM\} \tag{10.14}$$

Where variables are the amount of fertilizer applied (*NIT*), residual nitrogen in soil, nitrogen uptake by crops, properties of soil (*Si*, i.e. soil texture, and hydrologic group), soil temperature (*TEM*), water inputs (*PPT*, i.e. irrigation/precipitation), water table height, and land management practices (*LM*).

The quantity of leached nitrate in a given year depends upon the activities and properties of an area contributing its pollution in the given year. That area is hereafter termed as the catchment zone of a well. Earlier research on groundwater pollution defined its catchment zone as a circular area around a well. For example, Nolan (2002) used a circular area with diameter 500 m to model nitrate contamination in groundwater in given period. One of the ways to determine catchment zone would be to use an analytical element method (AEM). The AEM is capable of determining catchment zones for number of time periods using hydrogeologic properties such as aquifer base elevation, aquifer thickness, porosity and hydraulic conductivity of the geologic layers, flow gradient, and net extraction from wells. The quantity of nitrate contributed to a well by certain land use within a catchment zone for each year is used in estimating the amount of nitrate accumulated over a period of time.

Suppose a decision must be made in each t about whether to regulate nitrogen use at the land surface because of an increase in the concentration of nitrate in well water. Does the decision maker wait until the groundwater in a well is contaminated when sampled or does she anticipate the exceedance of the nitrate standard and intervene, through regulation, before a contamination incident occurs? We assert that a point in time exists when the regulator will have estimated that the probability of exceeding the standard at some point in the future will be great enough to take action to mitigate an adverse effect in the present.

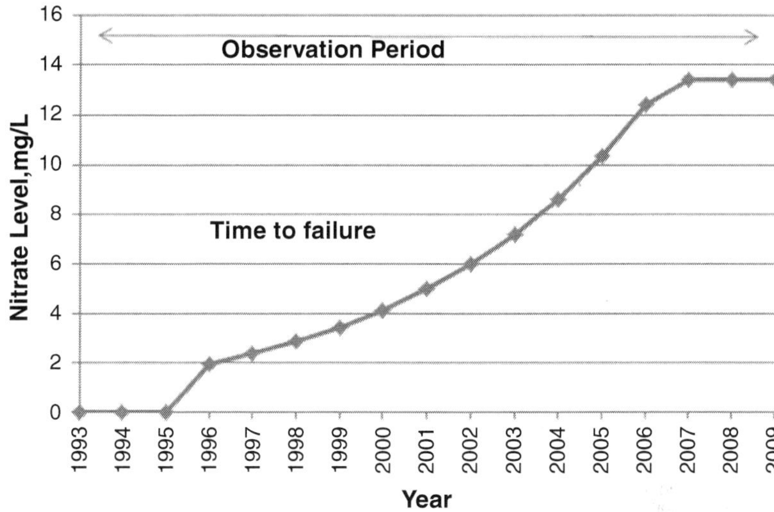

Fig. 10.2 Hypothetical illustration of cumulative nitrate indicator (CNI_t) of nitrate level

To capture the dynamic nature of the nitrate contamination problem, we introduce the cumulative nitrate indicator (CNI) shown in Fig. 10.2. Cumulative nitrate pollution in a well can be modeled with a difference equation (Yadav 1997). Equation (10.15) is the difference between the accumulated nitrates in the previous time period since time period i plus the addition of nitrate to the pool in the year the nitrogenous fertilizer was applied:

$$CNI_t = CNI_{t-1} + (\Delta NO_3)_t \qquad (10.15)$$

where CNI_t is the value of the cumulative nitrate indicator in year t, $CNI_t - 1$ is the value of the cumulative nitrate indicator in year $t - 1$, and (ΔNO_3) is the nitrate concentration change over the course of 1 year.

The nitrate concentration change in a year depends upon the surficial activities and weather variables in previous years and the soil characteristics, and characteristics of the geology underneath. The *CNI* provides the foundation for estimating the probability of exceedance of the regulatory standard. The *CNI* captures the spatially and temporally cumulative exposure of the aquifer to nitrate contamination while accounting for nitrate degradation during transport to the aquifer. The *CNI* can be expressed as:

$$CNI_t = CNI_{t-1} + \beta^0 \mathbf{e}_t^0 + \sum_{g \in G} c^g \left(\mathbf{q}_{t-g}, \mathbf{v}_{t-g}, \mathbf{z}_{t-g}, \mathbf{e}_{t-g} \right) \qquad (10.16)$$

where $\boldsymbol{\beta}$ represents the coefficients for explanatory variables, g designates the groundwater catchment zone that affects the well bottom, g years after a surface activity by the function $c^g\left(\mathbf{q}_{t-g}, \mathbf{v}_{t-g}, \mathbf{z}_{t-g}, \mathbf{e}_{t-g},\right)$. Note that g is both a label that designates the catchment zone and a parameter—the average travel time from the zone to the well bottom. G is the set of groundwater zones identified for a well or potential well and some recovery is possible depending on the properties of the well bottom, $e^{0}{}_{t}$.

The CNI_t, because it is calibrated to use all available information to estimate groundwater nitrate concentration, also brings together all information needed to assess the likelihood that contamination standards will be exceeded. Nitrate accumulation is determined by fertilizer application and other nitrogen sources, leaching of deposited nitrogen, and its transportation to groundwater strata. Transportation of leached nitrate to an aquifer depends on the travel time through geologic layers underneath the soil surface to the aquifer and sinks. There is a lag time in surface water infiltration through the hydrogeologic system, which depends on factors such as the thickness and infiltration rates of the unsaturated zone (Oakes 1982). Estimation of travel time to an aquifer must include the variability of the surficial geology and its hydraulic conductivity, dispersion, advection, permeability and oxidation zones, pump and rates of extraction, and the age and recharge rate of the ground water system (Canter 1997; Marsily 1986; Bear 1979).The CNI_t is an input to the conditional probability of exceeding a concentration level k, threshold of nitrate contamination that adversely affects humans. A survival (1 − failure, F) analysis (Kleinbaum 1996; Lancaster 1990; Kalbfleisch and Prentice 1980) is applied by the regulator using α.[6] The exceedance probability of a given loss is the combination of the loss L and the exceedance probability α. The first step is to estimate the instantaneous potential per unit time for a contamination event to occur as the hazard rate of transitioning from an uncontaminated state to a contaminated state of the groundwater. The nitrate concentration hazard function is the instantaneous rate of leaving the current state to destination k per unit time period at t:

$$h(t) = \sum_{k=1}^{k} h_k(t) \tag{10.17}$$

where
$h_k(t)dt = \Pr(\text{departure to state } k \text{ in the short interval } (t, t+dt), \text{given survival to } t$

$$h_k(t) = \lim_{dt\to 0} \frac{\Pr(t \le T \le t+dt, D_k = 1 | T \ge t)}{dt} = \frac{p_k f_k(t)}{\bar{F}(t)} \tag{10.18}$$

where T is the duration time of staying in state k, D_k is a set of K dummy variables with a value of 1 if state k is entered and 0 otherwise, p_k = Pr(when departure occurs to destination k) and can be estimated with a rank-ordered or

[6] It is assumed that good regulatory policy reduces or eliminates the adverse health effects of nitrates on humans.

mixed logit statistical regression estimated with the variables in Eq. (10.12), $\bar{F}(t)h_k(t)dt = p_k f_k(t)dt$,

$$\begin{aligned}\bar{F}(t)h_k(t)dt &= \Pr(\text{survival to } t) \times \Pr(\text{departure to } k \text{ in } (t, t+dt))|\text{given survival to } t \\ &= \Pr(\text{departure to } k \text{ in } t, t+dt) \qquad , \text{ and}\end{aligned}$$

$$\begin{aligned}\bar{F}(t) &= \sum_{k=1}^{K} p_k \bar{F}_k(t),\ \bar{F}_k(t) \\ &= \Pr(\text{survival to } t, \text{ given that when departure occurs it is to } k)\end{aligned}$$

The specific model used in the application is the Weibull proportional intensities hazard function (Lancaster 1990):

$$h_k(t) = a\exp\left\{x_k'\beta_k\right\}t^{a-1} \tag{10.19}$$

and

$$\alpha = \bar{F}(t) = \exp\left\{-t^a \sum_{k=1}^{K} \exp\{\mathbf{x}_k'\beta_k; CNI_t\}\right\} \tag{10.20}$$

Where $\mathbf{x}'\beta_k$ here are explanatory covariates, $\bar{F}(t)$ is a cumulative distribution function that is the survivor function (Lancaster 1990) and is synonymous with the exceedance probability α, a is the Weibull parameter, and $p_k f_k(t)$ is the probability of an event that is defined as the nitrate concentration level in a number of wells in a region that exceeds a maximum contamination level (MCL) for specific health effects as a result of land use j. The greater the number of wells exceeding MCL, the more damaging the event—or loss L—is to the regulator. There are two components to estimating the loss, L: the size and extent of the natural resource damage to the ground water and the lost economic benefits of agricultural production for the area causing the natural resource damage, L_i.

Regulation of agricultural production requires the estimation of the economic loss associated with a loss event L_i, $i = 0, \ldots, I$, possible events that cause a given amount of resource damage. The economic loss is the cost to producers of regulating nitrogenous fertilizer application at the intensive margin as a tax on fertilizer inputs or as a standard limiting use, and/or at the extensive margin as an incentive to reduce the amount of crop acreage (e.g., the USDA Conservation Reserve Program). This calculation applies in our region of interest above the social risk threshold. By definition, the optimal loss is the loss in agricultural production in this region. The optimal loss will be the value of the resource at risk below the risk threshold. However, because the decision affects the producer and the public differently, there is an asymmetry of loss in the decision. See the appendix for a description of a Bayesian decision approach to the problem. Here we

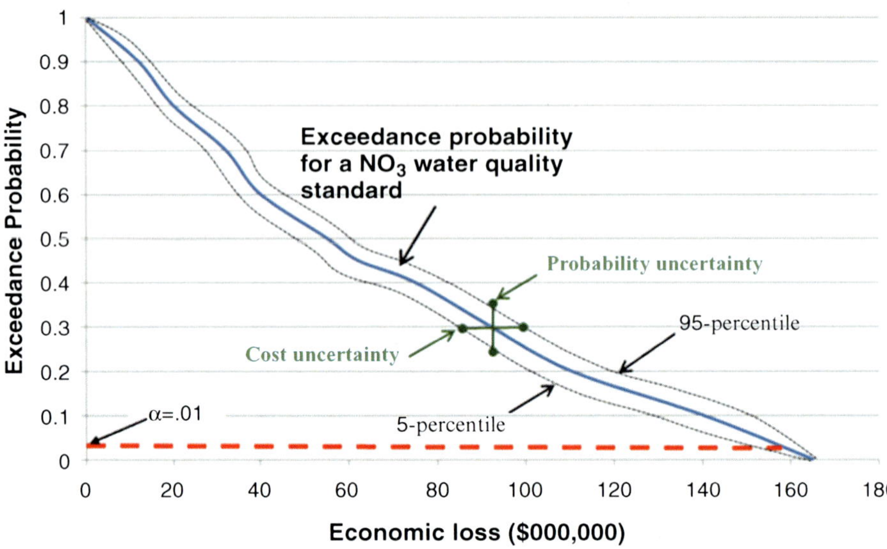

Fig. 10.3 Hypothetical exceedance probability curve for nitrate water quality standard threshold

estimate the economic loss associated with the regulatory constraint in Eq. (10.3). The loss to the producers L_i is the economic loss from event i:

$$L_i = \mathbf{P}(\mathbf{Q}(\mathbf{R}^*) - \mathbf{Q}) \tag{10.21}$$

Combining the survivor function in Eq. (10.20) with the expected loss in Eqs. (10.21) and (10.22), yields an estimate of the exceedance probability of a given economic loss $\alpha(L_i)$ (see Grossi and Kunreuther 2005, for an application of a survivor function to catastrophe modeling that is termed a probability of exceedance) is based on nitrate health standards. The exceedance probability for a given level of loss from event i is:

$$\alpha(L_i) = \Pr(L > L_i) \tag{10.22}$$

The risk to the decision maker of a groundwater resource failure can now be determined. To conduct a risk analysis we set a tolerance level at a specific exceedance probability for a regulatory standard (i.e. nitrate concentration in groundwater). The risk tolerance level for $\alpha = 0.01$ is shown in Fig. 10.3. The dashed line shows for that exceedance probability an economic loss (cost to the producers) if the health standard is exceeded. The vertical axis represents the exceedance probability and the horizontal axis is the economic cost due to the regulation associated with an estimated amount of natural resource damage.

Alternatively, we can construct a marginal groundwater protection supply curve based on the marginal loss in production. This creates a more direct—and less difficult to appropriately apply—measure of the value of the protected resource.

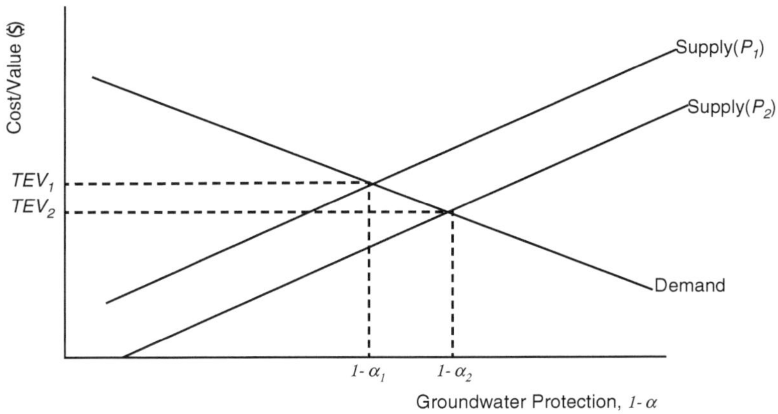

Fig. 10.4 Supply and demand for groundwater protection

This supply curve shifts as crop prices change from, say, $\boldsymbol{P_1}$ to $\boldsymbol{P_2}$ (Fig. 10.4). At the observed levels of groundwater protection, α_1 and α_2, the marginal cost of this protection level equals the marginal value of the groundwater protected, TEV_1 and TEV_2. From the observed history of variation in crop prices and the corresponding variation in groundwater protection, we can estimate a demand curve for groundwater protection that aggregates all of the values of the resource, for total economic value.

10.3.3 VOI: Comparison of Sensor Data Sets

Implementation of the IAA assumes that individual producers use prior information about markets, prior production, and regulations to decide whether to plant corn or do something else with the land. MRLI is used to observe those decisions. Thus, a policy could either target specific sensitive parcels of land, or, more likely, adjust general rules or incentives that change the pattern of planting and fertilizer application across the region's landscape. The estimates are based on the characteristics and derived classification products of each distinct sensor carried on a specific satellite platform and their imagery archives. For the National Aeronautics and Space Administration's (NASA) and the USGS's Landsat program, a 38-year archival history of observations exists while the Indian Space Research Organization's Advance Wide Field Sensor (AWiFS) has a 5-year history. The sensors' attributes and their expected cost savings can be compared.

The sensor and resolution characteristics of each type of MRLI provide data at a given error rate (i.e. typical remote sensing user and producer accuracy assessments) in detecting the land use practices of individual farmers that is central to the VOI determination. Calculation of VOI involves a two-phase analysis: (1) estimate VOI derived from remotely sensed land imaging versus traditional sampling methods[7] and (2) estimate VOI derived from a satellite sensor aboard Landsat versus another sensor on a different satellite (e. g., AWiFS). The VOI is the incremental cost savings in an application to demonstrate an economic return to the investment in the imagery. The $VOI_{\boldsymbol{\omega}(1)}$ is the benefit B possible given information from some source $\boldsymbol{\omega}(1)$ minus the information available without that source $\boldsymbol{\omega}(0)$:

$$VOI_{\omega(1)} = B_{\omega(1)} - B_{\omega(0)} \tag{10.23}$$

Alternatively stated, the difference between the predicted loss of value of production as a result of more stringent regulation pertaining to prevent groundwater resources loss using $\boldsymbol{\omega}(1)$ and $\boldsymbol{\omega}(0)$ is the proxy for VOI. Thus, some important VOI questions follow: which information source ($\boldsymbol{\omega}$), is more accurate and comprehensive? Which MRLI source has fewer errors of commission and omission? Does the estimation error affect the decision? How does an information archive improve prediction of resource loss from stock pollutants? Because an MRLI archive is critical for estimating the historical use of the land and its associated legacy of effects on common pool resources, does the temporal extent of the archive make any difference? Can MRLI sensors from different satellites like Landsat and AWiFS be coupled to improve the LULC classification probability distribution and associated estimates?

To summarize aspects of the previous sections and to frame the following example, the model is implemented using the following steps:

1. Identify an operational use of remotely sensed data that has a quantifiable economic value and policy relevance, in this case, the effects of land use practices on groundwater quality. Highlight the intersection of policies that encourage biofuels from corn production, incentivize a reduction of agricultural production for protecting resources via USDA conservation programs, and conflict with provisions of the Safe Drinking Water Act.
2. Apply Landsat or other MRLI to monitor changes in LULC that affect ecosystem functions, goods and services to estimate how the use of MRLI data brings tangible economic benefits to users.
3. Use annual MRLI observations as the basis for estimating crop yields and nutrient loading into the soil. Estimate the joint outputs of agricultural production and pollution. Couple observed land uses over time with water quality test

[7] Traditional sampling methods have included the following trajectory over time: prior to 1945, crop area estimates were not consistently available; from 1954 to 1978, area sampling frames with aerial photography were determined and field-surveys conducted; from 1978 to 1999, Landsat supplemented aerial photography for sampling stratification, but field surveys still included regression estimators for major crop acreages, harvest by region, state and county as well as livestock numbers, economic variables and farm demographics; from 2000 to present, the remotely sensed and classified Cropland Data Layer provides wall-to-wall crop types and areas, yet it still requires the June Agricultural Survey to collect ~11,000 field-based samples nation-wide (Hale et al. 1999; Lubowski et al. 2005).

data at different depths in a statistical survivor analysis. Use the conditional probability of exceeding a nitrate regulatory standard as the threshold for regulating agricultural production.
4. Measure the economic consequences for farm income and potential effects on private and public well water supplies resulting from repeated nitrogen usage.

10.4 Example

Groundwater has both market and nonmarket values. These include use values that are sometimes partially paid for, such as drinking water and irrigation supply, and non-market use values such as wetlands and spring water sources for in-stream flow. Groundwater also has non-use values that are related to quality, such as option value (Weisbrod 1964), and existence value (Krutilla 1967). The nonmarket value can be estimated using revealed-preference methods such as travel cost and hedonic pricing, or stated-preference methods such as contingent valuation and conjoint analysis. The overall purpose of the various techniques is to determine the willingness to pay (WTP), in this case WTP for protecting groundwater quality. WTP estimated using contingent valuation ranges from $57.37 to greater than $1434.37 per household per year (Crutch-field et al. 1997; Poe 1999). Jordan and Elnagheeb (1993) used a contingent valuation payment card to estimate WTP specifically for nitrates reduction at $204.89 per household per year for public wells, and $237.17 for private wells. Poe and Bishop (1999) estimated WTP for a one-unit improvement in nitrate contamination at $136.64 per household per year in a study conducted in Wisconsin, and in Delaware it was estimated at $150.79 by Sparco (1995). Loomis et al. (2009) used conjoint analysis to estimate WTP for reducing health risks in infants by reducing nitrate in drinking water using actual and hypothetical markets. WTP estimates for reduction in risk of shock, brain damage and mortality in infants was $2, $3.70 and $9.43, respectively, in actual markets, and $14, $26, and $66, respectively, in hypothetical markets, indicating bias in the hypothetical situation. Contingent valuation and conjoint analysis also can be used to value groundwater protection. The range in values here, however, suggests that the application of these methods contains variability in outcomes, and uncertainty about the correct, absolute value.

Complete and correct aggregation of all the use and non-use, market and non-market values of clean groundwater to a total economic value demand curve[8] is fraught with uncertainty and methodological difficulties. Care is needed to avoid (1) double accounting of some values contributing to more than one component value and (2) omission of unidentified, yet significant, values. Furthermore, as

[8] We will express demand as the as the schedule for the price, *TEV*, that society is willing to pay to ensure groundwater will be protected with certainty $1 - \alpha$. Thus, *TEV* is dollar value of marginally increasing $1 - \alpha$.

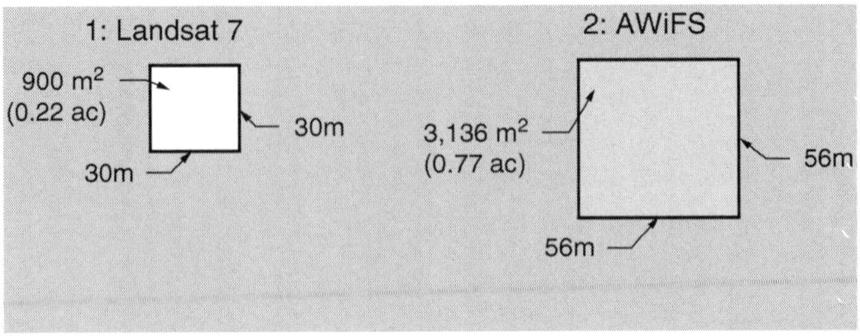

Fig. 10.5 Spatial characteristics of two MRLI sensors. The thermal band (10.6) has a resolution of 60 m, and the panchromatic band (10.8) has a resolution of 15 m (Note: Representations of spatial sensor resolutions are to relative scale)

suggested above, the accuracy of the component values is limited by the valuation techniques used.

In the example, MRLI is used to observe (screen) the land uses of a population of land parcels in Iowa that vary in their biophysical, ownership, and location characteristics. The land parcels are assigned a land use j, (j = corn, soybeans, other agriculture, or developed). An observation is processed for locations across a landscape relevant to a groundwater resource that results in a label for that land parcel; this label assigns a land use j for all land associated with a groundwater resource. Using spatial autocorrelation of well nitrate measurements can determine the possible spatial extent, lag distance, and direction of the influence that other well nitrate observations may have on a particular well, which can relate to the delineation of the catchment zone. After receiving the land-use signal and updating the exceedance probability for the regulatory standard, the regulator makes the decision to either regulate or not regulate the use of nitrogen given the crop rotation pattern covering the groundwater recharge area.

10.4.1 Data

Two primary sensors are compared in the example: Landsat 7 Enhanced Thematic Mapper plus ETM + = ($\boldsymbol{\omega}(1)$), and AWiFS = ($\boldsymbol{\omega}(0)$). Some basic characteristics of the two sensors are detailed in Fig. 10.5. The return intervals and number of bands for the two sensors are as follows:

1. Landsat 7: 16 days,[9] 8 bands (3 visible, 2 short-wave infrared, 1 thermal infrared, 1 mid-wave infrared, and 1 panchromatic).
2. AWiFS: 5 days, 4 bands (3 visible, 1 short-wave infrared).

[9] Please note, with Landsat 5 and 7 operating concurrently with polar-opposite orbits, the revisit rate is 8 days.

Among other factors, the ability of sensors to detect ground features accurately is a function of pixel size, spectral signatures, band combinations, frequency of overflights, cloud cover, and image processing, classification, and analysis techniques. Assumptions about the scale of certain features important to the integrated assessment (e. g., field, parcel, farm, census block, ZIP code, county) are implicit to the scale to which a sensor can resolve. For example, a typical center-pivot irrigation system on a large, factory farm can cover 128 acres, an area resolvable by both sensors. Some family and organic farms, however, are much smaller and may be only one field on a portion of a parcel. These scales may be resolvable only by AWiFS or Landsat, just Landsat, or not by MRLI at all. Furthermore, the return intervals of the different sensors, especially when considered in conjunction with the potential for cloud cover, can create an incentive or disincentive for using one sensor over another. This is further complicated by the number of bands and potential band combinations available from each sensor for classification purposes.

Returning to the economic model with the MRLI capabilities and sensor characteristics in mind, $\boldsymbol{\omega}(1)$ is the information with continuing Landsat data from the 38 year history of Landsat. $\boldsymbol{\omega}(0)$ is the data available if Landsat were discontinued and includes the best available alternative data, which we presume to be from the AWiFS remote-sensing platform. $\boldsymbol{\omega}(0)$ includes the Landsat historical archive, but these archival data are less compatible with the AWiFS data so calibrating risk models and assessing the cumulative condition of the sites will be compromised. $\boldsymbol{\omega}(1)$ includes the continuation of spatially—explicit LULC data derived from Landsat observations, specifically whether the land is used to grow corn, soybeans, or other cover types, or is developed. The primary difference in the information structure and availability between $\boldsymbol{\omega}(1)$ and $\boldsymbol{\omega}(0)$ is that no new Landsat observations are available.

10.4.1.1 MRLI Observation and Classification

Figure 10.6 displays a time series of MRLI observations and classification of LULC from 2000 to 2008. The land is classified each year according to what the sensor detects, the ability of the classifier, and the availability of ancillary data. The 9 years of observations depicted in Fig. 10.6 indicate that: (1) crop rotations, although persistent, are not perfectly repetitive as evidenced by the field outlined in blue; (2) apparent differences exist in the resolution of ground features in the earlier years with Landsat relative to the later years with AWiFS; and (3) the area that is neither corn nor soybean is consistent over time but less prevalent with AWiFS. Each of these points leads to a different amount of actual and modeled nitrogen use. For example, with year-over-year blanket coverage of the area, there is less chance of missing disruptions to the crop rotation patterns and the associated changes in nitrogen application than compared to traditional sampling techniques.

Next, the classified land uses are input to the agricultural production function in Eqs. (10.6), (10.7), (10.8), (10.9), (10.10), and (10.11) to estimate nitrogen use. If the

Fig. 10.6 Cropland data layer: Landsat (2000–2005) and AWiFS (2006–2008)

land has been classified as corn, based on the county and state agriculture records, we calculate the amount of nitrogen applied for a given range of yields in the area. If not corn, the amount of nitrogen for soybeans is less and for fallow it is zero.

10.4.1.2 Historical Agriculture and the Hydrogeologic System

This section provides agricultural and hydrogeologic context for the application of the conceptual framework to estimate CNI_t, α, and p_k in Iowa. Figures 10.7 and 10.8 provide historical context for the acreages and yields of corn and soy harvested in Iowa. It is interesting to note the general linear trends (with some outliers) in yield

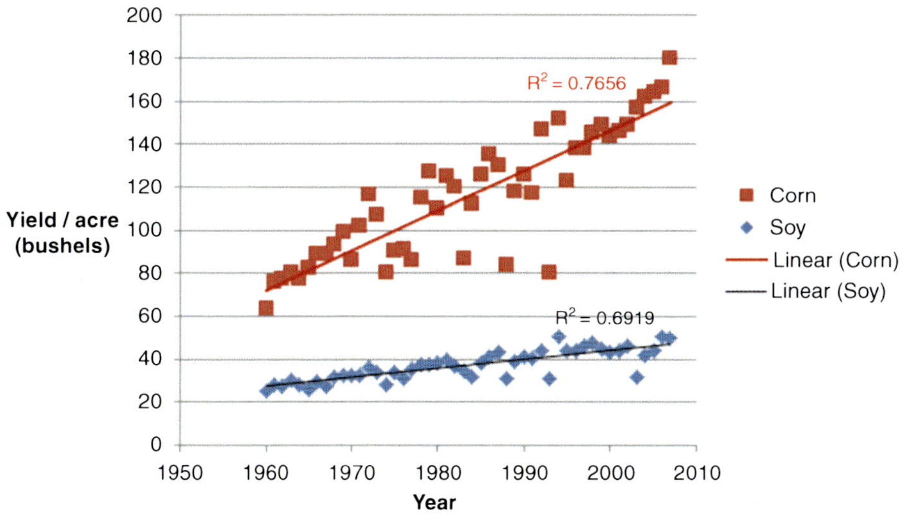

Fig. 10.7 Corn and soybean trends in yield in Iowa (Source: Iowa State University Extension Service 2010)

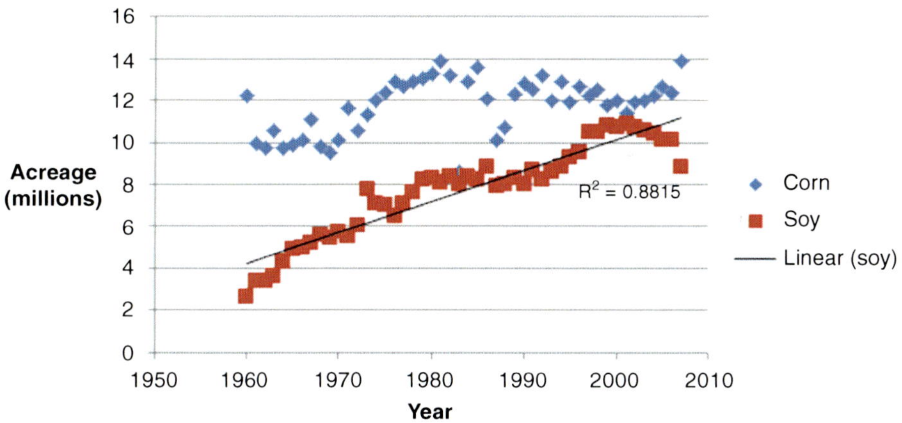

Fig. 10.8 Corn and soybean trends in acreage in Iowa (Source: Iowa State University Extension Service 2010)

per acre for corn and soybeans in Iowa from 1960 to 2007 (Fig. 10.7), which suggests that agricultural production technology and techniques became more sophisticated, industrialized, and efficient. One of the influencing agricultural production technologies is increased use of nitrogenous fertilizer, which results in an increase in nitrate concentration in groundwater. The general trends, which are less linear, for acreage in production for corn and soybeans in Iowa from 1960 to 2007 are shown in Fig. 10.8. In both Figs. 10.7 and 10.8, the R^2 values are provided

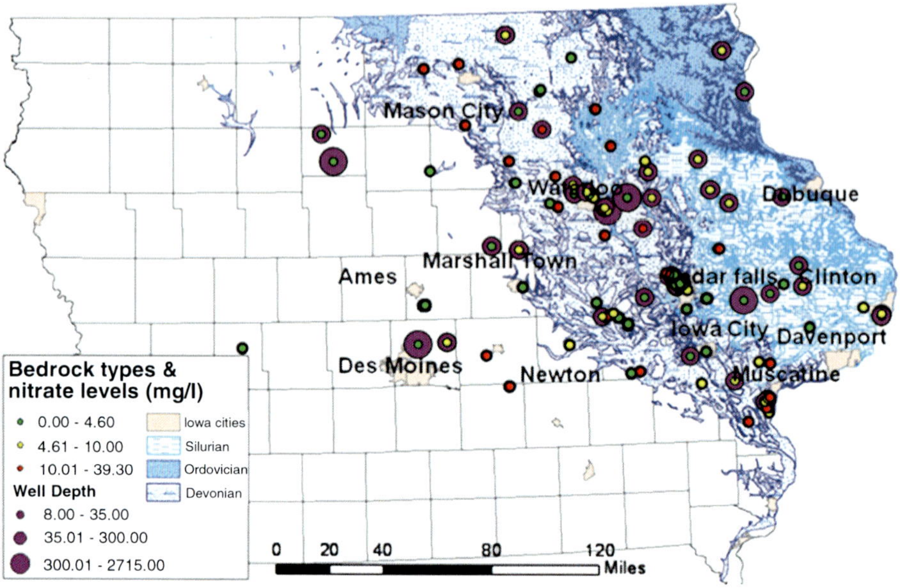

Fig. 10.9 USGS monitored sites and groundwater provinces. Approximately 20,000 Iowa Department of Natural Resources wells are monitored, but not shown

(where linearity is assumed), and the explanatory power of the variance in the trend lines are high.

In comparing the correlation of national and state trends—as well as in-state trends—in acreage, given the apparent indication that the trends are non-linear, the significance was tested with Kendall's tau and Spearman's rank tests instead of Pearson's product–moment correlation test. For corn, correlations were 0.745 and 0.880 (both p-values < 0.01), respectively, suggesting a statistically significant correlation in acreage dedicated to production over time at national and state levels. For soybeans, correlations were 0.809 and 0.928 (both p-values < 0.01), respectively, also suggesting a statically significant correlation over time in acreage dedicated to production at national and state levels. In Iowa, the correlation between corn and soybean acreage for the Kendall's tau test was not found to be significant at the p-value < 0.01 level, but the Spearman's rank test was found to be significant at that level, thereby suggesting an inconclusive result that the annual rotation patterns between crops are not necessarily regular and definitively correlated.

Turning to the groundwater resources information, well data and hydrogeologic properties of well-locations are available from the USGS and the Iowa Department of Natural Resources. Locations, depths, and nitrate levels of the USGS National Water Quality Assessment database and the groundwater provinces of northeast Iowa are shown in Fig. 10.9. The driving factors of the hydrogeologic system, depending on the hydraulic conductivity and permeability, groundwater moves at rates from a few inches a year to several feet per minute. Groundwater resources occur at various depths and in a variety of materials (Prior et al. 2003). Surficial aquifers occur in relatively loose granular sediments that lie between the land surface and deeper bedrock. The surficial aquifers and aquitards consist of alluvium (water-deposited sand and gravel), loess (wind-deposited silt) and glacial till (pebbly or sandy clay deposited by ice). Thickness of these materials ranges from 0 m in parts of northeastern Iowa to 180 m in west-central Iowa. Loess and glacial till are fine textured and have moderate to low permeability. Alluvial aquifers are unconfined. Some alluvial deposits consist of fine grained-silts and clays; others are coarse thick and extensive. If coarse and permeable materials, alluvial aquifers may have high yield and occur at depths of less than 30 m. Significant alluvial aquifers occur along the Mississippi River corridor in eastern Iowa.

Bedrock aquifers consist of solid rock layers such as limestone, dolomite, and sandstone. The groundwater in deep aquifers in Iowa can be older than 10,000 years. Bedrock consists of sedimentary rock layers, limestone and dolomite (carbonate rocks) as well as shale, siltstone, and sandstone. Thickness ranges from 1,580 m in southwest Iowa and about 240 m in the northeast. Some of the more prominent deep aquifers are the: Dakota aquifer, Mississippian aquifer, Silurian aquifer, Devonian aquifer, and Cambrian Ordovician aquifer.

10.4.2 Comparing Loss Estimates of Two Alternative MRLIs

This section presents an example of a hypothetical application of the integrated assessment approach with alternative data sets, Landsat and AWiFS. It includes the calculation of an estimate of the economic loss associated with the regulation of the maximum contamination level standards using the remotely sensed observations as well as the difference in the estimated economic loss between them. This difference can have an important effect on the decision to regulate.

We begin by identifying the relevant decision time increment of 1 year for t and setting α for a risk tolerance level. These two components are combined to provide the estimate of the expected loss $\alpha(L_i)$. To calculate $\alpha(L_i)$, we estimate the conditional exceedance probabilities for MCL regulatory standards of 4 and 10 mg/l as specified by the U.S. Environmental Protection Agency (EPA). The current MCL of 10 mg/l is more than double that of Germany and South Africa, where the standard is 4.4 ppm (Kross et al. 1995). Other nitrate concentrations may be relevant for a parallel analysis, since only below 0.2 mg/l is the risk considered low, while above 4.8 mg/l carries known risks (Nolan and Hitt 2006). The nitrosamines and nitrosamides that result when nitrates react with organic compounds are associated with 15 types of cancers including tumors in the bladder, stomach, brain, esophagus, bone and skin, kidney, liver, lung, oral and nasal cavities, pancreas, peripheral nervous system, thyroid, trachea, acute myelocytic leukemia, and T and B cell lymphoma (Mirvish 1995).

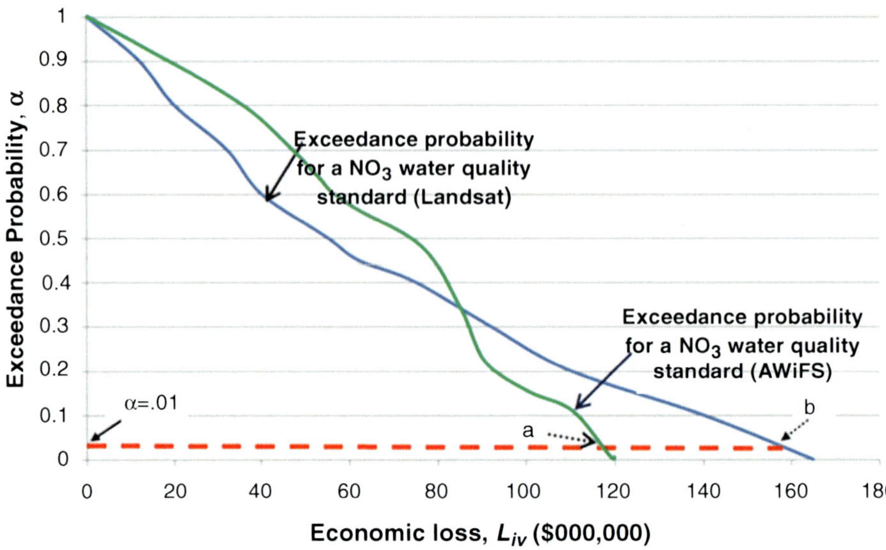

Fig. 10.10 Hypothetical exceedance probability curves for Landsat and AWiFS in northeastern Iowa

For the example, consider a hazard rate for exceeding the nitrate standard (based on the MCL analyzed in the wells sampled) in a region of Iowa of 0.05 as a result of land use *j*. The greater the number of wells that exceed MCL in the region, the more severe event *i* would be. In our hypothetical example, $(p_k) = 0.0005$ is the regional probability for an event that has an $\alpha_{\omega n} = 0.01$ averaged over the number of wells, *n* in the region. If regulation occurs, the economic loss, L_i, is the total loss in value of agricultural production due to the regulatory change (e.g., changing the tax on nitrogenous fertilizer).

The total production loss can cover a wide range depending on event size in a major agricultural region that is also heavily dependent on ground water resources for potable water as in Iowa. Consider the region described by Fig. 10.9. For our hypothetical severe event ($p_k = 0.0005$), the horizontal dashed line located at $\alpha = 0.01$ identifies the expected loss estimate based on each type of MRLI data. The two MRLI information types yield different loss estimates. The economic loss at $\alpha = 0.01$ could be either a little under $120 million for Landsat (point a) or almost $160 million for AWiFS (point b) in Fig. 10.10. Depending on which MRLI information is available, the cost of preventing the low probability, severe-contamination event is either $160 million or $120 million.

The VOI for this example is the estimated difference in the economic value of agricultural production lost if the MRLI information of the lower cost regulation (fewer acres in corn production) is implemented. That is, in this illustration of the method, if Landsat is available and corn production is regulated, the benefit (savings) of avoiding the loss of the groundwater is $40 million less in income loss to the farmer than is the case with AWiFS.

10.5 Summary

In this chapter, we have assembled models from a variety of scientific disciplines into a general framework of economic decision making. The integrated assessment approach links MRLI data to models of productive land uses, emissions and environmental damage caused by the land uses and risks to natural resources to better assess the joint production of market goods and environmental degradation. The VOI for MRLI in this particular case is based on the more efficient and accurate estimation of the joint production of market and environmental (i.e. non-market) goods made possible by better information.

In our application, we link the MRLI data to agricultural production, nitrate loading and groundwater vulnerability models to estimate the joint output of agricultural production and nitrate groundwater pollution. The VOI of MRLI for this case is the increased value of agricultural production that is possible because of better-calibrated regulatory measures for the protection of groundwater. Given that this is only one case study involving estimates of a limited set of goods, the VOI of MRLI should be considered a lower bound of its overall value to society.

We illustrate the method and application with a hypothetical example in which the VOI derives from: (1) providing cost effective information on the population of land activities across space and over time to analyze a particular harm to ecosystem services, and (2) reducing the risk of a regulatory decision error in cases where groundwater pollution is a likely problem. The use of MRLI information must show an incremental cost savings or result in a new application to demonstrate an economic return.

This research provides a method to demonstrate the value of MRLI information in an operational application and to assist in estimating the cost effectiveness of investment in space-based remote sensing that informs congressional policy makers and other stakeholders about the potential environmental risks associated with specific agricultural and health policies and regulations. Empirical application of the conceptual framework would involve a two-phase analysis: first, estimate VOI derived from remotely sensed land imaging vs. traditional sampling methods, and then determine VOI derived from Landsat versus. other satellite sensors. VOI pertaining to other regulatory decisions can be similarly estimated; for example:

- In the agricultural sector, nitrogen application and surface water protection, pesticide application and protection of surface water and groundwater resources, and water quality monitoring of impaired waterways that fall under the jurisdiction of EPA's TMDLs;
- In the mineral sector, mining for coal and downstream pollution; and
- LULC changes that affect patterns and processes of landscapes such as wildlife corridors and habitat connectivity for listed species, and indices of biodiversity, ecosystem integrity, and forest health.

The value of information arises from the spatial data or temporal archive or both provided by MRLI. The framework developed in this chapter has many potential applications.

Appendix

Previous studies on the economics of information and uncertainty established that information becomes valuable when the information can be profitably employed in a decision making process by reducing the risk facing decision makers, especially when the consequences to the decision maker are uneven (Morgan and Henrion 1990). In this application where the consequences are uneven, the underlying behavior is the use of nitrate fertilizers on fields. Of critical interest to the regulator is when to intervene to control the rate of change in the groundwater quality to avoid the costly treatment of water wells. To intervene either too early or too late relative to a regulatory threshold is to affect individuals and firms, e.g., agricultural production, negatively or to contaminate either or both shallow and deep groundwater. On the other hand, to intervene too late allows a greater potential of contamination, i.e., the regulator's risk. Bayesian Decision Analysis can be applied as a regulator's decision problem involving uncertainty.

The loss function we face is asymmetric (Fig. 10.A.1) since groundwater resource damage is very costly to mitigate and thus the total economic value of diminished current and potential uses and existence of a pristine resource is large relative to the marginal loss of agricultural output. Edwards (1988) found the willingness to pay to prevent groundwater damage increased linearly with risk. The marginal risk to the groundwater increases as nitrate fertilizer application increases, we model the social loss of over application with an exponential functional form.

The modified linex loss function in Fig. 10.A.1 is adapted to the risk analysis in the hypothetical example (van Noortwijk and van Gelder):

$$L(\Delta) = -c_v \mathrm{In}(\alpha) \cdot (\lambda^* - \lambda) + \frac{d}{1-d} L_{i(j)} \alpha \left[\exp\left\{ \mathrm{In}(\alpha) \cdot \frac{\lambda^* - \lambda}{\lambda} \right\} - 1 \right] \quad (10.\mathrm{A}.1)$$

where c_v is the additional variable cost to reduce nitrogen use, $\mathrm{In}(\alpha)$ is the q-quantile (level of risk tolerance) of the nitrate concentration distribution, and λ is the parameter of the nitrate loading distribution.

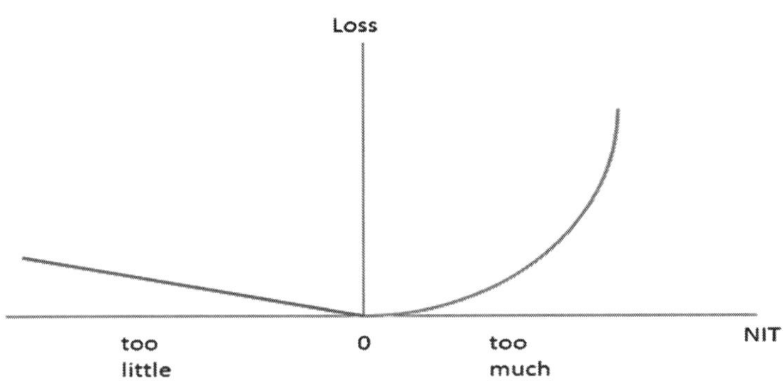

Fig. 10.A.1 Decision risk faced by regulators of societal cost of groundwater damage (*red*) vs. loss of agricultural production (*blue*). Asymmetric loss to regulator: too little NIT = crop income/profit loss; too much NIT = risk of aquifer loss

10. Commentary: Satellite Observations and Policy Improvements for Agriculture and the Environment

Catherine Shelley Norman

Bernknopf, Forney, Raunikar and Mishra (BFRM 2010) consider the value of medium-resolution land imagery (MRLI) to a regulator focused on keeping water quality risks to an acceptable level. They present a general model of the biological, physical and economic processes at work, and then outline a specific example focusing on corn and soybean rotations and the fate and transport of associated nonpoint source pollution in a nitrogen-limited region of Iowa. MRLI allows the regulator to monitor changes in land use and optimize regulations to reflect nutrient burdens on the system, economic costs, and the value of mitigation plans.

In offering a comprehensive, integrated approach to valuing the information provided by programs like Landsat, the authors' focus is on valuing the decisions that hinge on the information provided by a given set of observations. Their model is unusual in that it takes multiple disciplinary perspectives—ecological processes, agricultural science, economics, and hydrology, among others—seriously and simultaneously. Regulation of land uses and management practices is well suited to medium-resolution land imagery. Monitors can use simultaneously updated, geographically complete information rather than relying solely on sampling programs that require regulators to select representative sites over time and across space. Space-based data can be coordinated with direct sampling to improve inferences from images and to support enforcement efforts.

Although the full data required to reach conclusions are not available to BFRM, this work is a strong and ambitious initial step toward a framework supporting improved program evaluation. I consider some clarifications, limitations, and possible extensions of this work, with a focus on producing credible estimates to inform decisions about the use of MRLI.

10.C.1. Major Contributions

Satellite imagery supports multiple overlapping objectives in the regulatory and political arena. BFRM focus on one agricultural regulatory application, working to quantify the benefits from that use alone. In theory, one could aggregate up to a total value for a space-based observation program by considering all users, though as we can see from this relatively straightforward application, the informational and

C.S. Norman (✉)
Department of Geography and Environmental Engineering and Department of Economics, The Johns Hopkins University, Baltimore, MD, USA
e-mail: norman@jhu.edu

technical demands associated with such an effort would be very high. Additionally, for many earth observations satellites, information on all uses is likely to be confidential for reasons of national security, and some part of the value will be strategic or political—and thus more challenging to quantify than ecosystem services. It is perhaps better to think of VOI in this framework as valuing data access or specific data uses rather than a program as a whole. If this kind of use is the primary driver of value for the program in question, it should still be emphasized that estimates produced in this way, no matter how comprehensive, will be lower bounds of value rather than estimates of total value. Even rough information about the scale of a given use relative to uses for the satellite information as a whole would provide useful context for decisionmaking.

Some VOI quantification of the sort proposed is critical to maintaining and instrumenting costly space-based information resources. Cost-benefit analysis is mandatory in many public decisionmaking processes and a common, well-understood framework in the remainder. If those of us who use this information cannot usefully answer questions about what it enables us to do, it is difficult to justify maintaining either the systems themselves or access to the information. BFRM present a model to capture the benefit to a policymaker focused on risk management (for avoiding threshold environmental effects) and thus assess the value, to regulators and the public good, of operating with the additional clarity and scope offered by space observation systems.

I attempt to lay out some of the most interesting questions this work posed below, focusing on the economics of regulation. Moving this model from the general to the specific requires the authors to confront political and public choices that are difficult to observe or theorize about in a way that yields usable numbers. It will be important in applications that the assumptions used to develop the regulator's objectives and constraints are made transparent, and perhaps subjected to sensitivity analyses.

10.C.2. Social Preferences and Risk

In Sect. 10.2.3, the authors note that we can infer the optimal risk of failure (in a whole matrix of failure points reflecting reduced water quality due to pollution) of a given regulatory regime for the regulator and link this to social risk preferences. The regulator takes those social preferences as given from a 'higher level of authority.' In the illustrating example, the analyst must infer the political level of tolerance for a multitude of risks of reaching various pollution thresholds in various environmental media and locations. I would be interested in much more detail on how this matrix is populated; there is a small but longstanding literature on identifying the preferences of government bodies (McFadden 1975, 1976; Ross 1984 are identified in the chapter), but recent efforts (e.g., Ahlroth et al. 2010; DeCanio and Norman 2005) are in much narrower applications, and even in such settings, inferences based on political choice are met with considerable skepticism.

Social decisions should yield some sort of information about the willingness of society to make trade-offs and incur costs, but methods for determining political willingness to pay (or similar) are by no means well established. BFRM view the regulator's problem as one of reducing risk to an acceptable level without providing information on social risk aversion; individual and firm risk aversion are difficult to infer outside highly controlled settings (Chetty 2006 describes some of the complexities involved), and moving from individual choices to social choices involves a fairly fundamental determination about the relationship between a government and the citizenry: ought a government represent the median voter? Or should the state, with a (potentially) much longer lifespan and broader area of influence, worry more about the future than individuals?

It might be more feasible to go backward from costs incurred for specific environmental efforts rather than to look at political decisions directly: those probabilities can be translated into expected payouts to replace ecosystem services if they are reduced or eliminated by insufficiently stringent policy choices—and it may be that that's what BFRM will do to quantify the requirements of the higher governmental authority—but there is insufficient information for the reader on the methods envisioned for general applications of this framework at present.

10.C.3. Regulator's Objectives

In the BFRM example, the regulator's objective is to maximize the total value of agricultural output, given the constraints established by the willingness of society to accept the ecological risks outlined above. The farmer's objective is to maximize profit. Thus, a farmer would prefer a lower yield method if costs were sufficiently reduced to preserve profits, but the regulator cannot support this. In the agricultural sector it is often the case that polluting inputs can be replaced with less polluting alternatives or increased handwork; as regulatory environments and water and land quality evolve over time, this may be a profit-maximizing, output-reducing solution, which appears to be excluded in this analysis.

It is not clear to me that EPA or USDA, the primary agencies in this environment, value total output over total profitability in this context. Readers would benefit from efforts to explain the motivation behind this modeling choice and its implications for VOI calculations. In particular, this choice seems closely tied to the choice to value MRLI information as some fraction of total farm revenues.

The language of the chapter seems to suggest that there is a positive value for the information from the satellite system only if the information allows the regulator to relax restrictions on farmers. I don't believe this is required in the model; if it is, the authors should explain their motivation in more detail or perhaps consider relaxing this assumption. Given the level of geophysical detail in the example, it seems plausible that relaxing nutrient loading restrictions in some places and tightening them in others might provide aggregate benefits. Additionally, there is value in improved information that leads to greater restrictions: if we've gotten the regulations wrong because a sampling regime or alternate set of instruments led us to underestimate aquifer risks, for example, we gain from the changes even if they reduce output or profits.

As an extension, a more dynamic perspective might allow adaptive management to be built in to policy choices. VOI would thus derive from a more flexible system as well as from a system that is better at a specific moment in time. Given the slow pace of the regulatory and legislative process, this would not entail changes in middle of a crop cycle; rather, a set of observations could mean that restrictions were automatically loosened or tightened according to a preset schedule, obviating the need for ongoing legislative action.

Also on a longer scale, it is worth constraining "optimal" policy choices in the model to those that are politically feasible and legally defensible. Detailed spatial data may reveal significantly different optimal restrictions on farmers in the same jurisdiction growing the same crop using the same technology. It would be disingenuous to suggest that the VOI is contingent on such policies' being enacted.

That said, given a broader array of policy choices, MRLI data combined with hydrologic and other spatial data could be used to improve environmental quality (or reduce risks to environmental services) at minimum costs by treating neighboring parcels differently. We might use these data to identify land values that may be declining to a point where the parcels are appropriate for conservation easements, for example, and payments could be based on forgone income opportunities.

It is also worth noting that in the longer term, VOI is very sensitive to the national or international policy regime in place. In addition to the gains from improved management of water resources, monitoring of agricultural land uses and changes will provide credible baselines for measuring carbon sequestration and perhaps granting offsets. This is true globally, of course, and in an environment where offsets are valuable, the entire community of nations included in the carbon dioxide regime would gain from the sum total of land imagery available. Establishing a baseline now, assuming monitoring of changes will be needed in the future, creates an option value for farmers and regulators anticipating policy changes of this nature.

10.C.4. Monitoring and Enforcement

Some of the value associated with MRLI in this context will come from the broad applicability to monitoring and enforcement of existing law. Monitoring and enforcement are currently exogenous in this model, but repeated imagery of all farmers in a region will affect compliance behavior. Although 17-day satellite sweeps and missed observations due to weather mean that enforcement opportunities from satellite information cannot be perfect, MRLI offers more observation of practices and outcomes (with less awareness of the specifics of observation than an inspector arriving at the farm gate) than conventional practice. Over time, habitual offenders should be identifiable, and for some rules even a single pass can provide evidence of noncompliance with watershed protection law. Even limited enforcement actions taken based on this information could have significant spillovers in compliance behavior for a region (Shimshack and Ward 2005).

Agricultural nonpoint sources are enormously significant nutrient sources in strained watersheds; resource limitations and the difficulty of tracing a given pollutant load to a specific plot of land mean that enforcement is difficult and inspections relatively rare. Improved compliance as a result of farmers' expectations about the use of satellite imagery (perhaps to direct site visits) will offer reduced uncertainty not only about watershed conditions but also about land use and growers' behavior as regulations change.

10.C.5. Baselines

When fuller data are available to complete the analysis envisioned in BFRM, clarity about the specific baseline considered (and alternative baselines) will be essential. If we are considering a proposed Landsat 10 mission, for example, it will have a value relative to no use of satellite imagery in regulatory decisionmaking, a value relative to the other satellites whose data will be used if the mission does not take place, and plausibly a value relative to whatever other alternatives may be available. Each of these alternatives also has costs, either to build and run or to purchase data from. We can then estimate how much of the value of the management plan derives from the improved information associated with the MRLI source. We will never make decisions about regulating agriculture for water quality protection in an information vacuum, so the marginal value is always contingent on our expectations about the information that will be available without the mission (or without a specific instrument's inclusion in the mission).

If, as the authors suggest, the next-best information is from the Advanced Wide Field Sensor (AWiFS), which is under the charge of the Indian Space Research Organization, the VOI in the proposed Landsat mission is likely to be increased by the greater control U.S. agencies would have over Landsat. This is harder to quantify than the ecosystem services approach outlined in BFRM, but a simplified approach might rely on estimates of the probability of having to move from AWiFS to another next-best solution in a given year.

10.C.6. VOI and Time-Series Continuity

In the example given by BFRM, we are concerned with the value of information associated with Landsat imaging. One important area that is not yet extensively developed is the value of continuity in the time series. On a short-term basis, the value of one satellite mission relative to another may be roughly comparable, but most researchers would prefer a mission that offers greater continuity with past and future missions over one that does not. In particular, scientists and policymakers interested in climate change and in carbon sequestration rely on comparability of current and older imagery. Direct comparisons are especially useful for looking at trends over large scales when medium- and low-resolution imagery does not allow detailed observation of the phenomenon of interest: we may not be able to count the trees on a plot of land, but seeing how the same instrument records changing levels of things associated with trees (height of ground cover, albedo, etc.) allows us to make inferences about changing ecology, land use, and productivity over time.

Valuing this is more complex than valuing a given piece of data over its entire use cycle, which is itself not straightforward. The switch from SIC codes to NAICS codes for the collection and provision of industrial data by the U.S. government prompted some discussion of the value of continuous time-series information and of the processes used to connect different time series; the Census Bureau's Economic Classification Policy Committee (1993) report presents the core concerns of data collectors and users. It details millions spent to construct "bridge" data to help users get some of the benefits of the long series even after classification schemes has been changed to reflect new patterns of production and provide improved comparability across nations.

What value there is in continuity will not automatically accrue to ongoing Landsat missions. Each mission has differing instrumentation and data characteristics. The additional value of information that is part of a 40-year stream of information of the same sort is thus contingent on efforts to merge the varying Landsat series (or to merge Landsat data usefully with alternative space-based observations). Information about those costs or about the number and type of users who would benefit from more determined efforts in this direction is not readily available. It does, however, seem likely that efforts to connect data from various Landsat missions will be easier, cheaper, and more likely to be pursued by a single entity responsible for all the data than if Landsat observations were replaced by AWiFS, as in the illustrative case.

10.C.7. Additional Considerations

This work covers a very large area, and space precludes consideration of everything that is important or interesting in this context. An extended analysis might take into account some or all of the considerations in this section.

In BFRM, financials are allowed to fluctuate weekly in the model. This makes sense for inputs to production but is less appropriate for revenues, since farmers can typically store corn and soybeans for a time if prices are not favorable. This storage decision could be added to the model, or annual output price figures could be used to approximate outcomes of storage decisions and costs.

Consideration of farm- and community-level spatial interactions would be interesting. Do we allow the production decisions of farmers to affect those of other producers spatially near them? If one farmer changes her crop rotation, do the neighbors respond? There may be thin local markets in processing or storing facilities or farm labor, or limited transportation or storage infrastructure that vitiate incentives toward more homogenized crop patterns in a region. This could operate through increased input costs or input cost volatility, affecting expected profit and income volatility.

Analysts would also need fuller details on the actual instrumentation considered for a given mission and how it could be used. For those unfamiliar with the details of the program, information about what spectra will be measured at what level of detail, and which will be used for the regulatory purposes envisioned, would clarify the application of the methodology to the illustration. Is the regulator hiring staff to visually inspect images of farms? Are heat and albedo sensors informing estimates of growth rates or nitrogen concentrations in specific locations? VOI will be very sensitive to the specific instrumentation and uses envisioned.

Lastly, broader consideration of ecosystem services, including carbon sequestration, biodiversity support, and erosion outcomes, would provide more opportunities for enhanced regulatory decisionmaking, thus increasing the VOI to environmental management based on MRLI.

10.C.8. Conclusion

The authors integrate a complex set of hydrogeologic, biophysical, social, and economic models to provide an estimate of the state of water quality and agricultural products in a given region with regulators with identical preferences but two possible data sources. The value to society of better information in this setting is estimated by examining the effects of changes in regulations associated with higher-quality MRLI. Although applying this approach will be data and time intensive, and explaining it to decisionmakers will need to done very carefully, this is an important step toward estimating the value of satellite information in a broad array of uses.

References

American Society for Photogrammetry and Remote Sensing Survey (ASPRS). (2006). *Moderate resolution imagery survey.*

Antle, J., & Just, R. J. (1991). Effects of commodity program structure on resource use and the environment. In R. Just & N. Bockstael (Eds.), *Commodity and resource policies in agricultural systems* (pp. 97–128). Berlin: Springer.

Antle, J., & McGuckin, T. (1993). Technological innovation, agricultural productivity, and environmental quality. In G. Carlson, D. Zilberman, & J. Miranowski (Eds.), *Agricultural and environmental resource economics* (pp. 175–220). New York: Oxford University Press.

Antle, J., & Valdivia, R. (2006). Modeling the supply of ecosystem services from agriculture: A minimum-data approach. *The Australian Journal of Agricultural and Resource Economics, 50*, 1–15.

Ahlroth, S., M. Nilsson, G. Finnveden, O. Hjelm, and E. Hochschorner, 2010. "Weighting and valuation in selected environmental systems analysis tools - suggestions for further developments." *Journal of Cleaner Production*, In Press, Available online, DOI: 10.1016/j.jclepro.2010.04.016.

Bear, J. (1979). *Hydraulics of groundwater*. New York: McGraw–Hill.

Bouma, J. A., van der Woerd, H. J., & Kuik, O. J. (2009). Assessing the value of information for water quality management in the North Sea. *Journal of Environmental Management, 90*, 1280–1288.

Brady, N. C., & Weil, R. R. (2002). *The nature and properties of soils* (13th ed.). Upper Saddle River: Prentice Hall.

Canter, L. W. (1997). *Nitrates in groundwater*. Boca Raton: CRC Press.

Chetty, R., 2006. "A New Method of Estimating Risk Aversion." *American Economic Review* 96: 1821-1834.

Crutch-field, S. R., Cooper, J. C., & Hellerstein, D. (1997). *Benefits of safer drinking water: The value of nitrate reduction* (Agricultural Economic Report No. 752). U.S. Department of Agriculture, Economic Research Service, Food and Consumer Economics Division.

Daily, G. C. (Ed.). (1997). *Nature's services: Societal dependence on natural ecosystems*. Washington, DC: Island Press.

de Marsily, G. (1986). *Quantitative hydrogeology*. Orlando: Academic.

DeCanio, S.J. and C.S. Norman, 2005. "Economics of the 'Critical use' of Methyl Bromide Under the Montreal Protocol," *Contemporary Economic Policy* 23(3): 376-393.

Earth Satellite Corporation. (1974). *Earth resources survey benefit-cost study* (Contract No. 14-08-0001-13519, 6 Vols and Appendices). Washington, DC: Earth Satellite Corporation.

ECON, Incorporated. (1974). *The economic value remote sensing of earth resources from space: An ERTS overview and the value of continuity of service*. Princeton: ECON, Inc.

Economic Classification Policy Committee (ECPC). 1993. Issues Paper No. 5, "The Impact of Classification Revisions on Time Series." Available at: http://www.census.gov/epcd/naics/issues5

Edwards, S. (1988). Option prices for groundwater protection. *Journal of Environmental Economics and Management* 15:475–487

Grossi, P., & Kunreuther, H. (2005). *Catastrophic modeling: A new approach to managing risk*. New York: Springer.

Hale, R. C., Hanuschak, G. A., & Craig, M. E. (1999). *The appropriate role of remote sensing in U.S. agricultural statistics*. National Agricultural Statistics Service (USDA), FAO regional project: Improvement of agricultural statistics in Asia and pacific countries, GCP/RAS/171/JPN.

Hall, H. (1998). *Choosing an empirical production function: Theory, nonnested hypotheses, costs of specifications* (Agricultural Economics Research Report No. 59). Lexington: Department of Agricultural Economics, University of Kentucky.

Heal, G. (1991). Economy and climate: A preliminary framework for microeconomic analysis. In R. Just & N. Bockstael (Eds.), *Commodity and resource policies in agricultural systems* (pp. 196–212). Berlin: Springer.

Hong, N., White, J. G., Weisz, R., Gumpertz, M. L., Duffera, M. G., & Cassell, D. K. (2007). Groundwater nitrate spatial and temporal patterns and correlations: Influence of natural controls and nitrogen management. *Vadose Zone Journal, 6*, 53–66.

Iowa State University Extension Service. (2010). *Chartbook corn tables*. Available at: http://www2.econ.iastate.edu/outrea.ch/agriculture/periodicals/chartbook/files/corn.htm. Accessed on 14 Sept 2010.

Isik, M., Hudson, D., & Coble, K. H. (2005). The value of site-specific information and the environment: Technology adoption and pesticide use under uncertainty. *Journal of Environmental Management, 76*, 245–254.

Jordan, J. L., & Elnagheeb, A. H. (1993). Willingness to pay for improvements in drinking water quality. *Water Resources Research, 29*, 237–245.

Kalbfleisch, J., & Prentice, R. (1980). *The statistical analysis of failure time data*. New York: Wiley. 321p.

Kalluri, S., Gilruth, P., & Bergman, R. (2003). The potential of remote sensing data for decision makers at the state, local and tribal level: Experiences from NASA's synergy program. *Environmental Science & Policy, 6*, 487–500.

Kleinbaum, D. (1996). *Survival analysis: A self-learning text*. New York: Springer.

Kross, B. C., Olson, M. L., Ayebo, A., & Johnson, J. K. (1995). Ameliorating effects of alternative agriculture. In J. Rechchigl (Ed.), *Soil amendments: Impacts on biotic systems* (pp. 153–214). Boca Raton: Lewis Publishers.

Krutilla, J. (1967). Conservation reconsidered. *American Economic Review, 57*, 787–796.

Lambert, D., Lowenberg-DeBoer, J., & Malzer, G. (2006). Economic analysis of spatial-temporal patterns in corn and soybean response to nitrogen and phosphorus. *Agronomy Journal, 98*, 43–54.

Lancaster, T. (1990). *The econometric analysis of transition data*. Cambridge: Cambridge University Press.

Lauer, J., Porter, P., & Opplinger, E. (1997). *The corn and soybean rotation effect*. Agronomy advice, Agronomy Department, University of Wisconsin. http://corn.agronomy.wisc.edu/aa/a014.aspx

Lence, S., & Hayes, D. (1995). Land allocation in the presence of estimation risk. *Journal of Agricultural and Resource Economics, 20*, 49–63.

Lichtenberg, E. (1991). Determination of regional environmental policy under uncertainty: Theory and case studies. In A. Dinar & D. Zimmerman (Eds.), *The economics and management of water and drainage in agriculture* (pp. 700–716). Boston: Kluwer Academic Publishers.

Livanis, G., Salois, M., & Moss, C. (2009). *A nonparametric kernel representation of the agricultural production function: Implications for economic measures of technology*. Paper presented at the 83rd Annual Conference of the Agricultural Economics Society, Dublin.

Loomis, J., Bell, P., Cooney, H., & Asmus, C. (2009). A comparison of actual and hypothetical willingness to pay of parents and non-parents for protecting infant health: The case of nitrates in drinking water. *Journal of Agricultural and Applied Economics, 4*, 697–712.

Lubowski, R. N., Vesterby, M., Bucholtz, S., Baez, A., & Roberts, M. J. (2005). *Major uses of land in the United States* (USDA Economic Information Bulletin, No. 14).

Macauley, M. K. (2006). The value of information: Measuring the contribution of space-derived earth science data to resource management. *Space Policy, 22*, 274–282.

Macauley, M. (2007). *The role of earth observations in revolutionizing management of natural resources and the environment: Identifying the Landsat contribution, report to the United States Geological Survey*. Washington, DC: Resources for the Future.

Macauley, M. (2010). *Climate adaptation policy: The role and value of information* (Resources for the Future, Issues Brief 10-10).

Martinez, Y., & Albiac, J. (2006). Nitrate pollution control under soil heterogeneity. *Land Use Policy, 23*, 521–532.

McFadden, Daniel, 1975. "The Revealed Preferences of a Government Bureaucracy: Theory." *The Bell Journal of Economics* 6(2): 401-416.

McFadden, Daniel, 1976. "The Revealed Preferences of a Government Bureaucracy: Empirical Evidence." *The Bell Journal of Economics,* 7(1): 55-72.

Meals, D. W., Dressing, S. A., & Davenport, T. E. (2010). Lag time in water quality response to best management practices: A review. *Journal of Environmental Quality, 39*, 85–96.

Miller, H., Sexton, N., Koontz, L., Loomis, J., Koontz, S., & Hermans, C. (2011). *The users, uses, and value of Landsat and other moderate-resolution satellite imagery in the United States Executive Report* (U.S. Geological Survey Open-File Report 2011-1031).

Mirvish, S. S. (1995). Role of N-nitroso compounds (NOC) and N-nitrosation in etiology of gastric, esophageal, nasopharyngeal and bladder cancer and contribution to cancer of known exposures to NOC. *Cancer Letters, 97*, 271.

Morgan, M. G., & Henrion, M. (1990). *Uncertainty: A guide to dealing with uncertainty in quantitative risk and policy analysis*. New York: Cambridge University Press.

Mortensen, J. R., & Beattie, B. R. (2005). *Does choice of response function matter in setting maximum allowable N-application rates in Danish agriculture?* (Cardon Research Papers, 2005-01). Available at: http://ag.arizona.edu/arec/pubs/researchpapers/2005-01mortensenbeattie.pdf

Nelson, R., & Winter, S. (1964). A case study in the economics of information and coordination: The weather forecasting system. *Quarterly Journal of Economics, 78*, 420–441.

Nelson, G., Schimmelpfennig, D., & Sumner, D. (2007). *Can satellite-based land imaging data be made more valuable for agriculture?* (Report to the US Department of Agriculture, Cooperative Agreement 58-6000-6-0047).

Nolan, B. T., & Hitt, K. J. (2006). Vulnerability of shallow groundwater and drinking-water wells to nitrate in the United States. *Environmental Science and Technology, 40*, 7834–7840.

Nolan, B. T., Ruddy, B. C., Hitt, K. J., & Helsel, D. R. (1997). Risk of nitrate in groundwaters of the Unites States—A national perspective. *Environmental Science and Technology, 31*, 2229–2236.

Nolan, B. T., Hitt, K. J., & Ruddy, B. C. (2002). Probability of nitrate contamination of recently recharged ground waters in the conterminous United States. *Environmental Science and Technology, 36*, 2138–2145.

Nolan, B. T., Puckett, L. J., Ma, L., Green, C. T., Bayless, E. R., & Malone, R. W. (2010). Predicting unsaturated zone nitrogen mass balances in agricultural settings in the United States. *Journal of Environmental Quality, 39*, 1051–1065.

North Dakota State University (in conjunction with North Dakota Corn Utilization Council). (1997). *Corn production guide*. Available at http://www.ag.ndsu.edu/pubs/plantsci/rowcrops/a1130-2.htm#Index. Accessed on 12 Jan 2009.

Oakes, D. B. (1982). Nitrate pollution of groundwater resources: Mechanisms and modeling. In: *Proceedings of an IIASA Task Force Meeting on nonpoint nitrate pollution of municipal water supply sources* (Issues of Analysis and Control, Series No. CP-82-S4). Laxenburg: International Institute of Applied Systems Analysis.

Philips, L. (1988). *The economics of imperfect information*. Cambridge: Cambridge University Press.

Poe, G. L. (1999). Maximizing the environmental benefits per dollar expended: An economic interpretation and review of agricultural environmental benefits and costs. *Society and Natural Resources, 12*, 571–598.

Poe, G. L., & Bishop, R. C. (1999). Valuing the incremental benefits of groundwater protection when exposure levels are known. *Environmental and Resource Economics, 13*, 341–367.

Prior, J., Howes, J. M., Libra, R., & VanDorpe, P. (2003). *Iowa groundwater basics* (Iowa Geological Survey Educational Series 6). Iowa Department of Natural Resources.

Ross, T. W. (1984). Uncovering regulators' social welfare weights. *The RAND Journal of Economics, 15*, 152–155.

Sheriff, G. (2005). Efficient waste? Why farmers over-apply nutrients and the implications for policy design. *Review of Agricultural Economics, 27*, 542–557.

Shimshack, J. and M.B. Ward, 2005. "Regulator Reputation, Enforcement, and Environmental Compliance," *Journal of Environmental Economics and Management* 50: 519-540.

Sparco, J. (1995). *Marginal valuation of health related attributes of ground water using conjoint analysis*. Northeastern Agricultural and resource economics association meetings, Burlington, VT.

The Financials. (2010). www.commodities.thefinancials.com. Verified June 15, 2010.

Tomer, M. D., & Burkart, M. R. (2003). Long-term effects of nitrogen fertilizer use on ground water nitrate in two small watersheds. *Journal of Environmental Quality, 32*, 2158–2171.

Van Noortwijk, J.M., and Van Gelder, P.H.A.J.M., 1998. Bayesian estimation of quantiles for the purpose of flood prevention. In: Coastal Engineering, *27*, 3529–3541.

Weisbrod, B. A. (1964). Collective consumption services of individual consumption goods. *Quarterly Journal of Economics, 78*, 471–477.

Wilkie, D. S., & Finn, J. T. (1996). *Remote sensing imagery for natural resources monitoring*. New York: Columbia University Press.

Williamson, R. A., Hertzfeld, H. R., Cordes, J., & Logsdon, J. M. (2002). The socioeconomic benefits of earth science and applications research: reducing the risks and costs of natural disasters in the USA. *Space Policy, 18*, 57–65.

Yadav, S.N. 1997. Dynamic optimization if nitrogen use when groundwater contamination is internalized at the standard in the long run Amercian Journal of Agricultural Economics, Vol. 79, No. 3, pp. 931–945.

Index

R. Laxminarayan and M.K. Macauley (eds.), *The Value of Information: Methodological Frontiers and New Applications in Environment and Health*, DOI 10.1007/978-94-007-4839-2, © Springer Science+Business Media Dordrecht 2012

Printed by Publishers' Graphics LLC
MO20120927-217-528